T0201387

# Introduction to Chemicals from Biomass

# Wiley Series in Renewable Resources

## Series Editor

*Christian V. Stevens – Faculty of Bioscience Engineering, Ghent University, Ghent, Belgium*

## Titles in the Series

**Wood Modification – Chemical, Thermal and Other Processes**
Callum A.S. Hill

**Renewables-Based Technology – Sustainability Assessment**
Jo Dewulf and Herman Van Langenhove

**Introduction to Chemicals from Biomass**
James Clark and Fabien Deswarte

**Biofuels**
Wim Soetaert and Erick Vandamme

**Handbook of Natural Colorants**
Thomas Bechtold and Rita Mussak

**Surfactants from Renewable Resources**
Mikael Kjellin and Ingegärd Johansson

**Industrial Application of Natural Fibres – Structure, Properties and Technical Applications**
Jörg Müssig

**Thermochemical Processing of Biomass – Conversion into Fuels, Chemicals and Power**
Robert C. Brown

**Biorefinery Co-Products: Phytochemicals, Primary Metabolites and Value-Added Biomass Processing**
Chantal Bergeron, Danielle Julie Carrier and Shri Ramaswamy

**Aqueous Pretreatment of Plant Biomass for Biological and Chemical Conversion to Fuels and Chemicals**
Charles E. Wyman

**Bio-Based Plastics: Materials and Applications**
Stephan Kabasci

**Introduction to Wood and Natural Fiber Composites**
Douglas Stokke, Qinglin Wu and Guangping Han

**Cellulosic Energy Cropping Systems**
Douglas L. Karlen

## Forthcoming Titles

**Cellulose Nanocrystals: Properties, Production and Applications**
Wadood Hamad

**Lignin and Lignans as Renewable Raw Materials: Chemistry, Technology and Applications**
Francisco García Calvo-Flores, José A. Dobado, Joaquín Isac García and Francisco J. Martin-Martinez

**Sustainability Assessment of Renewables-Based Products: Methods and Case Studies**
Jo Dewulf, Steven De Meester and Rodrigo Alvarenga

**Biorefinery of Inorganics: Recovering Mineral Nutrients from Biomass and Organic Waste**
Erik Meers and Gerard Velthof

**Bio-Based Solvents**
François Jerome and Rafael Luque

# Introduction to Chemicals from Biomass

SECOND EDITION

Edited by

JAMES CLARK
*Department of Chemistry, Green Chemistry Centre of Excellence,
University of York, UK*

FABIEN DESWARTE
*Biorenewables Development Centre, The Biocentre,
York Science Park, UK*

WILEY

This edition first published 2015
© 2015 John Wiley & Sons, Ltd

*Registered Office*
John Wiley & Sons, Ltd, The Atrium, Southern Gate, Chichester, West Sussex,
PO19 8SQ, United Kingdom

For details of our global editorial offices, for customer services and for information about how
to apply for permission to reuse the copyright material in this book please see our website at
www.wiley.com.

*Library of Congress Cataloging-in-Publication Data*

Introduction to chemicals from biomass / edited by James Clark, Fabien Deswarte. – Second edition.
    pages cm
  Includes bibliographical references and index.
  ISBN 978-1-118-71448-5 (cloth)
1. Biomass chemicals.   2. Organic compounds.   I. Clark, James H., editor.
II. Deswarte, Fabien E. I., editor.   III. Title: Chemicals from biomass.
  TP248.B55I68 2015
  662′.88–dc23

                                                                    2014045943

A catalogue record for this book is available from the British Library.

Set in 10/12pt Times by SPi Publisher Services, Pondicherry, India

Printed in Singapore by C.O.S. Printers Pte Ltd

1   2015

# Contents

# List of Contributors

**Mehrdad Arshadi** Department of Forest Biomaterials and Technology, Swedish University of Agricultural Sciences (SLU), Sweden

**Thomas M. Attard** Department of Chemistry, Green Chemistry Centre of Excellence, University of York, UK

**James A. Bergman** Materials Science and Engineering Department, Iowa State University, USA

**Vitaliy L. Budarin** Department of Chemistry, Green Chemistry Centre of Excellence, University of York, UK

**James Clark** Department of Chemistry, Green Chemistry Centre of Excellence, University of York, UK

**Fabien Deswarte** The Biorenewables Development Centre, The Biocentre, York Science Park, UK

**Jiajun Fan** Department of Chemistry, Green Chemistry Centre of Excellence, University of York, UK

**Thomas J. Farmer** Department of Chemistry, Green Chemistry Centre of Excellence, University of York, UK

**Joseph A. Houghton** Department of Chemistry, Green Chemistry Centre of Excellence, University of York, UK

**Andrew J. Hunt** Department of Chemistry, Green Chemistry Centre of Excellence, University of York, UK

**Michael R. Kessler** School of Mechanical and Materials Engineering, Washington State University, USA

**Tsz Him Kwan** School of Energy and Environment, City University of Hong Kong, Hong Kong

**Wan Chi Lam** School of Energy and Environment, City University of Hong Kong, Hong Kong

**Carol Sze Ki Lin** School of Energy and Environment, City University of Hong Kong, Hong Kong

**Mark Mascal** Department of Chemistry, University of California Davis, USA

**Avtar S. Matharu** Department of Chemistry, Green Chemistry Centre of Excellence, University of York, UK

**Egid B. Mubofu** Department of Chemistry, University of Dar es Salaam, Tanzania

**Igor Polikarpov** Grupo de Biotecnologia Molecular, Instituto de Física de São Carlos, Universidade de São Paulo, Brazil

**Antoine Rouilly** National Polytechnic Institute of Toulouse, France

**Anita Sellstedt** Department of Plant Physiology, UPSC, Umeå University, Sweden

**David Turley** NNFCC – The Bioeconomy Consultants, The Biocentre, York Science Park, UK

**Carlos Vaca-Garcia** National Polytechnic Institute of Toulouse, France; King Abdulaziz University, Center of Excellence for Advanced Materials Research, Saudi Arabia

# Series Preface

A multitude of important processes which have a major influence on our everyday lives involve the use and modification of renewable resources. Applications can be found in the energy sector, chemistry, pharmacy, the textile industry, paints and coatings, to name but a few.

The area interconnects several scientific disciplines (agriculture, biochemistry, chemistry, technology, environmental sciences, forestry, etc.), which makes it very difficult to have an expert view on the complicated interaction. The idea to create a series of scientific books, focusing on specific topics concerning renewable resources, has therefore been very opportune and can help to clarify some of the underlying connections in this area.

In a very fast-changing world, trends are not only characteristic of fashion and political standpoints; science is not free of its hypes and buzzwords either. The use of renewable resources is however much more important than a fad or fashion. As the lively discussions among scientists continue about how many years we will still be able to use fossil fuels – opinions range from 50 to 500 years – they do agree that the reserve is limited and that it is essential to search for new energy carriers and for new material sources.

In this respect, renewable resources are a crucial area in the search for alternatives for fossil-based raw materials and energy. In the field of energy supply, biomass- and renewable-based resources are part of the solution alongside other alternatives such as solar energy, wind energy, hydraulic power, hydrogen technology and nuclear energy.

In the field of material sciences, the impact of renewable resources will probably be even bigger. Integral utilisation of crops and the use of waste streams in certain industries will grow in importance, leading to more sustainable methods of producing materials.

Although our society was much more (almost exclusively) based on renewable resources centuries ago, this disappeared in the Western world in the nineteenth century. It is now time to focus again on this field of research. This does not mean '*retour à la nature*' however, but should be a multidisciplinary effort on a highly technological level to conduct research into new opportunities and to develop new crops and products from renewable resources. This will be essential to guarantee a certain standard of living for the growing number of people living on our planet. *The* challenge for the coming generations of scientists is to develop more sustainable ways to create prosperity and to fight poverty and hunger in the world. A global approach is certainly favoured.

This challenge can only be dealt with if scientists are attracted to this area and are recognised for their efforts in this interdisciplinary field. It is therefore also essential that consumers recognise the fate of renewable resources in a number of products. Furthermore, scientists need to communicate and discuss the relevance of their work.

The use and modification of renewable resources may not follow the path of the genetic engineering concept in terms of consumer acceptance in Europe. Related to this aspect, this series will certainly help to increase the visibility of the importance of renewable resources.

Being convinced of the value of the renewables approach for the industrial world as well as for developing countries, I was myself delighted to collaborate on this series of books focusing on different aspects of renewable resources. I hope that readers become aware of the complexity, the interconnections and the challenges of this field and that they will help to communicate the importance of renewable resources.

I sincerely thank the people of Wiley's Chichester office, especially David Hughes, Jenny Cossham and Lyn Roberts, for recognising the need for such a series of books on renewable resources, for initiating and supporting it and for helping to carry the project to the end. I also thank my family, especially my wife Hilde and children Paulien and Pieter-Jan, for their patience and for allowing me the time to work on the series when other activities seemed to be more inviting.

Christian V. Stevens, Faculty of Bioscience Engineering
Ghent University, Belgium
Series Editor 'Renewable Resources'
June 2005

# Preface

The first decade of the twenty-first century saw the emergence of biofuels as a major, international and, as it developed, complex industry. It is quite likely that the second decade will not only see the maturing of the biofuels industry but also the emergence of a biochemicals industry that will hopefully learn from the strengths and weaknesses of biofuels. Key areas in biofuels that we can learn from include the need to avoid any competition with food (with a few possible exceptions for highly valuable and necessary non-food products such as speciality pharmaceuticals), the value of wastes as feedstocks and the importance of valorising the whole crop including by-products. Second-generation biofuels, including biodiesel from food waste and bio-alcohols from sugarcane bagasse, are already available and value chains have been developed for some by-products, notably for the glycerine produced in most biodiesel production processes. Consumer concerns over the food versus (bio)fuel issue has also helped encourage the development of standards that will soon cover all biobased products, at least in Europe.

Biobased chemicals have been slower to emerge than biofuels. While the chemical industries are as dependent on petroleum as the fuel industries, there has been less political and public pressure to create alternatives to liquid petroleum fuels partly because the public does not connect chemicals to (diminishing) fossil reserves in the same way that it does for fuels. We have however seen strong activity with biosuccinic acid for example, and several companies – established and new – are showing activity in the biobased space. There is a strong view that biobased chemicals should enjoy the same government incentives as biobased fuels, and that without this their market penetration will be slow. The biopreferred program in the US is also sometimes cited as an example of how governments can help.

Fiscal incentives alongside proper standards will certainly help, but the biggest drivers will be: (1) increasing demand from end-users of chemicals for products derived from renewables and with lower environmental footprints; and (2) the availability of more efficient technologies to maximise the chemical potential of biomass. In this second edition of the successful and increasingly topical book *Introduction to Chemicals from Biomass*, we discuss the state-of-the art in technologies, products and resources, investigate the overall life-cycle and perform a techno-economic assessment of the area, including its role in future biorefineries. With the latter point in mind, we include one chapter on biobased energy production.

While we seek to ultimately supersede petroleum-based industries, we must learn from the co-production of fuels and chemicals in the highly cost-effective petroleum refineries created in the twentieth century. We should aim for similar, if not better, levels of efficiency and strive for zero-waste biorefineries in the twenty-first century. An integrated approach to future biorefineries is described in Chapter 1.

What will be the feedstocks of future biorefineries? Food-grade resources are unlikely to feature to any significant extent but food supply chain wastes, from farm to fork, are expected to become more and more important. These and other important renewable resources are discussed in Chapter 2.

How can we get chemical value from biomass? In recent years we have seen thermochemical methods gain popularity, which now complement the more established biochemical methods. These processes, alongside pretreatment technologies (necessary as biomass comes in many, and often awkward, shapes and forms), will be the workhorses of the future biorefineries; such methods are described and compared in Chapter 3.

In Chapters 4–7 we look at the chemical product types that can be made in the biorefinery. Platform molecules will be the building blocks of the future bio-economy; we can expect many future industries to be dependent on these in the same way that they are currently reliant on petrochemicals such as benzene, ethane and butadiene. Products from these platform molecules will include solvents, paints and coatings, agrochemicals, pharmaceuticals, adhesives, dyes and many others. The chemical industry is currently built on about 100,000 chemicals, over 90% of which are based on non-renewable resources. Switching a good proportion of these to biobased chemicals is an enormous but vital challenge.

About half of the chemical value of petroleum ends up in polymers and materials. Modern society is heavily dependent on these; we use plastics, fibres and composites in many industries from automobile construction to aerospace. Biomass is a natural source of some of these biomaterials, especially if we can learn to make better use of nature's largest macromolecules including cellulose. However, in other cases we need to manufacture biobased materials using small molecules obtained from biomass. Commercial success has already been demonstrated in this field from well-established polylactic acid (PLA) to the new biobased polyethylene (PE); many more materials will follow!

Of course, energy needs will continue to dominate the overall resources picture. The US and EU will place energy as their highest priority and will aim to move towards both sustainability but also independence of supply. By learning from current refineries and adding value to the biomass harvested for energy through higher-value chemicals production, we should make biorefineries more cost-effective and resilient to the highly dynamic energy situation.

In Chapter 8 we take a look at the 'big picture': how can we deliver a self-sufficient bio-economy? There are few that dispute our need to move in this direction, but making the new economy work at the same level of efficiency as the well-established petro-economy is incredibly challenging. This is a challenge we all need to share, as chemical production and use will surely continue to be at the heart of the future bio-economy.

James Clark
York
June 2014

# 1

# The Biorefinery Concept: An Integrated Approach

**James Clark[1] and Fabien Deswarte[2]**

[1] *Department of Chemistry, Green Chemistry Centre of Excellence, University of York, UK*
[2] *The Biorenewables Development Centre, The Biocentre, York Science Park, UK*

## 1.1 Sustainability for the Twenty-First Century

The greatest challenge we face in the twenty-first century is to reconcile our desires as a society to live lives based on consumption of a wider range of articles both essential (e.g. food) and luxury (e.g. mobile phones) with the fact that we live on a single planet with limited resources (to make the articles) and limited capacity to absorb our wastes (spent articles). While some will argue that we should not be limited by our own planet and instead seek to exploit extra-terrestrial resources (e.g. mining the asteroids), most of us believe it makes more sense to match our lifestyles with the planet we live on.

We can express this in the form of an equation whereby the Earth's capacity (EC) is defined as the product of world population $P$, the economic activity of an individual $C$ and a conversion factor between activity and environmental burden $B$:

$$EC = P \times C \times B.$$

Since we live in a time of growing $P$ and $C$ (through the rapid economic development of the mega-states of the East in particular), and if we assume that all the indicators of environmental stress (including climate change, full landfill sites,

*Introduction to Chemicals from Biomass*, Second Edition. Edited by James Clark and Fabien Deswarte.

pollution and global warming) are at least partly correct, then to be sustainable we must reduce *B*. There are two ways to do this:

1. dematerialisation: use less resources per person and hence produce less waste; and
2. transmaterialisation: use different materials and have a different attitude to 'waste'.

While many argue for dematerialisation, this is a dangerous route to go down as it typically requires that the developing nations listen to the developed nations and 'learn from their mistakes'. While many of our manufacturing processes in regions such as Europe and North America are becoming increasingly more efficient, we continue to treat most of our waste with contempt, focusing on disposal and an 'out of sight, out of mind' attitude. We also have to face the unavoidable truth that people in developing countries want to enjoy the same standard of living we have benefited from in the developed world; pontificating academics and politicians in the West talking about the need to reduce consumption will have little impact on the habits of the rest of the world!

Transmaterialisation, as it would apply to a sustainable society based on consumer goods, is more fundamental. It makes no assumption about limits of consumption other than the need to fit in with natural cycles such as using biomass at no more than the rate nature can produce it. Transmaterialisation also avoids clearly environmentally incompatible practices (such as using short-lifetime articles that linger unproductively in the environment for long periods of time, e.g. non-biodegradable polyolefin plastic bags) and bases our consumption pattern on the circular economy model, with spent articles becoming a resource for other manufacturing [1]. This model is essentially the same as the green chemistry concept, at least in terms of the chemical processes and products that dominate consumer goods, described in more detail in Section 1.4.

## 1.2    Renewable Resources: Nature and Availability

We need to find new ways of generating the chemicals, energy and materials as well as food that a growing world population (increasing *P*) and growing individual expectations (increasing *C*) needs, while limiting environmental damage. At the beginning of transmaterialisation is the feedstock or primary resource and this needs to be made renewable (see Figure 1.1). An ideal renewable resource is one that can be replenished over a relatively short timescale or is essentially limitless in supply. Resources such as coal, natural gas and crude oil come from carbon dioxide, 'fixed' by nature through photosynthesis many millions of years ago. They are of limited supply, cannot be replaced and are therefore non-renewable. In contrast, resources such as solar radiation, wind, tides and biomass can be considered as renewable resources, which are (if appropriately managed) in no danger of being over-exploited. However, it is important to note that while the first three resources can be used as a renewable source of energy, biomass can be used to produce not only energy but also chemicals and materials, the focus of this book.

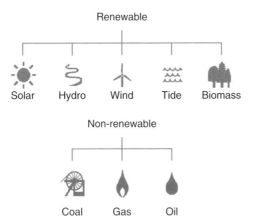

**Figure 1.1** *Different types of renewable and non-renewable resources.*

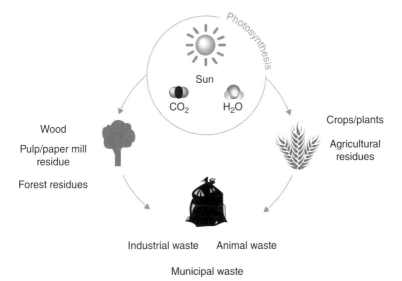

**Figure 1.2** *Different types of biomass.*

By definition, biomass corresponds to any organic matter available on a recurring basis (see Figure 1.2). The two most obvious types of biomass are wood and crops (e.g. wheat, maize and rice). Another very important type of biomass we tend to forget about is waste (e.g. food waste, manure, etc.), which is the focus of Section 1.3. These resources are generally considered to be renewable as they can be continually re-grown/regenerated. They take up carbon dioxide from the air while they are growing (through photosynthesis) and then return it to the air at the end of life, thereby creating a closed loop [2].

**Table 1.1**  *Biomass potential in the EU [5].*

| | Biomass potential (MTonnes oil equivalent) | | |
| --- | --- | --- | --- |
| | 2010 | 2020 | 2030 |
| Organic wastes | 100 | 100 | 102 |
| Energy crops | 43–46 | 76–94 | 102–142 |
| Forest products | 43 | 39–45 | 39–72 |
| Total | 186 | 215–239 | 243–316 |

Food crops can indeed be used to produce energy (e.g. biodiesel from vegetable oil), materials (e.g. polylactic acid from corn) and chemicals (e.g. polyols from wheat). However, it is becoming widely recognised by governments and scientists that waste and lignocellulosic materials (e.g. wood, straw and energy crops) provide a much better energy production opportunity than food crops since they avoid competition with the food sector and often do not require as much land and fertilisers to grow. In fact, only 3% of the 170 million tonnes of biomass produced yearly by photosynthesis is currently being cultivated, harvested and used (food and non-food applications) [3]. Indeed, according to a report published by the USDOE and the USDA [4], the US alone could sustainably supply more than one billion dry tons of biomass annually by 2030. As seen in Table 1.1, the biomass potential in Europe is also enormous.

## 1.3   The Challenge of Waste

Waste is a major global issue and is becoming more important in developing countries, as well as in the West. According to the World Bank, world cities generate about 1.3 billion tonnes (Gt) of solid waste per year, and this is expected to increase to 2.2 Gt by 2025 [6]. Globally, solid waste management costs will increase from today's $200 billion per year to about $375 billion per year in 2025. Cost increases will be most severe in low-income countries (more than five-fold increases) and lower–middle income countries (more than four-fold increases). Global governments need to put in place programmes to reduce, reuse, recycle or valorise as much waste as possible before burning it (and recovering the energy) or otherwise disposing of it.

Few countries have a constructive waste management policy whereby a significant proportion of the waste is used in some way (see Figure 1.3); reliable data are however not easily available from developing countries, other than anecdotal evidence such as from India where many people apparently make a living from waste [7]. The increasing costs of traditional fossil reserves, along with concerns over security of supply and the identification of critical raw mineral materials by the European Union (EU) is beginning to make people realise that the traditional linear economy model of extract-process-consume-dispose is unsustainable [8].

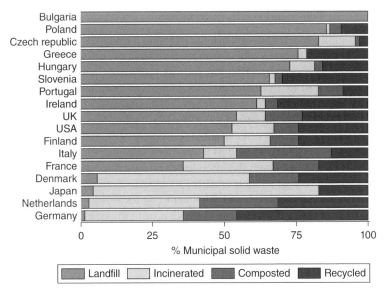

***Figure 1.3***  *The fate of waste in different countries.*

Rather, we must move towards a circular economy whereby we continue to make use of the resources in articles when they are no longer required in their current form. This is *waste valorisation.*

Waste produced in the food supply chain is a good example of a pre-consumer type of waste generated on a large scale all over the world. Sixty percent of this is organic matter, which can represent up to 50% of all the waste produced in a country. Food waste is ranked third of 15 identified resource productivity opportunities in the McKinsey report '*Resource Revolution: Meeting the World's Energy, Material, Food and Water Needs*' [9]. But there are few examples that take us away from the totally wasteful and polluting landfilling or first-generation and limited-value recycling practices such as composting and animal feed production. We need to design and apply advanced methods to process food waste residues in order to produce high-value-added products including chemicals and materials which can be used in existing and future markets. Society also needs a paradigm shift on our attitude towards waste, and this needs to be steered by governments (and trans-government agencies such as the EU) worldwide. One government-driven incentive is the increasingly expensive waste disposal costs. The EU Landfill Directive, for example, has caused landfill gate fees to increase from £40–74 to £68–111 between 2009 and 2011 [10]. Improved resource utilisation can positively influence the profits of industry as well as enable new companies to start up, produce new growth and expand innovation opportunities by moving towards the ultimate sustainability goal of a zero-waste circular economy [11, 12].

### 1.3.1    Waste Policy and Waste Valorisation

The first significant waste policies in the EU were introduced in the early 1970s. These were aimed at developing a uniform definition of 'waste' on the basis of a range of policies and laws aimed at regulating production, handling, storage and movement, as well as treatment and disposal of waste. Their objective was to reduce the negative effects of waste generation on human health and the environment [13]. Essentially, the definition of waste is 'substances or objects that the holder discards or intends or is required to discard'. Differentiating waste from by-products and residues, as well as waste from substances that have been fully recovered, are constant issues that need to be resolved if valorisation routes additional to current first-generation practices (composting, animal feed and anaerobic digestion) are to be developed. The hierarchy for waste management places priority on preventing waste arising in the first instance, consistent with the philosophy of green chemistry (the best way to deal with waste is to avoid its formation in the first place), and relegates disposal or landfilling to the worst waste management option [14]. Among the intermediate waste management options, re-use and recycling (e.g. to make chemicals) is preferred to energy recovery; this seems sensible given the greater resource consumption and pollution associated with the production of chemicals, although the value of energy continues to grow. Significantly, a new policy approach to waste management that takes account of the whole life-cycle of products was introduced in 2008, along with an emphasis on managing waste to preserve natural resources and strengthen the economic value of waste [13].

EU Member States are required to draw up waste prevention programmes that help to break the link between economic growth and waste generation, an important development on the road to zero waste. The EU guidelines identify two main approaches to food waste prevention:

1. behavioural change; and
2. sectoral-based approaches aimed at companies, households, institutions, etc.

There is a significant directive to shift biodegradable municipal waste away from landfill by imposing stringent reduction targets on EU Member States (65% by weight by 2016 against 1995 levels, with intermediate reduction targets). Food waste is considered as biodegradable waste for the purpose of the Directive. Another factor driving the diversion of biodegradable food waste from landfill towards other waste management options is the widely recognised importance of reducing greenhouse gas (GHG) emissions to the atmosphere.

Where international transportation is contemplated for the treatment of waste, including transportation to other EU Member States, trans-frontier shipment of waste rules will also need to be considered further to the Basel Convention [15]. Shipments of waste for disposal are generally prohibited, but the rules applicable to shipments of waste for recovery depend on the classification of the waste concerned and on the destination of the waste. Waste from agro-food industries is generally found on the 'green list', subject to the condition that it is not infectious, which should enable much useful food waste to be transported. A potential policy and regulatory disincentive to the reprocessing of food wastes into chemical substances is the dovetailing of end-of-waste status and chemical substances legislation, most notably through the major new REACH (Registration, Evaluation, Authorisation and restriction of CHemicals) legislation affecting chemicals manufactured or used in the EU (see Section 1.4). The testing and administrative costs of achieving a registration under REACH are considerable; cost sharing by co-registration is only partially successful as many companies are reluctant to collaborate in areas where they are competing (e.g. over the sale of the same substance). This is a major disincentive for industry and producers in the EU, especially small and medium enterprises (SMEs) producing novel substances resulting from food waste reprocessing who may find the compliance costs of REACH legislation a major barrier to commercialising the process. With other major economies outside the EU also showing interest in adopting similar legislation to REACH (including the US, where current legislation is variable from state to state, and China), manufacturing and distribution outside of the EU will have to overcome this potential barrier.

### 1.3.2    The Food Supply Chain Waste Opportunity

Alternative feedstocks to conventional fossil raw materials have attracted increasing interest over recent years for the manufacture of chemicals, fuels and materials [16]. In the case of biomass as a renewable source of carbon, feedstocks including agricultural and forestry residues are converted into valuable marketable products, ideally by using a series of sustainable and low-environmental-impact technologies, so that the resulting products are genuinely green and sustainable. The facilities where such transformations take place are often referred to as biorefineries, the focus of this book [17].

Food supply chain waste (FSCW) is emerging as a biomass resource with significant potential to be employed as a raw material for the production of fuels and chemicals, given the abundant volumes globally generated and its inherent diversity of functionalised chemical components [18].

Several motivating factors for the development of advanced valorisation practices on residues and by-products of food waste are available, such as the abundance, ready availability, under-utilisation and renewable nature of the significant quantities of functionalised molecules including carbohydrates, proteins, triglycerides, fatty acids and phenolics. Various waste streams also contain valuable compounds including antioxidants, which could be recovered, concentrated and re-used in applications such as food and lubricants additives. Examples of such types of wastes and associated 'corresponding target ingredient for recovery' have been used to highlight the potential of FSCW as a source of valuable chemical components [19].

The development of such valorisation routes may address the main weakness of the food processing industry and aim to develop a more sustainable supply chain and waste management system. Such routes can solve both resource and waste management problems. The important issues associated with agro-food waste include:

1. decreasing landfill options;
2. uncontrolled greenhouse gas emissions;
3. contamination of water supplies through leaching of inorganic matter; and
4. low efficiency of conventional waste management methods, notably incineration and composting.

Up-to-date and accurate data on the production of food waste (FW) at every stage of the food supply chain are difficult to obtain, but food waste is being mapped in Europe as part of the new COST (European Cooperation in Science and Technology) Action TD1203. There are strong drivers for stakeholders and public organisations in food processing and other sectors to reduce costs and develop suitable strategies for the conversion and valorisation of side streams. The development of knowledge-based strategies to realise the potential of food waste should also help to satisfy an increasing demand for bio-derived chemicals, fuels and materials, and probably affect waste management regulations over the years to come. The valorisation of FSCW is necessary in order to improve the sustainability and cost-effectiveness of food supply and the manufacture of chemicals. Together with the associated ethical and environmental issues and the drivers for utilising waste, the pressures for such changes are becoming huge.

### 1.3.3   Case Study: Citrus Waste

Citrus fruits are grown in many regions of the earth, including Latin and Central America and the southern USA, southern Europe, northern and southern Africa, China and India. Of the various fruit types orange is the largest in volume, representing about 95 million tonnes (Mt) annually. Major producers include Brazil, USA, China, India and southern Europe, particularly Italy and Spain. After extraction of the juice, the residual peel accounts for 50 wt% of fruit that is costly to treat and is highly regulated. However, with the high volumes of citrus production and processing, there is a real opportunity to better utilise this resource for animal feed (although it has low protein content) and essential oil extraction. Simple calculations show that the amount of organic carbon available in the peel and other residues from juicing corresponds to over 5 Mt, similar in weight to the total amount of (mostly non-renewable, typically oil-derived) carbon used by the UK for the manufacture of all of its chemicals [20].

Major components of wet orange peel are water (80% by weight), soluble sugars cellulose and hemicellulose, pectin and D-limonene. The demand for pectin (a valuable food thickener and cosmetic ingredient) and limonene (a flavour and fragrance additive for many household products and a 'green' solvent, e.g. for cleaning electronics where it replaces atmospherically harmful halogenated solvents) is increasing.

The production of chemical products from wet orange peel has had very limited commercial success: limonene and other oils are extracted and sold but only for a small proportion of the peel, while pectin is generally sourced from other fruit (apples and lemons). The current methods for the production of pectin requires a two-stage process involving the use of mineral acids, which generates large amounts of contaminated wastewaters (from neutralising the waste acid) adding to the cost of the final product, although there is a high demand for pectin for food and non-food applications.

One way to improve the economics of this process of is to employ an integrated technology that yields multiple products. Recent work has demonstrated that low-temperature microwave processing of citrus peel such as orange yields limonene and pectin as well as porous cellulose and other products in one process, thus offering the real possibility of developing a microwave biorefinery that could be employed wherever citrus waste is concentrated [21]. New uses for limonene have also been reported recently, notably as a solvent for organic chemical manufacturing processes where there is growing pressure to reduce the process environmental footprint and use more renewable compounds [22].

## 1.4   Green Chemistry

Green chemistry emerged in the 1990s as a movement dedicated to the development of more environmentally benign alternatives to hazardous and wasteful chemical processes as a result of the increased awareness in industry of the costs of waste and of government regulations requiring cleaner chemical manufacturing. Through a combination of meetings, research funding, awards for best practice and tougher legislation, the green chemistry movement gained momentum through the 1990s and into the twenty-first century. New technologies which addressed key process chemistry issues such as wasteful separations (e.g. through the use of easy-to-separate supercritical $CO_2$), atmospherically damaging volatile organic solvents (e.g. through the use of involatile ionic liquids), hazardous and difficult-to-separate process auxiliaries (e.g. by using heterogeneous reagents and catalysts) and poor energy utilisation (e.g. through alternative reactors such as microwave heating) were developed and promoted. The importance of metrics for measuring process greenness also became recognised and was championed by the pharmaceutical industry as well as by academics [23]. The pharmaceutical industry has led the way in many examples of green chemistry metrics in practice, including solvent selection guides and assessment of the environmental impacts of different processes [24].

The legislative, economic and social drivers for change impact all of the main chemical product life-cycle stages, resources, production and products.

Diminishing reserves and dramatic fluctuations in the price of oil, the most important raw material for chemicals, have been highlighted. However, the wider reality is of resource depletion of many key minerals and price increases for commodities affecting almost all chemical manufacturing as well as other important industries, notably electronics [25]. There has also been an exponential growth in product-focused legislation and non-governmental organisation (NGO) pressure, threatening the continued use of countless chemicals. The most important legislative driver is REACH [26]. This powerful legislation requires the thorough testing of all chemicals used at quantities of more than 1 tonne per year in Europe (including those manufactured outside of the EU and imported in). Persistence, bioaccumulation and toxicity (PBT) are the key assessments.

While there has been considerable debate on the impact of REACH on the European chemical industry due to the high costs of assessment and testing and the inevitable bureaucracy, the biggest results will ultimately be the identification of chemicals that require authorisation or restricted use. At the time of writing, the list of chemicals effectively black-listed or at least highlighted as being of serious concern (making their use very difficult due to NGO and consumer pressure) is growing and already causing alarm in industries whose own processes or supply chains rely on the same chemicals. Solvents are an area of great concern as they are widely used in many industries; several polar (e.g. N-methylpyrolloidone), polarisable (e.g. chlorinated aliphatics) and non-polar (e.g. hexane) solvents are likely to fall foul of REACH. In such cases a critical issue is suitable alternatives. While replacing some chemicals may not prove too difficult, in many cases there are no suitable alternatives. This certainly applies to solvents where currently used compounds may well have a complex set of desirable properties (liquid range, boiling point, polarity/polarisability, water miscibility, etc.); finding a suitable alternative that also has better PBT characteristics can be very difficult. This is an area where renewable products may prove to be very important. New solvents with a diverse range of properties (e.g. terpenes, esters, polyethers) can be tailored for some problematic processes, such as limonene as referred to earlier [22].

By using the low-environmental-impact technologies developed in the 1990s to obtain safe 'REACH-proof' chemicals from large-volume bioresources, we can take a major step towards the creation of a new generation of green and sustainable chemicals as well as tackle the escalating waste problems faced by modern society. Through the use of green chemistry techniques to obtain organic chemicals and materials from biomass and materials and metals from waste electronics and other consumer waste, we can help establish a life-cycle for many products that is sustainable on a sensible timescale within the human lifespan.

We must make better use of the primary metabolites in biomass. Cellulose, starches and chitin need to be used to make new macromolecular materials and not simply act as a source of small molecules; this can include composites and blends with synthetic polymers as we move towards a sustainable chemical industry. The small molecules that we obtain in this way need to become the building blocks of that industry: compounds such as lactic acid, succinic acid and fatty acids, glycerol and sugars, as well as ethanol and butanol are all needed to feed the industry, using green chemistry methods to convert them into replacements for the very large number of organic chemicals in current use. This includes developing synthetic pathways starting from oxygenated, hydrophilic molecules, but we must avoid wasteful and costly separations from dilute aqueous fermentation broths. A wider range of chemistry in water including more water-tolerant catalysts is needed, as are other important synthetic strategies such as the reduction in the number of process steps through telescoped reactions. The future green chemistry toolkit needs to be flexible and versatile as well as clean, safe and efficient [27, 28].

## 1.5   The Biorefinery Concept

### 1.5.1   Definition

A biorefinery is a facility or a network of facilities that converts biomass including waste (Chapter 2) into a variety of chemicals (Chapters 4 and 5), biomaterials (Chapter 6) and energy (Chapter 7), maximising the value of the biomass and minimising waste. This integrated approach is gaining increased commercial and academic attention in many parts of the world [29, 30]. As illustrated in Figure 1.4, advanced biorefineries are analogous in many ways to today's petrorefineries [31].

Similarly to oil-based refineries, where many energy and chemical products are produced from crude oil, biorefineries produce many different industrial products from biomass. These include low-value high-volume products such as transportation fuels (e.g. biodiesel, bioethanol), commodity chemicals and materials and high-value low-volume products or specialty chemicals such as cosmetics or nutraceuticals [32]. Energy is the driver for developments in this

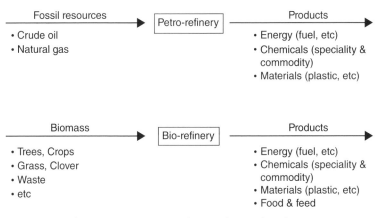

**Figure 1.4**   *Comparison of petrorefinery v. biorefinery.*

area, but as biorefineries become more and more sophisticated with time, other products will be developed. In some types of biorefinery, food and feed production may also be incorporated.

According to the Joint European Biorefinery Vision for 2030 [33], a significant proportion of the overall European demand for chemicals, energy and materials will be met using biomass as a feedstock by 2030:

- 30% of overall chemical production is expected to be bio-based in nature by this date (for high-added-value chemicals and polymers, the proportion might even be >50%);
- 25% of Europe's transport energy needs will be supplied by biofuels, with advanced fuels (and in particular bio-based jet fuels) taking an increasing share; and
- 30% of Europe's heat and power generation will be derived from biomass.

### 1.5.2    Different Types of Biorefinery

Three different types of biorefinery have been described in the literature [34, 35]:

- Phase I biorefinery (single feedstock, single process and single major product);
- Phase II biorefinery (single feedstock, multiple processes and multiple major products); and
- Phase III biorefinery (multiple feedstocks, multiple processes and multiple major products).

#### *1.5.2.1    Phase I Biorefinery*

Phase I biorefineries use only one feedstock, have fixed processing capabilities (single process) and have a single major product. They are already in operation and have proven to be economically viable. In Europe, there are now many phase I biorefineries producing biodiesel [36]. They use vegetable oil (mainly rapeseed oil in the EU) as a feedstock and produce fixed amounts of biodiesel and glycerine through a single process called transesterification (see Figure 1.5). They have almost no flexibility to recover investment and operating costs. Other examples of phase I biorefinery include today's pulp and paper mills and corn grain-to-ethanol plants.

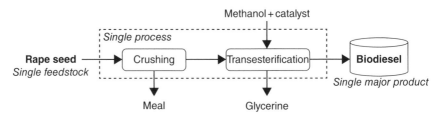

***Figure 1.5***    *The biodiesel process: an example of a phase I biorefinery.*

### 1.5.2.2 Phase II Biorefinery

Similarly to phase I biorefineries, phase II biorefineries can only process one feedstock. However, they are capable of producing various end-products (energy, chemicals and materials) and can therefore respond to market demand, prices, contract obligation and the operating limits of the plant.

Recent studies have revealed that a biorefinery integrating biofuels and chemicals offers a much higher return on investment and meets its energy and economic goals simultaneously [37]. For instance, Wageningen University performed a study in 2010 in which 12 full biofuel value chains – both single-product processes and biorefinery processes co-producing value-added products – were technically, economically and ecologically assessed. The main overall conclusion was that the production costs of the biofuels could be reduced by about 30% using the biorefinery approach [38].

One example of a phase II biorefinery is the Novamont plant in Italy, which uses corn starch to produce a range of chemical products including biodegradable polyesters (Origi-Bi) and starch-derived thermoplastics (Mater-Bi). Another example of this type of biorefinery is the Roquette site of Lestrem in France that produces a multitude of products including polyols, native and modified starches, proteins and derivatives, cyclodextrins, organic acids and resins (see Figure 1.6).

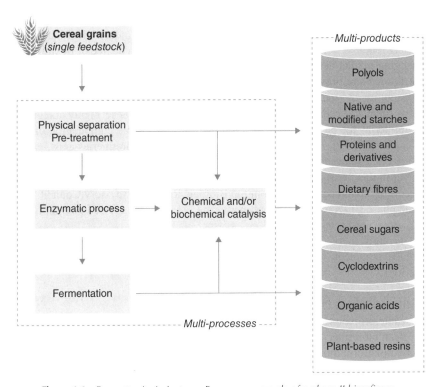

**Figure 1.6** *Roquette site in Lestrem, France: an example of a phase II biorefinery.*

Roquette produces more than 600 carbohydrate derivatives worldwide and is now leading a major programme (the BioHub™ programme) aiming to develop cereal-based biorefineries and a portfolio of cereals-based platform chemicals (e.g. isosorbide) for biopolymers as well as specialty and commodity chemicals production.

Ultimately, all phase I biorefineries could be converted into phase II biorefineries if methods of upgrading the various side streams could be identified. For example, a phase I biodiesel processing plant could be turned into a phase II biorefinery if the operator started to convert the (crude) glycerol into valuable energy and/or chemical products (see Chapter 4 for potential chemical products from glycerol). It is in fact recognised that energy or biofuel generation will probably form the initial backbone of numerous phase II biorefineries. Indeed, crude oil refining also started with the production of energy; it now employs sophisticated process chemistry and engineering to produce complex materials and chemicals that 'squeeze every ounce of value' from each barrel of oil [31].

### *1.5.2.3   Phase III Biorefinery*

Phase III biorefineries correspond to the most developed/advanced type of biorefinery. They are not only able to produce a variety of energy and chemical products (as phase II biorefineries do), but can also use various types of feedstocks and processing technologies to produce the multiplicity of industrial products our society requires. The diversity of the products gives a high degree of flexibility to changing market demands (a current by-product might become a key product in the future) and provides phase III biorefineries with various options to achieve profitability and maximise returns [33]. In addition, their multi-feedstock nature helps them to secure feedstock availability and offers these highly integrated biorefineries the possibility of selecting the most profitable combination of raw materials [39, 40]. Although no commercial phase III biorefineries exist, extensive work is currently being carried out in the EU, the US (the present leading player in this field) and elsewhere on the design and feasibility of such facilities. According to a recent report from the Biofuels Research Advisory Council, full-scale phase III (zero-waste) biorefineries are not expected to become established in Europe until around 2020 [5].

Currently, there are five phase III biorefinery systems being pursued in research and development, which will be discussed in more detail in the following sections:

1. lignocellulosic feedstock biorefinery;
2. whole-crop biorefinery;
3. green biorefinery;
4. two-platform biorefinery; and
5. marine biorefinery.

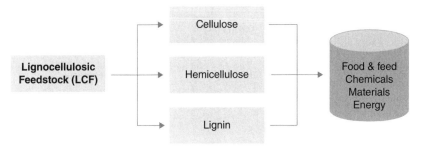

**Figure 1.7**    *Simplified diagram of a lignocellulosic feedstock biorefinery.*

#### 1.5.2.3.1    Lignocellulose Feedstock Biorefinery

A lignocellulose feedstock biorefinery will typically use lignocellulosic biomass such as wood, straw and corn stover. The lignocellulosic raw material (consisting primarily of polysaccharides and lignin) will enter the biorefinery and, through an array of processes, will be fractionated and converted into a variety of energy and chemical products (see Figure 1.7).

In the US, ZeaChem is currently developing its first commercial lignocellulose feedstock biorefinery at the Port of Morrow in Boardman, Oregon. Located adjacent to their demonstration facility, the 25 million gallons per year integrated biorefinery is expected to produce bio-based fuels, C2 chemicals (acetic acid, ethyl acetate, ethanol and ethylene) and C3 chemicals (propionic acid, propanol and propylene) from nearby woody biomass and agricultural residues using a hybrid process of biochemical and thermochemical processing.

Another example of an imminent lignocellulosic feedstock biorefinery is SP Processum in Sweden, which corresponds to an integrated cluster of industries converting wood into energy, chemicals and materials (see Figure 1.8). This is probably one of the best examples of industrial symbiosis in the world, with one industry using the waste or by-product of another as a raw material [41]. Among the member companies are AkzoNobel Surface Chemistry (production of thickeners for water-based paints and the construction industry), Domsjo Fabriker (production of dissolving pulp and paper pulp), Ovik Energy (energy production and distribution) and Sekab (production of ethanol, ethanol derivatives and ethanol as fuel).

In reality, while the sole products of existing pulp and paper manufacturing facilities today are pulp and paper (phase I biorefinery), these facilities are geared to collect and process substantial amounts of lignocellulosic biomass. They therefore provide an ideal foundation on which to develop advanced lignocellulose feedstock biorefineries. Additional processes could be built around pulp mills, either as an extension or as an 'across-the-fence'-type company.

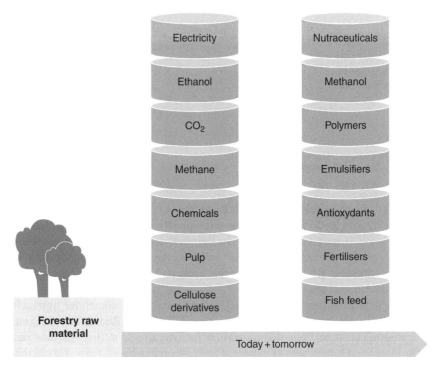

**Figure 1.8**    *Processum biorefinery in Sweden: an example of a lignocellulosic feedstock biorefinery.*

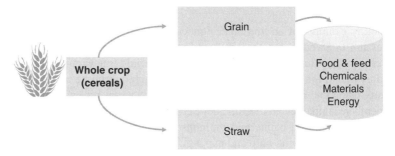

**Figure 1.9**    *Simplified diagram of a whole-crop biorefinery.*

### 1.5.2.3.2    Whole-Crop Biorefinery

A whole-crop biorefinery will employ cereals (e.g. wheat, maize, rape, etc.) and convert the entire plant (straw and grain) into energy, chemicals and materials (see Figure 1.9).

The first step will involve the separation of the seed from the straw (collection will obviously occur simultaneously to minimise energy use and labour cost). The seeds may then be processed to produce starch and a wide variety of products

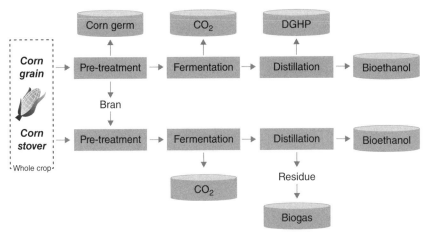

**Figure 1.10** *POET-DSM advanced biofuels biorefinery, US:an example of a whole-crop biorefinery.*

including ethanol and bioplastics, as it is currently done in phase II biorefineries. In parallel, the straw may undergo various conversion processes as described in the previous section.

POET (formerly known as Broin Companies), one of the largest producers of ethanol in the world, has built a commercial whole-crop biorefinery in Iowa which started operations in 2014. Through the Liberty project (jointly funded by POET, DSM Advanced Biofuels and the US Department of Energy), a corn grain-to-ethanol plant was converted into a commercial-scale whole-crop biorefinery designed to utilise advanced corn fractionation and lignocellulosic conversion technology to produce ethanol from corn cobs, leaves, husk and some stalk (see Figure 1.10). The facility also produces a number of valuable products including corn germ and a protein-rich dried distillers grains (Dakota Gold® HP or DGHP), which can be used as animal feed.

### 1.5.2.3.3    Green Biorefinery

Green biorefinery is another form of phase III biorefinery which has been extensively studied in the EU (especially Germany, Austria and Denmark) over the last 10–20 years [42, 43]. It takes green biomass (such as green grass, lucerne, clover, immature cereals, etc.) and converts it into useful products including energy, chemicals, materials and feed through the use of different technologies including fermentation (see Figure 1.11). Rich in water, green biomass is typically separated into a fibre-rich press cake and a nutrient-rich green juice. The green juice contains a number of useful chemicals such as amino acids, organic acids and dyes. The press cake can be used for fodder or to produce energy, insulation materials, construction panels and biocomposites, etc.

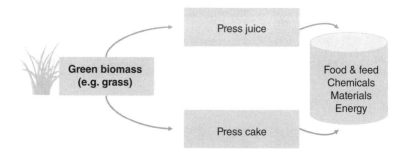

**Figure 1.11**    *Simplified diagram of a green biorefinery.*

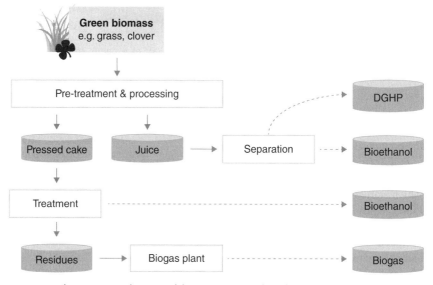

**Figure 1.12**    *Schematic of the Austrian green biorefinery. Based on [44].*

A green biorefinery demonstration plant has recently been set up in Brandenburg (Germany) and produces high-value proteins, lactic acid and fodder from 20,000 tonnes per year of alfalfa and wild mix grass. In this facility, insulating material, reinforced composites for production of plastics and biogas for heat and power are all produced from grass in an integrated process. Another example of this form of phase III biorefinery is in Austria, where the processing of green biomass from silage ensures a decentralised and seasonally independent feedstock process (see Figure 1.12). A demonstration facility based in Utzenaich currently produces a range of chemicals (e.g. lactic acid, amino acids) and fibre-derived products (e.g. animal feed, boards, insulation materials, etc.) as well as electricity and heat from grass silage.

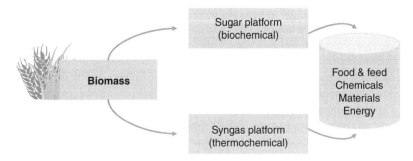

**Figure 1.13**   *Simplified diagram of a two-platform biorefinery.*

#### 1.5.2.3.4   Two-Platform Biorefinery

Another form of biorefinery recently defined by the National Renewable Energy Laboratory (NREL) is the two-platform biorefinery. As depicted in Figure 1.13, the feedstock is separated into a sugar platform (biochemical) and a syngas platform (thermochemical). Both platforms can offer energy, chemicals, materials and potentially food and feed, therefore making use of the entire feedstock(s) (see Chapter 3). The sugar platform is based on biochemical conversion processes and focuses on the fermentation of sugars extracted from biomass feedstocks. The syngas platform thermolytically transforms biomass into gaseous or liquid intermediate chemicals that can be upgraded to transportation fuels as well as commodity and speciality chemicals [45].

No biorefinery of this type currently exists in Europe but sugar conversion technologies (e.g. wheat grain-to-ethanol fermentation) and gasification approach (e.g. Linde's Carbo-V® process) are independently used. Opinions vary widely on the best strategy to combine these two platforms. However, it is most likely that, as new technologies are developed, multiple biorefinery designs will emerge commercially depending on the location of the plant and the feedstock(s) used.

Importantly, the sugar platform and many other (non-thermochemical) processes likely to be incorporated into a biorefinery will almost certainly generate some waste products that will be difficult to convert into value-added materials and chemicals. Such wastes and residues could potentially represent an important source of energy within the biorefinery, and are an ideal candidate for thermochemical conversion [46].

#### 1.5.2.3.5   Marine Biorefinery

There is considerable interest in the use of micro- and macro-algae (seaweed) as a biorefinery feedstock to produce food, feed, biofuels and chemicals. Although their potential is considerable, marine biorefineries are in their infancy compared to the other types of phase III biorefineries (based on terrestrial crops) described in the preceding sections [47].

**Table 1.2**    *Comparison of petrorefinery and biorefinery in terms of feedstock, conversion processes and products.*

|          |                      | Petro-refinery                        | Biorefinery                                      |
|----------|----------------------|---------------------------------------|--------------------------------------------------|
| Feedstock | Location            | Rich deposits in some areas           | Widely distributed                               |
|          | Density              | High                                  | Low                                              |
|          | Availability         | Continuous but finite                 | Seasonal but renewable                           |
|          | Chemical composition | Hydrocarbons; not functionalised      | Highly oxygenated and functionalised             |
| Conversion processes | | Optimised over 100 years              | Require further research and technological development |
| Products |                      | On the market and to high specification | Quality needs to be standardised               |

### 1.5.3    Challenges and Opportunities

Biorefinery products (energy, chemicals and materials) will unavoidably have to compete with existing and future petroleum-derived products. As seen in Table 1.2 (comparison of biorefinery and petrorefinery characteristics in terms of feedstock, process and products), the two types of refinery display major differences, which translate into a number of challenges to and opportunities for the rapid and widespread deployment of biorefineries.

#### 1.5.3.1    Feedstock

Biorefinery feedstocks mainly consist of existing arable crops (e.g. cereals, oilseeds), dedicated biomass crops (e.g. perennial lignocellulosic crops), biowaste (e.g. agricultural and forestry residues, food and municipal wastes) and algae.

In contrast to fossil resources which are found in rich deposits ('mine mouth' or 'well head'), biomass is widely distributed geographically (multiplicity of 'farm gate' or 'waste sources') [41, 48]. As such, the economic viability of biorefineries largely depends on the availability of a reliable supply of appropriate quality biomass at fair prices [49].

Biomass is clearly not an unlimited resource. Consequently, it is critical to ensure that additional uses of biomass do not compromise the ability to produce food and feed in sufficient quality and quantity [49]. Due to limited land availability for biomass production for both food and non-food applications, the following requirements are now widely recognised.

• We must increase the productivity and output of biomass from forest and agricultural land through the development of high-yielding arable and perennial crop varieties with greater value for industrial processing (e.g. higher oil or starch content, more digestible lignocellulose) and increased tolerance to pests and diseases under changing climatic conditions [50].

- We must unlock the potential of greatly untapped resources such as waste and algae [51]. In the EU alone, unused biowaste (e.g. agriculture and forestry residues, waste water treatment sludge, organic household waste, food processing waste, debarking waste) amounts to a total of 2.8 billions tonnes per year [49].

In addition, biomass typically exhibits a low bulk density and a relatively high water content (up to 90% for grass), which makes its transport in its raw state much more expensive than the transfer of natural gas or petroleum. Reducing the cost of collection, transportation and storage through pre-processing biomass into a higher-density, aerobically stable, easily transportable material is therefore critical to developing a sustainable infrastructure capable of working with significant quantities of raw material [52].

The most common approach used to increase biomass density is grinding. By chopping bailed straw, for example, a 10-fold densification can typically be achieved. An alternative strategy that can provide a material of even higher density is pelletisation. Through conversion of ground straw into pellets, the density of the material could be further increased by a factor of three [53]. This pre-treatment also provides the added benefit of providing a much more uniform material (in size, shape, moisture, density and energy content) which can be much more easily handled (see Chapter 3). Pre-processing might be performed onsite but can also be done during harvesting. An example of technology recently developed to address the engineering challenge presented by low-bulk-density biomass such as wheat straw is a multi-component harvester that can simultaneously and selectively harvest wheat grain and the desired parts of wheat straw in a single pass [52].

Another issue associated with the use of (fresh) biomass is its perishable character or susceptibility to degradation. Taking straw as an example once more, fermentation will begin if the moisture content of baled straw is kept above 25% for a prolonged period of time, resulting in a dramatic reduction of the quality of the raw material. In some cases, spontaneous combustion in the stacks can even take place [54]. This issue is particularly important given that, in contrast to fossil resources (which are of permanent availability and are continuously pumped and mined), the availability of biomass is seasonal. In order to ensure a continuous year-round operation of the biorefinery, biomass may have to be stabilised (e.g. dried) prior to (long-term) storage. For example, the Austrian green biorefinery tackles this problem by processing not only direct-cut grasses but also silage, which can be prepared in the growing season and stored in a silo [55, 56].

In summary, it is essential that we develop versatile and sustainable biomass supply chains and cost-effective infrastructures for production, collection, storage and pre-treatment of biomass. As highlighted by Nilsson and Kadam, the economic success of large biorefineries will greatly depend upon the fundamental logistics of a consistent and orderly flow of feedstocks [54, 57]. Localised small-scale (and perhaps mobile) pre-treatment units will be necessary to minimise transportation costs and supply the biorefinery with a stabilised feedstock (e.g. in the form of a dry solid or a liquid such as pyrolysis oil) which can be stored, allowing the biorefinery to run continuously all year long [58]. Such an approach will yield the added benefits of reducing the environmental impact of transportation [55], allowing farmers to gain a greater share of the total added value of the supply chain.

### *1.5.3.2　Conversion Processes*

The major impediment to biomass use is the development of economically viable methods (physical, chemical, thermochemical and biochemical) to separate, refine and transform it into energy, chemicals and materials [29, 59]. Biorefining technologies (most of which require further research and technological development) have to compete with well-developed and very efficient processes which have been continuously improved by petrorefineries over the last 100 years; the latter demonstrate a very high degree of technical and cost optimisation. Large investments will therefore be required from the public and private sector to bring these technologies to maturity through research, development, demonstration and deployment [60, 61]. The EU and a consortium of bio-based industries have recently committed to jointly invest over €2.8 billion in research and innovation between 2014 and 2020 [62].

Priorities for research and development include the following topics.

- *Pre-processing*: there is currently no effective way to separate the major components of biomass (i.e. cellulose, hemicellulose and lignin). More sophisticated and milder pre-treatment methods therefore need to be developed [63].
- *Chemical catalysis and biochemical processes*: in contrast to fossil resources, biomass feedstocks are composed of highly oxygenated and/or highly functionalised chemicals (see Table 1.3). This means that we must apply significantly different chemistries (e.g. reduction instead of oxidation) to convert them into the valuable chemical products our society is built on and, in particular, develop new catalysts that are able to work in aqueous media [46, 64]. These include new biocatalysts (microorganisms and enzymes) being developed through the novel field of synthetic biology.
- *Thermochemical processes*: research should focus on scaling-up and integrating these processes into existing production units as well as end-product quality improvement [50].

**Table 1.3**　*General chemical composition of selected biomass components and petroleum. Reproduced with permission from [65]. Copyright © 2008, John Wiley & Sons, Ltd.*

| Cellulose/starch | $[C_6(H_2O)_5]_n$ | Gasoline | $C_6H_{14}$–$C_{12}H_{26}$ |
| Hemicellulose | $[C_5(H_2O)_4]_n$ | Diesel | $C_{10}H_{22}$–$C_{15}H_{32}$ |
| Lignin | $[C_{10}H_{12}O_4]_n$ | | |

- *Downstream processes*: biorefineries need to develop intelligent process engineering to deal with separation and separation, by far the most wasteful and expensive stage of biomass conversion and currently accounting for 60–80% of the process cost of most mature chemical processes [66]. For example, the production of chemicals (e.g. succinic acid) and fuels (e.g. bioethanol) through fermentation processes generates very dilute and complex aqueous solutions which will have to be dealt with using clean and low-energy techniques [64].

For biorefineries to flourish, enhanced cooperation between academia and industry needs to be supported and new unconventional partnerships between traditionally separate industry sectors need to be developed (e.g. agri-food businesses and chemical companies). A number of regional bioeconomy clusters bringing together industry (large and small companies), academia, investors and policy-makers have emerged across the world over the last decade to encourage these necessary collaborations (e.g. IAR in France, BioVale in UK).

Last but not least, all the processes employed in future biorefineries will have to be environmentally friendly. It is essential that we use clean technologies and apply green chemistry principles throughout the biorefinery in order to minimise the environmental footprints of its products and ensure its sustainability, as discussed in Section 1.4 and described in more detail in Chapter 3. Future biorefineries will have to be highly energy-efficient and make use of mostly zero-waste production processes [33].

### 1.5.3.3   Products

There are currently a number of factors driving the development and commercialisation of bio-based products. These include high oil prices, consumer preference for and corporate commitment to sustainable products, and government mandates and support for the bio-economy [38].

However, a competitive price and an equal or superior level of performance compared to their fossil-based counterparts are central to the viability of bio-based products. Although some bioplastics are cost-competitive, most are 2–4 times more expensive than conventional plastics, limiting their uptake to date [49]. Importantly, biorefineries should not limit themselves to producing existing products but instead should aim to develop new families of products, taking full advantage of the native properties of biomass and its components [29].

A major issue for biomass as a raw material for industrial product manufacture is variability. Standards and certifications therefore need to be established as new biofuels, biomaterials and bioproducts are introduced to the market to assure end-users of a bio-based product's quality, its performance and its bio-based content (see Chapter 8).

A secure and long-term policy and regulatory framework is also needed to provide certainty for companies and investors seeking to exploit biomass/biowaste as a feedstock. As with many emerging industries, new incentives will be required to stimulate market uptake and attract the necessary private investment required in the development of new bio-based products and the large-scale deployment of integrated biorefineries [36, 37]. For example, the US Federal Government set up the BioPreferred programme in 2002 to increase the purchasing of bio-based products [49].

One of the main drivers for the use of bio-based products (energy, chemicals and materials) is their potential environmental benefits compared to petroleum-derived products (e.g. carbon dioxide emission reduction, biodegradability). It is therefore essential that we assess the environmental impact of all the energy and chemical products manufactured by biorefineries (across their life cycle) to ensure that they are truly sustainable and represent real (environmental and societal) advantages compared to their petroleum-derived analogues [67]. In particular, the impacts of direct and indirect land-use change and biomass production on regional biodiversity need to be evaluated as part of this assessment [50]. Work is currently underway through a number of initiatives to reach agreement at an international level on sustainability principles, criteria and indicators [36].

### 1.5.4   Biorefinery Size

Biorefineries are emerging around the world in a variety of different forms and sizes and encompass a combination of large-scale facilities (which can take full advantage of the economies of scale, enjoying greater buying power when acquiring feedstocks) and small-scale plants (which can keep transport costs to an absolute minimum and take full advantage of available process integration technologies). Their optimal size, which will obviously depend upon the nature of the feedstock(s) processed, the location of the plant, the technologies employed and the demand for given products (not 'one size fits all'), will correspond to a balance between the increasing cost of transporting pre-treated biomass and the decreasing cost of processing as the size of the biorefinery increases [33, 36].

Many of these integrated biorefineries are expected to be located in rural or densely forested areas in close proximity to the biomass (e.g. POET in US, Processum in Sweden), while some are likely to emerge in large ports and refinery complexes (e.g. Nestle Oil in Rotterdam, ENI in Venice), making the most of existing infrastructures and easy access to key markets and customers.

## 1.6   Conclusions

Current industrial economies are largely dependent on oil, which provides the basis of most of our energy and chemical feedstocks; in fact, over 90% (by weight) of all organic chemicals are derived from petroleum [29]. However, crude oil reserves are finite and world demand is growing. In the meantime, there is increasing concern over the impact of these traditional manufacturing processes on the environment (i.e. effect of $CO_2$ emissions on climate change). In order to maintain the world population in terms of food, fuel and organic chemicals and tackle climate change, it has been recognised by a number of governments and companies that we need to substantially reduce our dependence on petroleum feedstock by establishing a bio-based economy [68].

For this purpose, long-term strategies that recognise the potential of local renewable resources including waste should be developed. Of paramount importance will be the deployment of biorefineries (of various size and shape) which can convert a variety of biofeedstocks into power, heat, chemicals and other valuable materials, maximising the value of the biomass and minimising waste. These integrated facilities will most likely employ a combination of physical, chemical, biotechnological and thermochemical technologies; they must be efficient and adopt the green chemistry principles in order to minimise environmental footprints and ensure the sustainability of all products generated (cradle-to-grave approach). Local pre-treatment of low-bulk-density and often wet biomass will be critical to the development of a sustainable infrastructure capable of working with significant quantities of raw material. Specific attention should therefore be paid to the development of these (local) processes. The challenge of the next decade will be to develop demonstration plants, which will require cross-sector collaborations and major investment in the construction of full-scale advanced biorefineries.

## 1.7   Acknowledgement

The authors would like to thank Juliet Burns for the illustrations included in this chapter.

## References

1. WRAP (2014) What is the Circular Economy? Available at http://www.wrap.org.uk/content/wrap-and-circular-economy (accessed 25 August 2014).
2. Deswarte, F.E.I. (2008) Can biomass save the planet? *Chemistry Review*, **17**, 17–20.
3. Sanders, J., Scott, E. and Mooibroek, H. (2005) Biorefinery, the bridge between agriculture and chemistry. 14th European Biomass Conference, Paris.
4. USDOE & USDA (2005) Biomass as feedstock for a bioenergy and bioproducts industry: the technical feasibility of a billion-ton annual supply. Available at http://www1.eere.energy.gov/biomass/pdfs/final_billionton_vision_report2.pdf (accessed 25 August 2014).
5. European Commission (2006) Biofuels in the European Union, a vision for 2030 and beyond. Available at http://ec.europa.eu/research/energy/pdf/biofuels_vision_2030_en.pdf (accessed 25 August 2014).
6. Hoornweg, D. and Bhada-Tata, P. (2012) *What a Waste: A Global Review of Solid Waste Management. Urban Development Series; knowledge Papers No. 15*, The World Bank, Washington DC. Available at http://documents.worldbank.org/curated/en/2012/03/16537275/waste-global-review-solid-waste-management (accessed 25 August 2014).

7. Dodson, J.R., Hunt, A.J. and Parker, H.L. (2012) Elemental sustainability: towards the total recovery of scarce metals. *Chemical Engineering and Processing*, **51**, 69–78.

8. Europa (2010) Report Lists of 14 Critical Mineral Raw Materials. Available at http://europa. eu/rapid/press-release_MEMO-10-263_en.htm (accessed 25 August 2014).

9. Dobbs, R., Oppenheim, J. and Thompson, F. (2011) Resource Revolution: Meeting the world's energy, materials, food and water. McKinsey & Company. Available at http://www. mckinsey.com/features/resource_revolution (accessed 25 August 2014).

10. WRAP (2011) Gate Fee Report. Available at http://www.wrap.org.uk/recycling_industry/ publications/wrap_gate_fees.html (accessed 25 August 2014).

11. Bio Intelligence Service (2010) Preparatory Study on Food Waste Across EU-27 for the European Commission. Available at http://ec.europa.eu/environment/eussd/pdf/bio_ foodwaste_report.pdf (accessed 25 August 2014).

12. Ellen MacArthur Foundation (2014) Circular Economy. Available at http://www.ellenmac arthurfoundation.org/circular-economy (accessed 25 August 2014).

13. EC (2008) Directive 2008/98/EC of the European Parliament and of the council of 19 November 2008 on waste and repealing certain directives. Available at http://eur-lex.europa. eu/LexUriServ/LexUriServ.do?uri=OJ:L:2008:312:0003:0030:en:PDF    (accessed    4 September 2014).

14. EEC (1975) Directive 775/442/EEC on waste. Available at http://eur-lex.europa.eu/legal-content/EN/TXT/?uri=CELEX:31975L0442 (accessed 4 September 2014).

15. EC (2006) Regulation (EC) No. 1013/2006 of the European Parliament and the Council of 14 June 2006 on shipments of waste. Available at http://eur-lex.europa.eu/legal-content/EN/ TXT/?uri=CELEX:32006R1013 (accessed 4 Sepetmber 2014).

16. Clark, J.H. and Deswarte, F.E.I. (eds) (2008) *Introduction to Chemicals from Biomass*, John Wiley & Sons, Ltd, Chichester.

17. Luque, R., Herrero-Davila, L., Campelo, J.M. *et al.* (2008) Biofuels: a technological perspective. *Energy and Environmental Science*, **1**, 542–564.

18. Lin, S.K.C., Pfaltzgraff, L.A., Herrero-Davila, L. *et al.* (2013) Food waste as a valuable resource for the production of chemicals, materials and fuels. Current situation and global perspective. *Energy and Environmental Science*, **6**, 426–464.

19. Galanakis, C.M. (2012) Recovery of high added-value components from food wastes: conventional, emerging technologies and commercialized applications. *Trends in Food Science and Technology*, **26**, 68–87.

20. Pfaltzgraff, L.A., De Bruyn, M., Cooper, E.C. *et al.* (2013) Food waste biomass: a resource for high-value chemicals. *Green Chemistry*, **15**, 307–314.

21. Balu, A.M., Budarin, V.L., Shuttleworth, P.S. *et al.* (2012) Valorisation of orange peel residues: waste to biochemicals and nanoporous materials. *ChemSusChem*, **5**, 1694–1697.

22. Clark, J.H., Macquarrie, D.J. and Sherwood, J. (2012) A quantitative comparison between conventional and bio-derived solvents from citrus waste in esterification and amidation kinetic studies. *Green Chemistry*, **14**, 90–93.

23. Lapkin, A. and Constable, D. (2008) *Green Chemistry Metrics*, John Wiley & Sons, Ltd, Chichester.

24. Dunn, P.J., Wells, A. and Williams, M.T. (eds) (2010) *Green Chemistry in the Pharmaceutical Industry*, Wiley-VCH, Weinheim.

25. Hunt, A.J. (2013) Elemental sustainability and the importance of scarce element recovery, in *Element Recovery and Sustainability* (eds A.J. Hunt, T.J. Farmer and J.H. Clark), RSC Green Chemistry Book Series, RSC, Cambridge, pp. 1–28.

26. EC (2006) Regulation No. 1907/2006 of the European Parliament and of the Council of 18 December. Concerning the Registration, Evaluation, Authorisation and Restriction of Chemicals (REACH), Establishing a European Chemicals Agency, Amending Directive 1999/45/EC and Repealing Council Regulation (EEC) No. 793/93 and Commission Regulation (EC) No. 1488/94 as well as Council Directive 76/769/EEC and Commission Directives 91/155/EEC, 93/67/EEC, 93/105/EC and 2000/21/EC. Available at http://europa. eu/legislation_summaries/internal_market/single_market_for_goods/chemical_products/ l21282_en.htm (accessed 4 September 2014).

27. Clark, J.H. (2009) Chemistry goes green. *Nature Chemistry*, **1**, 12–13.

28. Clark, J.H., Kraus, G., Stankiewicz, A. *et al.* (eds) (2013) *Natural Product Extraction: Principles and Applications*, RSC Green Chemistry Book Series, RSC, Cambridge.

29. Bozell, J.J. (2008) Feedstocks for the future – biorefinery production of chemicals from renewable carbon. *CLEAN – Soil, Air, Water. Special Issue: Feedstocks for the Future: Renewables in Green Chemistry*, **36**, 641–647.

30. Halasz, L., Povoden, G. and Narodoslawsky, M. (2005) Sustainable process synthesis for renewable resources. *Resources, Conservation and Recycling*, **44**, 293–307.

31. Realff, M.J. and Abbas, C. (2004) Industrial symbiosis – refining the biorefinery. *Journal of Industrial Ecology*, **7**, 5–9.

32. Cherubini, F. (2010) The biorefinery concept: using biomass instead of oil for producing energy and chemicals. *Energy Conversion and Management*, **51**, 1412–1421.

33. Star-COLIBRI (2011) Joint European Biorefinery Vision for 2030. Available at http://www. star-colibri.eu/files/files/vision-web.pdf (accessed 25 August 2014).

34. Kamm, B. and Kamm, M. (2004) Principles of biorefineries. *Applied Microbiology and Biotechnology*, **64**, 137–145.

35. Fernando, S., Adhikari, S., Chandrapal, C. and Murali, N. (2006) Biorefineries: current status, challenges, and future direction. *Energy & Fuels*, **20**, 1727–1737.

36. Wellisch, G., Jungmeier, A., Karbowski, M.K.P. and Rogulska, M. (2010) Biorefinery systems – potential contributors to sustainable innovation. *Biofuels, Bioproducts and Biorefining*, **4**, 275–286.

37. Bozell, J.J. and Petersen, G.R. (2010) Technology development for the production of biobased products from biorefinery carbohydrates—the US Department of Energy's "Top 10" revisited. *Green Chemistry*, **12**, 539–554.

38. IEA Bioenergy (2012) Task 42 Biorefinery: Biobased Chemicals – value added products from biorefineries. Available at http://www.ieabioenergy.com/wp-content/uploads/2013/10/ Task-42-Biobased-Chemicals-value-added-products-from-biorefineries.pdf (accessed 25 Aug 2014).

39. Maung, T.A., Gustafson, C.R., Saxowsky, D.M. *et al.* (2013) The logistics of supplying single vs. multi-crop cellulosic feedstocks to a biorefinery in southeast North Dakota. *Applied Energy*, **109**, 229–238.

40. de Jong, E., van Ree, R., van Tuil, R. and Elbersen, W. (2006) Biorefineries for the chemical industry – a Dutch point of view, in *Biorefineries – Biobased Industrial Processes and Products. Status Quo and Future Directions* (eds B. Kamm, M. Kamm and P. Gruber), Wiley-VCH, Weinheim.

41. Gravitis, J. (2007) Zero techniques and systems – ZETS strength and weakness. *Journal of Cleaner Production*, **15**, 1190–1197.

42. Ecker, J., Schaffenberger, M., Koschuh, W. *et al.* (2012) Green biorefinery upper Austria – pilot plant operation. *Separation and Purification Technology*, **96**, 237–247.

43. Kamm, B., Hille, C., Schonicke, P. and Dautzenberg, G. (2010) Green biorefinery demonstration plant in Havelland (Germany). *Biofuels, Bioproducts and Biorefining*, **4**, 253–262.

44. Schnitzer, H. (2006) Agro-based Zero Emissions Systems. Environmentally Degradable Polymers from Renewable Resources Workshop, Bangkok. Available at http://www.unido.org/fileadmin/import/58371_Presentation_Keynote1_Hans_Schnitzer.pdf (accessed 4 September 2014).

45. Wright, M. and Brown, R.C. (2007) Comparative economics of biorefineries based on the biochemical and thermochemical platforms. *Biofuels, Bioproducts and Biorefining*, **1**, 49–56.

46. Ragauskas, A.J., Williams, C.K., Davison, B.H. *et al.* (2006) The path forward for biofuels and biomaterials. *Science*, **311**, 484–489.

47. Taylor, G. (2008) Biofuels and the biorefinery concept. *Energy Policy*, **36**, 4406–4409.

48. Wright, M. and Brown, R.C. (2007) Establishing the optimal sizes of different kinds of biorefineries. *Biofuels, Bioproducts and Biorefining*, **1**, 191–200.

49. BSI (2013) Bio-based products: Guide to standards and claims. Bio-based and Renewable Industries for Development and Growth in Europe, Strategic Innovation and Research Agenda. Available at http://shop.bsigroup.com/en/ProductDetail/?pid=000000000030262005 (accessed 4 September 2014).

50. Star-COLIBRI (2011) European Biorefinery Joint Strategic Research Roadmap. Available at http://beaconwales.org/uploads/resources/Vision_2020_-_European_Biorefinery_Joint_Strategic_Research_Roadmap.pdf (accessed 25 August 2014).

51. F. Fava, G. Totaro, L. Diels, *et al.* (2013) Biowaste biorefinery in Europe: opportunities and research and development needs. *New Biotechnology*, doi: 10.1016/j.nbt.2013.11.003.

52. Hess, J.R., Thompson, D.N., Hoskinson, R.L. *et al.* (2003) Physical separation of straw stem components to reduce silica. *Applied Biochemistry and Biotechnology*, **105–108**, 43–51.

53. Deswarte, F.E.I., Clark, J.H., Wilson, A.J. *et al.* (2007) Toward an integrated straw-based biorefinery. *Biofuels, Bioproducts and Biorefining*, **1**, 245–254.

54. Kadam, K.L., Forrest, L.H. and Jacobson, W.A. (2000) Rice straw as a lignocellulosic resource: collection, processing, transportation, and environmental aspects. *Biomass and Bioenergy*, **18**, 369–389.

55. Koschuh, W., Thang, V.H., Krasteva, S. *et al.* (2005) Flux and retention behaviour of nanofiltration and fine ultrafiltration membranes in filtrating juice from a green biorefinery: a membrane screening. *Journal of Membrane Science*, **261**, 121–128.

56. Thang, V.H. and Novalin, S. (2007) Green biorefinery: separation of lactic acid from grass silage juice by chromatography using neutral polymeric resin. *Bioresource Technology*, **99**, 4368–4379.

57. Nilsson, D. (1999) SHAM: a simulation model for designing straw fuel delivery systems. Part 2: model applications. *Biomass and Bioenergy*, **16**, 39–50.

58. Bruins, M.E. and Sanders, J.P.M. (2012) Small-scale processing of biomass for biorefinery. *Biofuels, Bioproducts and Biorefining*, **6**, 135–145.

59. Cherubini, F. and Strømman, A.H. (2011) Chemicals from lignocellulosic biomass: opportunities, perspectives, and potential of biorefinery systems. *Biofuels, Bioproducts and Biorefining*, **5**, 548–561.

60. Peck, M.P., Bennett, S.J., Bissett-Amess, R. *et al.* (2009) Examining understanding, acceptance, and support for the biorefinery concept among EU policy-makers. *Biofuels, Bioproducts and Biorefining*, **3**, 361–383.

61. Van Dael, M., Márquez, N. and Reumerman, P., *et al.* (2013) Development and techno-economic evaluation of a biorefinery based on biomass (waste) streams – case study in the Netherlands. *Biofuels, Bioproducts and Biorefining*, doi: 10.1002/bbb.1460.

62. European Public-Private Partnership on Biobased Industries (2012) Accelerating Innovation and Market Uptake of Biobased Products. Available at http://biconsortium.eu/sites/default/files/downloads/BIC_BBI_Vision_web.pdf (accessed 4 September 2014).

63. Gomez, L.D., Steele-King, C.G. and McQueen-Mason, S.J. (2008) Sustainable liquid biofuels from biomass: the writing's on the walls. *New Phytologist*, **178**, 473–485.

64. Clark, J.H. (2007) Green chemistry for the second generation biorefinery – sustainable chemical manufacturing based on biomass. *Journal of Chemical Technology and Biotechnology*, **82**, 603–609.

65. Pu, Y., Zhang, D., Singh, P.M. and Ragauskas, A.J. (2008) The new forestry biofuels sector. *Biofuels, Bioproducts and Biorefining*, **2**, 58–73.

66. Abels, C., Carstensen, F. and Wessling, M. (2013) Membrane processes in biorefinery applications. *Journal of Membrane Science*, **444**, 285–317.

67. Gallezot, P. (2007) Process options for converting renewable feedstocks to bioproducts. *Green Chemistry*, **9**, 295–302.

68. van Dam, J.E.G., de Klerk-Engels, B., Struik, P.C. and Rabbinge, R. (2005) Securing renewable resource supplies for changing market demands in a bio-based economy. *Industrial Crops and Products*, **21**, 129–144.

# 2
# Biomass as a Feedstock

Thomas M. Attard[1], Andrew J. Hunt[1], Avtar S. Matharu[1], Joseph A. Houghton[1] and Igor Polikarpov[2]

[1]*Department of Chemistry, Green Chemistry Centre of Excellence, University of York, UK*
[2]*Grupo de Biotecnologia Molecular, Instituto de Física de São Carlos, Universidade de São Paulo, Brazil*

## 2.1 Introduction

Since the 1990s, decreasing fossil reserves, rising oil prices, concerns over security of supply and environmental impacts have led to a global policy shift back towards the use of biomass as a local, renewable and low-carbon feedstock. The biorefinery concept is analogous to today's petroleum refineries, which convert biomass into multiple value-added products including energy, chemical and materials in an integrated facility [1]. However, many first-generation biorefineries utilise feedstocks that compete with food or feed [2]. The use of lignocellulosic agricultural residues (that do not compete with food) and food supply chain (FSC) wastes ('from farm to fork') can aid in the creation of flexible zero-waste networks, applicable to a variety of low-value local feedstocks. The utilisation of such wastes can lead to the development of novel interconnecting webs of products that can meet the demands of existing and new industries.

*Introduction to Chemicals from Biomass*, Second Edition. Edited by James Clark and Fabien Deswarte.
© 2015 John Wiley & Sons, Ltd. Published 2015 by John Wiley & Sons, Ltd.

## 2.2    Lignocellulosic Biomass

There are a large variety of lignocellulosic materials containing varying compositions of lignin, cellulose and hemicellulose that are ideal feedstocks for a biorefinery. These materials can also provide a complex mixture of phytochemicals which may find use in high-value applications. In order to maximise the potential of such feedstocks, a holistic approach with the use of innovative clean technologies should be used to obtain the maximum value and number of products from these renewable resources. Two agricultural residues which offer great potential as feedstocks for biorefineries are sugarcane bagasse and wheat straw. Here we will focus on how phytochemicals (waxes) from agricultural residues, such as sugarcane bagasse and wheat straw, can be utilised in higher-value applications.

Sugarcane (*Saccharumofficinarum*) is a tall, perennial, $C_4$ grass belonging to the Poaceae family [3]. The plant originated from south and southeast Asia, but is now widespread throughout various tropical and sub-tropical countries [3, 4]. It is approximately 5 cm thick and generally grows to a height of 2–6 m. The crop is cultivated in around 200 countries and a number of different horticultural varieties exist, differing in length and stem colour [3]. According to the world crop statistics that have been gathered by the Food and Agriculture Organization (FAO), the global harvest of sugarcane has drastically changed over the past six decades from 260 million tonnes in 1950, to 770 million tonnes in 1980 to 1,525 million tonnes in 2007. There has been nearly a six-fold increase in the global sugarcane harvest from 1950 to 2007 [4]. The top three countries (Brazil, India and China) contributed around 67% of the total sugarcane production in 2010; Brazil alone contributed approximately 43%. The global sugarcane production and expansion is therefore dominated by Brazil. From 2000 to 2007, 75% of sugarcane area increases was solely attributed to Brazil [4].

The sugarcane is processed in conventional sugar mills, which involves crushing the sugarcane to extract the juices followed by heating of the juices to form syrup and crystallising out the sugar from the syrup [4]. A fibrous residue remains after conventional milling, which is referred to as sugarcane bagasse [5]. This is a lignocellulosic material which is collected in large amounts following sugarcane processing. In a typical sugar mill, the processing of 1 metric tonne of sugarcane yields around 270 kg of bagasse (with 50% moisture), which produces approximately 135 kg of dry matter [6, 7].

Sugar and ethanol plants normally use around 50% of this dry matter to generate heat and power. The rest is normally stockpiled by sugar mills, which poses an environmental problem to both the sugar mills and surrounding districts since stockpiling for long periods of time could increase the risk of spontaneous combustion [8]. There have been various reports which highlight the use of this sugarcane bagasse in a variety of applications ranging from animal feed to the production of various industrial enzymes (cellulases, lipases, etc.), chemicals, pulp and paper [8]. As such, sugarcane bagasse is an ideal feedstock for utilisation as part of an integrated holistic biorefinery. Extraction of valuable phytochemicals including waxes, prior to biomass destruction during biochemical and thermal treatments, can significantly increase the overall financial returns.

In order to maintain a stable wax market, there is an ever-growing demand for synthetic waxes due to the decline in petroleum waxes. This demand has led to the continual search for natural resources that are readily available and found in sufficient amounts. This, together with the use of environmentally benign technology, will allow for sustainable development [9, 10].

When adhering to the strict chemical definition, the term 'waxes' refers to the ester products formed from the esterification of long-chain fatty acids with long-chain primary alcohols [11]. However, the term 'plant wax' is often used collectively to describe the complex mixture of surface lipids covering the aerial tissues of herbaceous plants [12].

Early morphological work, reviewed by Martin, Juniper and Baker, led to the first description of the terminology defining plant wax in which the 'cuticular membrane' was used to describe the entire waxy coating which lined the outer surface of the plant epidermal cells [13, 14]. It comprises three distinct regions: the exterior epicuticular wax; the cuticle proper; and the interior intracuticular wax [12]. The cuticle proper is made up of a biopolyester consisting of hydroxy and epoxy fatty acids, collectively known as cutin, which covers the epidermal cells forming an electrodense layer [12, 15, 16]. The intracuticular wax, which is embedded in the cutin, contains amorphous mixtures of lipids that link the cuticle to the cell wall matrix [12, 16]. The epicuticular wax refers to the complex mixture of surface lipids, comprising cyclic and aliphatic long-chain molecules, which cover the cuticle proper forming a smooth film exterior or crystalloids [11, 12]. It consists of a large variety of chemicals which may be subdivided on the basis of their structure, functional group type and their homologue distribution [11].

Plant wax consists of both cyclic and aliphatic long-chain components. The most common long-chain aliphatic compounds include hydrocarbons, primary alcohols, aldehydes, fatty acids and wax esters. Long-chain aliphatic compounds that are less common include ketones, β-diketones and secondary alcohols [11]. Cyclic compounds that are found in the plant cuticular wax include sterols, flavonoids and terpenoids. These are summarised in Tables 2.1 and 2.2 [12].

In the majority of plant waxes, the hydrocarbon fraction consists of a number of *n*-alkanes varying in chain length from $C_{25}$ to $C_{35}$. The predominant *n*-alkanes are those which contain an odd number of carbon atoms. In most cases, the predominant chain length (usually 90% or more of the total hydrocarbon fraction) is $C_{29}$ or $C_{31}$ [17]. This is a generalisation however, and there are a number of exceptions in relation to the number of carbon atoms and the predominant chain length. There have been some studies, in which alkanes having chain lengths less than $C_{25}$ have been reported. In addition, in the majority of algae the predominant chain length is $C_{17}$. There have also been reports in which alkanes of chain length $C_{62}$ have been identified in cane grass wax [17].

**Table 2.1** Various types of long-chain aliphatic compounds found in plant waxes.

| Aliphatic compounds | Structure | Chain length | Preferred no. C atoms |
|---|---|---|---|
| *n*-alkanes | | $C_{19}$–$C_{37}$ | Odd |
| Primary alcohols | | $C_{12}$–$C_{36}$ | Even |
| Aldehydes | | $C_{14}$–$C_{34}$ | Even |
| Fatty acids | | $C_{12}$–$C_{36}$ | Even |
| Wax esters | | $C_{30}$–$C_{60}$ | Even |
| Secondary alcohols | | $C_{21}$–$C_{33}$ | Odd |
| Ketones | | $C_{25}$–$C_{33}$ | Odd |
| β-Diketones | | $C_{27}$–$C_{35}$ | Odd |

**Table 2.2** *Major cyclic compounds found in plant waxes.*

| Cyclic compounds | Structure |
| --- | --- |
| Sterols | |
| Flavonoids | |
| Terpenoids | |

On the other hand, in the case of primary alcohols, aldehydes and fatty acids, the predominant compounds are those which possess an even number of carbon atoms [11]. In a number of plant waxes, the chain length of ketones is related to the hydrocarbon chain length. Two examples are nonacosan-15-one and hentriacontan-16-one, which are normally present in plant waxes where the predominant hydrocarbons are nonacosane and hentriacontane, respectively [18–21]. In addition, in some plant tissues, mixtures of ketones occur where mixtures of alkanes are present [22]. However, in the case of β-diketones, the chain length of the compounds is not related to the hydrocarbon chain length of the same tissue. Tritriacontan-16,18-one appears to be the most common β-diketone [23, 24]. Occasionally, the type of secondary alcohol present is closely related to the type of ketone found in the plant wax [25, 26].

Wax esters generally comprise *n*-alkanoic acids and *n*-alkan-1-ols and often possess an even number of carbon atoms. In plant waxes, the presence of double bonds and branches in wax esters is rare, unlike microbial and animal waxes. The fatty acid and fatty alcohol portions of the ester usually correspond to the free fatty acid and free fatty alcohol in the plant wax [27].

The extraction of lipids molecules can traditionally be carried out using conventional organic solvents such as hexane, dichloromethane and chloroform, which pose a number of toxicological and environmental problems [28]. The use of alternative green solvents and technologies is necessary for the safe and sustainable extraction of waxes. One such solvent is supercritical carbon dioxide [29]. Carbon dioxide is an ideal supercritical solvent for a number of applications ranging from extraction processes to pharmaceutical applications [30]. This is because the critical temperature is only 31.1°C; the benefits of near-critical operation can therefore be exploited at temperatures below 35°C. In addition, carbon dioxide is non-flammable, has minimal toxicity and is widely available. It is relatively inexpensive, recyclable and is an unregulated solvent. One slight disadvantage is the relatively high critical pressure of carbon dioxide (73.8 bar). However, operating at such pressures has become fairly routine in industrial-scale extraction processes in which supercritical carbon dioxide is used, such as in the extraction of hops and decaffeination of coffee [30, 31].

The extraction of waxes from biomass using supercritical carbon dioxide has been previously carried out [28, 32–35]. Supercritical carbon dioxide was found to be an ideal solvent for the extraction of plant lipids. The advantage of using supercritical carbon dioxide as a solvent is that the extraction of non-polar compounds can be made selective by fine-tuning the solvent power, which is done by varying the temperature and pressure [31, 36–39]. Furthermore, the extraction yields can be improved by adding polar modifiers (e.g. methanol, ethanol) which increases the solvent polarity; however, this results in a decrease in selectivity towards plant lipids as a higher proportion of polar compounds are extracted [40, 41].

Successful extraction of lipids from bagasse wax has been carried out and the extraction process was optimised by carrying out a factorial experimental design, whereby the effect of temperature and pressure on the extraction yields was modelled using a dimensionless factor coordinate system (Figure 2.1). These optimised conditions yield a complex mixture of lipids that can be utilised in a wide variety of applications.

The range of interesting molecules that have been identified in the wax extracts of sugarcane bagasse are summarised in Table 2.3. Applications of these waxes include use in nutraceuticals and ingredients for cleaning products, flavours, degreasers, cosmetics and lubricants. Further discussion of potential applications is provided in the following section.

Each class of hydrophobic compounds present in the bagasse wax has several properties which make them ideal for a variety of applications.

Long-chain hydrocarbons have been shown to display semiochemical properties, where they play a role in plant–insect interactions [44]. Work has been carried out on the 'pseudocopulatory' behaviour of male bees, *Andrenanigroaena*, towards the flowers of *Ophryssphegodes*. Results have shown that this orchid synthesises a variety of chemical compounds which are present in the sex pheromone of

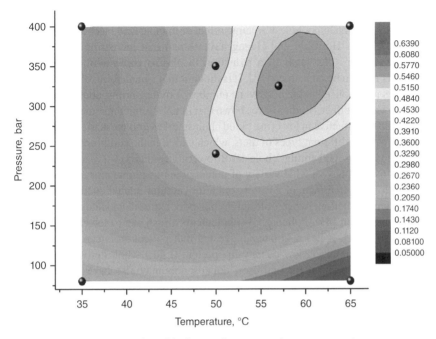

**Figure 2.1**  *Percentage crude yield of wax of sugarcane bagasse at varying temperatures and pressures.*

**Table 2.3**  *Group of compounds found in the wax of sugarcane bagasse and their applications [42–49].*

| Wax group | Compounds | Applications |
| --- | --- | --- |
| *n*-alkanes | $C_{23}$–$C_{31}$ | Semiochemicals |
| Primary alcohols | $C_{26}$, $C_{28}$ | Nutrient supplements, cosmetics |
| Long-chain aldehydes | $C_{26}$, $C_{28}$ | Food flavourings |
| Saturated fatty acids | $C_{16}$, $C_{18}$ | Cleaning compounds, lubricating oils/grease, detergents |
| Unsaturated fatty acids | $C_{18.2}$, $C_{18.3}$ | Nutrient supplements |
| Wax esters | $C_{44}$–$C_{54}$ | Cosmetics, hard wax polishes, lubricants, coatings, plasticers |
| Sterols | β-sitosterol, campesterol | Nutrient supplements |

*Andrenanigroaena* in similar abundances. Gas chromatography-electroantenno-graphic data indicate that a total of 14 compounds are present in the orchids which are found in the attractive odour sample of female bees and which cause an electroantennographic response in the antennae of males. Gas chromatography mass spectrometry (GC-MS) data indicates that these compounds are saturated and unsaturated long-chain hydrocarbons have chain lengths which vary from $C_{21}$ to $C_{29}$ [44].

Polyunsaturated fatty acids are known to have an effect on serum cholesterol in humans [43]. The hypocholesterolemic effect of linoleic acid has been well established [50–52]. Horrobin and Huang have shown that increasing the intake of linoleic in one's diet leads to a reduction in plasma cholesterol, though large amounts need to be consumed [53]. It is thought that a metabolite of linoleic acid, which is metabolised in the body via a number of routes, brings about this cholesterol-lowering effect. Studies have shown that in normolipidemic men, α-linolenic acid is just as effective in lowering blood cholesterol as linoleic acid [54]. Research has shown that different types of diets, which varied in the composition of unsaturated fatty acids, had similar cholesterol-lowering results.

Policosanols have a wide variety of potential applications, most notably in the prevention and treatment of a variety of cardiovascular-related conditions such as poor arterial function, hypercholesterolemia, poor antioxidant status and intermittent claudication [49]. Policosanol supplements are shown to lower the levels of cholesterol [54]. It has also been suggested that policosanols can act as potent antioxidants, inhibiting low-density-lipoprotein (LDL) -cholesterol peroxidation [49, 54]. Apart from medicinal uses, policosanols may also have potential cosmetic applications. Policosanols were found to be ideal components in cosmetics as anti-acne agents, as an emollient and for the control of sebum secretion [48].

Phytosterols are of particular interest as they have a variety of potential biological and physiological applications. The three main phytosterols that are present in the human diet are β-sitosterol, campesterol and stigmasterol [47]. They are widely known to act as efficient anticancer compounds. It has been estimated that the risk of cancer can be significantly decreased by as much as 20% with a phytosterol-enriched diet.

The high molecular weight of wax esters makes them one of the most significant groups of molecules found in plant waxes. This is because the high molecular weight allows for many valuable applications ranging from cosmetics to hard wax polishes, lubricants, coatings and plasticisers [46].

Sugarcane bagasse has therefore shown to be a promising feedstock for the generation of added-value products. Another significant agricultural residue in the world is wheat straw. The world production of wheat grain in 2010 was estimated to be 651.8 million tonnes by the FAO in 2011. Every tonne of wheat grain leads to the generation of 1.3 tonnes of wheat straw, which is of low economic value [55]. This by-product is extensively used in livestock feed and bedding; however, there is a growing interest in utilising wheat straw as a biomass feedstock for use in the production of fuel sources such as bioethanol. The main component of wheat straw that is used in biofuel production is the lignocellulosic component [56, 57]. However, also present in wheat straw are a large range of secondary metabolites that have potential economic value [58].

The epicuticular wax that coats the surface of the wheat straw (1% w/w) can generate a number of high-value products [28, 59]. The use of supercritical carbon

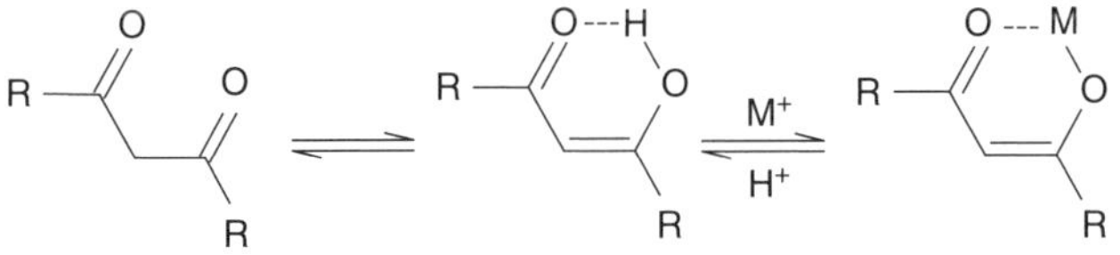

**Figure 2.2**   *Keto-enoltautomerism and metal chelation of a β-diketone.*

dioxide leaves the dewaxed wheat straw free of any residual solvent so that the lignocellulosic components can be recovered without any further treatment. This is ideal, especially if a holistic biorefinery approach is considered. The supercritical extraction of wax from wheat straw has been described in detail by Deswarte *et al.*, Sin *et al.* and Athukorala *et al.* [28, 35, 60].

The primary compounds found in the crude wheat straw wax include long-chain fatty acids, fatty alcohols, alkanes, sterols, hydroxyl-β-diketones and β-diketones [61]. The β-diketone constituents of the wax are of particular interest due to their potential industrial applications, including their use as a natural chelating agent for metal removal, and utilisation for their hydrophobic properties in coatings. Supercritical extractions of wheat straw wax have demonstrated that $1000\,\mu g\,g^{-1}$ of 14,16 hentriacontanedione can be obtained leading to a wax with a 35% purity of this β-diketone [35].

The ability of β-diketones to form metal chelates is of considerable interest and paves the way for a potential market. There is growing environmental concern regarding metal waste contamination [62] and also the long-term sustainability of elements [63, 64]. As shown in Figure 2.2, β-diketones are able to undergo keto-enoltautomerism; the enolic hydrogen can then be replaced by a metal ion, with the β-diketone then forming a chelate ring [65]. β-diketones have been shown to be effective chelators and can be easily isolated from the rest of the wax by chelating to copper (II) salts to form insoluble copper complexes [66].

The hydrophobic properties of β-diketones may also be utilised in industrial applications. Numerous plants display superhydrophobic and self-cleaning properties which reduce water loss and prevent the adhesion of debris or potential pathogens. These properties are a result of the hydrophobic cuticle (composed of cutin and epicuticular waxes) that covers the plant surface [67]. These hydrophobic components may be used in a wide range of bio-inspired applications including self-cleaning windows, textiles and packaging.

With the continual decline of petrochemical resources, there is an ever-growing demand for alternative feedstocks for the production of chemicals and fuel. There is a wide range of readily available low-value biomass feedstocks such as agricultural residues, which can be transformed into a variety of products by applying green chemical technologies. The use of agricultural residues in an integrated biorefinery will lead to the production of environmentally compatible and sustainable chemicals and materials for the twenty-first century. Extraction of secondary metabolites such as epicuticular waxes from biomass is an important first step in a holistic biorefinery, as it will lead to the generation of added-value products and also utilise as much of the biomass as possible, resulting in a close-to-zero-waste biorefinery.

## 2.3    Food Supply Chain Waste

Food supply chain waste (FSCW) is a serious problem which is increasing in parallel with global population. Roughly one-third of food produced for human consumption is wasted, amounting to roughly 1.3 billion tonnes annually. Within Europe and North America the food waste per capita is in the range of 280–300 kg per year [68]. Roughly 90 million tonnes of food waste is produced in the European Union (EU) annually, with around 39% of that from the manufacturing sector, 42% from households, 14% from the food service and 5% from retail/wholesale as depicted in Figure 2.3 [69].

In the UK alone 15 million tonnes of food waste (FW) is generated every year, 7.2 million tonnes of which is generated in UK households and 4.3 million tonnes from food supply chains (FSC), equating to roughly 42% and 29%, respectively [70]. With such a large amount of waste produced by the FSC, and with landfill gate fees having increased from £40–74 in 2009 to £80–121 in 2013 [71, 72], industries can no longer afford to disregard the financial burden that waste entails and are being forced to consider methods for eliminating or utilising it. FW has large scope for valorisation to convert into functional high-value chemicals, materials or energy as highlighted by the McKinsey Global Institute. FW was ranked 3rd of 15 identified resource productivity opportunities as part of its 2011 report, entitled 'Resource revolution: meeting the world's energy, material, food and water needs' [73].

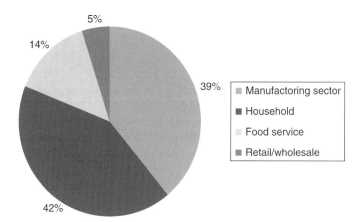

***Figure 2.3***   *Proportions of waste from different sectors generated in the EU.*

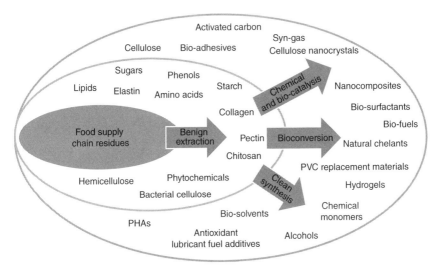

**Figure 2.4** *Components present in FW and their uses in common consumer applications.*

As shown in Figure 2.4, food waste is an eclectic mix of organic (predominantly C, H, O and N) and inorganic compounds, representing an exciting inventory of chemical compounds either as feedstocks or intermediaries for further processing [74–77].

For example, Clark and co-workers [78–80] have shown that citrus waste can be converted into chemicals for home and personal care industries such as limonene **1** and pectin **2** using microwave-assisted clean technologies (pectin is the generic term for a polymer comprising galacturonic acid units of at least 65%. The acid groups may either be free ($-CO_2H$), combined as a methyl ester ($-CO_2CH_3$), or as $Na^+$, $K^+$, $Ca^{2+}$ or ammonium salts.)

Lin *et al.* demonstrated the potential of bakery wastes (pastries and cakes) as a valuable source of (bio)succinic acid **3,** a useful platform molecule or biodegradable polymers (e.g. polyhydroxybutyrate **4**), through judicious selection of microbial strains in fermentation processes [81, 82].

**3**                    **4**

Fermentation of sugary wastes is a classical route to $C_2$ (bio)ethanol, but Atsumi *et al.* [83] have developed non-fermentative pathways for synthesis of branched-chain higher alcohols as biofuels using bio-engineered strains of *Escherichia coli* to successfully produce $C_4$ bioalcohols (butanol and isobutanol). Waste coffee grounds have been successfully exploited as a source of fatty acids and fatty acid esters suitable for conversion into biodiesel. Waste coffee grounds are rich in oils (10–15% by weight dry basis) with a profile amenable for conversion to biodiesel; this is very interesting as coffee is the second most-traded commodity in the world [84].

Proteins from food waste are an essential source of amino acids that can be converted into bulk organic chemicals. For example, L-phenylalanine **5** can be converted to styrene **7** via cinnamic acid **6**. Interestingly, the fermentative production of L-phenylalanine **5** from biomass has become so efficient that the most economical way to produce cinnamic acid **6** is probably from L-phenylalanine **5** by the reverse reaction [74].

**5**                    **6**                    **7**

PAL = L-phenylalanine ammonia lyase

Glutamic acid **8** and lysine **9** can be hydrolysed from proteinaceous food wastes and serve as a potentially useful feedstock chemical for a variety of commodity chemicals. For example, *N*-methylpyrrolidone (NMP) **10** and *N*-vinylpyrrolidone (NVP) **11** can be sourced from glutamic acid **8**, whereas lysine **9** is a source of 5-aminovaleric acid **12**, 1,5-diaminopentane **13** and caprolactam **14** [74].

Cashew nut shell liquid (CNSL), a by-product of the cashew industry, is an interesting feedstock chemical because it provides a source of *meta*-substituted phenolic compounds, namely anacardic acid **15,** cardol **16,** cardanol **17** and trace amounts of methylcardol **18.** CNSL is the dark reddish-brown viscous liquid (pericap fluid) found in the soft honeycomb structure of the cashew nut. CNSL has many applications in polymer-based industries, including friction linings, clutch disks, paints and varnishes and intermediates for the chemical industry [77].

## 2.4    Mango Waste: A Case Study

Mango waste represents an interesting and currently under-utilised resource for feedstocks chemicals. Mango is known as the 'king of fruits' in certain areas [85] and is becoming increasingly important as an exported crop to the western world due to its rising popularity [86]. It had a global production of around 39 million tonnes in 2010, which represents nearly a four-fold increase in production over the last 50 years (Figure 2.5) [87, 88].

The mango fruit is a large (2.5 to >30 cm) fleshy drupe weighing up to 1 kg, consisting of a woody endocarp, an edible mesocarp and a thick exocarp [85]. Most of the mango produced in the world is eaten fresh with 1–2% made into juices, concentrates, jams and so on [89]. The processing of mango yields 35–60% waste product by weight with peel being 16–30% and the kernel/bean being 10–30% of the total fruit weight [90, 91]. With current global production of over 40 million tonnes this translates to over 14 million tonnes of waste generated per annum, representing significant scope for the development of techniques for valorising mango waste.

Mango waste is a rich source of chemicals [92–94], for example mangiferin **19** [95] and quercetin 3-*O*-galactoside **20** [96], lipids and fatty acids, such as stearic **21,** oleic **22** and palmitic acid **23** [97].

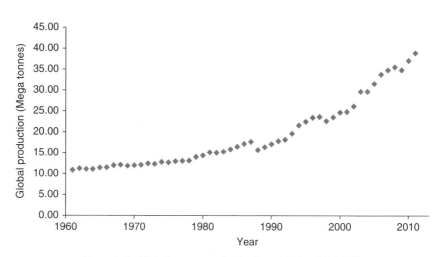

***Figure 2.5***   *Global mango production from 1961 to 2011 [88].*

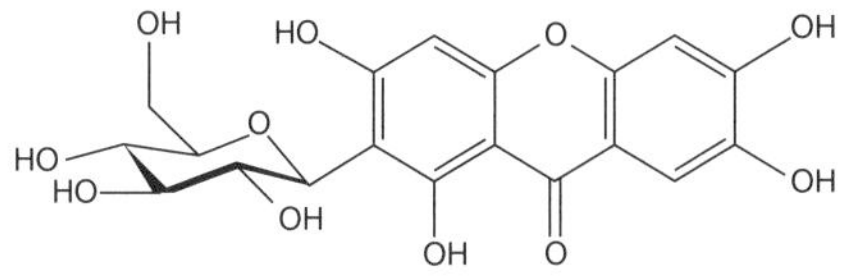

Mangiferin **19**

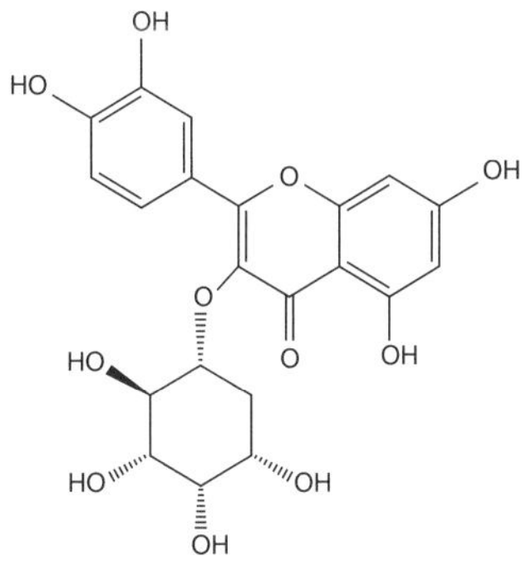

Quercetin 3-*O*-galactoside **20**

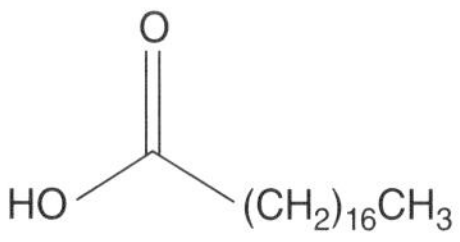

Stearic acid **21**

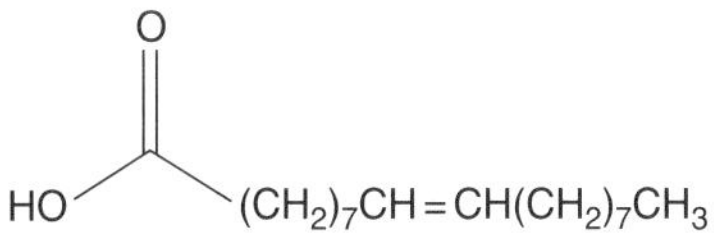

Oleic acid **22**

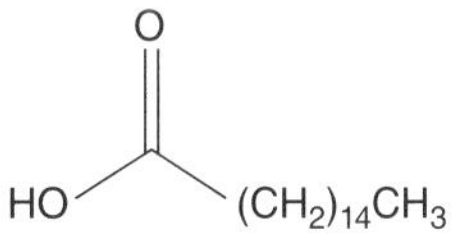

Palmitic acid **23**

The kernel has been traditionally ground to make flour in India and Nigeria [97] and the seed had been shown to have a very high lipid concentration. The mango bean butter obtained via extraction with either solvents or cold/hot pressing has been shown to be of very high quality with potential to be used as a replacement for cocoa butter in cosmetic and food applications. The fatty acid profiles of both cocoa butter and mango bean butter have been shown to be comparable with the same main four fatty acids present in each, these being oleic **22**, stearic **21**, palmitic **23** and linoleic, with similar ratios observed in both examples [97–99].

Mango waste, especially the peel, has great promise as a potential source of food-grade pectin with extraction yields of up to 21% and promising values for the galacturonic acid content and degree of esterification [100].

Along with the extraction of oil, there is evidence that phenolic compounds with high antioxidant activity could be extracted [101]. These phenolic compounds, including flavanols (e.g. **24**) and xanthones (e.g. **25**) [102], can be used in food manufacturing due to their antioxidant properties. Also found within the kernels are gallotannins **26** [103]; these are of special interest because they have been shown to have antimicrobial properties [104, 105].

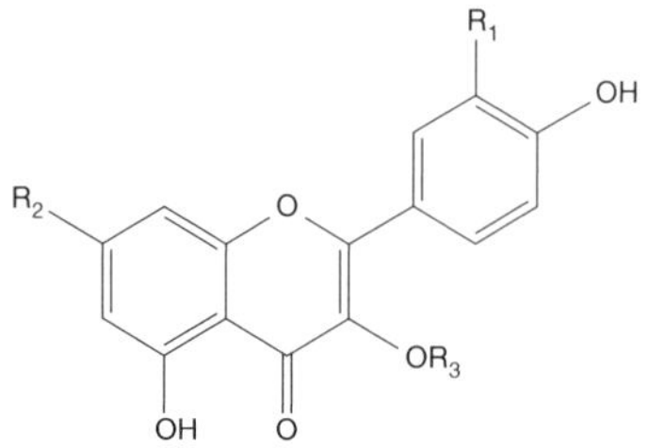

General structure of flavanol **24**

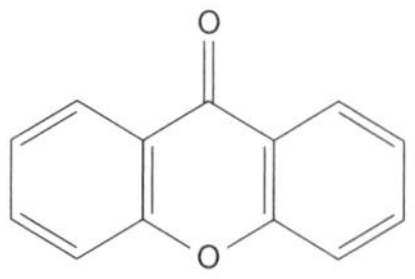

Xanthone **25**

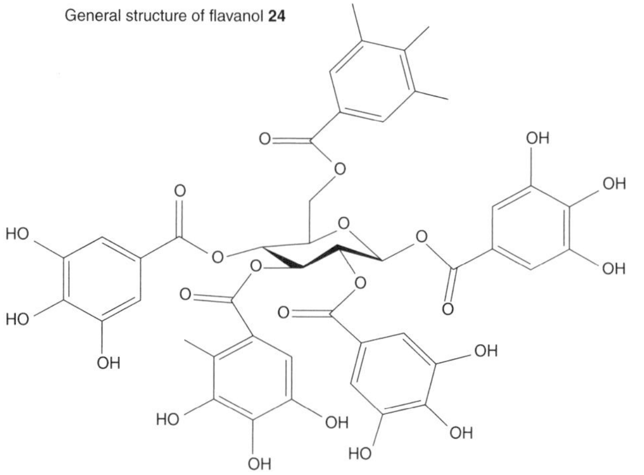

Gallotannin **26**

## 2.5   Concluding Remarks

Within the current state of industry, with non-renewable resources running out and the cost for waste disposal increasing, waste valorisation is an incredibly important concept that is beginning to be acknowledged as the future of industrial waste processing. Biomass including woody waste, marine waste and FSC wastes are rich in interesting molecules. The techniques for extraction of these chemicals is still undergoing extensive research to determine the most efficient, selective and green methods available, and the scaling-up process required to realise an industrial-scale biowaste biorefinery provides additional difficulties involving the combination of chemistry, economics and legislation. Although classical solvent extractions are the main quoted method for extraction, more modern and greener techniques such as microwave-assisted solvent-free supercritical $CO_2$ and subcritical water extractions are currently being explored as viable alternatives, with positive results indicating the retention of thermally sensitive and volatile components. The pyrolysis of biomass through microwave treatment has been extensively researched, especially the methodology and optimal conditions, but further scope is available for research into biowaste-specific torrefaction as the literature available within this field is sparse.

# References

1. Budarin, V.L., Shuttleworth, P.S., Dodson, J.R. *et al.* (2011) Use of green chemical technologies in an integrated biorefinery. *Energy and Environmental Science*, **4**, 471–479.
2. Clark, J.H., Budarin, V., Deswarte, F.E.I. *et al.* (2006) Green chemistry and the biorefinery: a partnership for a sustainable future. *Green Chemistry*, **8**, 853–860.
3. Nigam, P.S. and Pandey, A. (2009) *Biotechnology for Agro-Industrial Residues Utilisation: Utilisation of Agro-Residues*, Springer, Dordrecht, Netherlands.
4. Zuurbier, P. and van de Vooren, J. (2008) *Sugarcane Ethanol: Contributions to Climate Change Mitigation and the Environment*, Wageningen Academic Publishers, Wageningen.
5. Pandey, A., Soccol, C.R., Nigam, P. and Soccol, V.T. (2000) Biotechnological potential of agro-industrial residues. I: sugarcane bagasse. *Bioresource Technology*, **74**, 69–80.
6. Pimenta, M.d.A. and Frollini, E. (1997) Lignin: utilization as a "macromonomer" in the synthesis of phenolic type resins. *Anais Associação Brasileira de Química*, **46**, 43–49.
7. Baudel, H.M., Zaror, C. and de Abreu, C.A.M. (2005) Improving the value of sugarcane bagasse wastes via integrated chemical production systems: an environmentally friendly approach. *Industrial Crops and Products*, **21**, 309–315.
8. Lavarack, B.P., Griffin, G.J. and Rodman, D. (2000) Measured kinetics of the acid-catalysed hydrolysis of sugar cane bagasse to produce xylose. *Catalysis Today*, **63**, 257–265.
9. Clark, J.H. and Deswarte, F. (2011) *Introduction to Chemicals from Biomass*, John Wiley & Sons, Ltd, Chichester.
10. Attard, T.M., Watterson, B., Budarin, V.L. *et al.* (2014) Microwave assisted extraction as an important technology for valorising orange waste. *New Journal of Chemistry*, **38**, 2278–2283.
11. Barthlott, W., Neinhuis, C., Cutler, D. *et al.* (1998) Classification and terminology of plant epicuticular waxes. *Botanical Journal of the Linnean Society*, **126**, 237–260.
12. Kunst, L. and Samuels, A.L. (2003) Biosynthesis and secretion of plant cuticular wax. *Progress in Lipid Research*, **42**, 51–80.
13. Martin, J.T. and Juniper, B.E. (1970) *The Cuticles of Plants*, Edward Arnold, London.
14. Cutler, D. F., Alvin, K. L., Price, C. E. and London, L. S. o. (1982) *The Plant Cuticle*. Paper presented at an International Symposium organised by the Linnean Society of London, 8–11 September 1980, Burlington House, London. Academic Press, London.
15. Kolattukudy, P.E. (1980) Biopolyester membranes of plants: cutin and suberin. *Science*, **208**, 990–1000.
16. Harwood, J.L. and Bowyer, J.R. (1990) *Lipids, Membranes and Aspects of Photobiology*, Academic Press, London.
17. Kolattukudy, P.E. and Liu, T.-Y.J. (1970) Direct evidence for biosynthetic relationships among hydrocarbons, secondary alcohols and ketones in Brassica oleracea. *Biochemical and Biophysical Research Communications*, **41**, 1369–1374.
18. Channon, H.J. and Chibnall, A.C. (1929) The ether-soluble substances of cabbage leaf cytoplasm. *Biochemical Journal*, **23**, 168–175.

19. Sahai, P.N.S. and Chibnall, A.C. (1932) Wax metabolism in the leaves of Brussels sprout. *Biochemical Journal*, **26**, 403–412.

20. Chibnall, A.C., Piper, S.H., El Mangouri, H.A. *et al.* (1937) The wax from the leaves of sandal (Santalum album Linn). *Biochemical Journal*, **31**, 1981–1986.

21. MacKie, A. and Misra, A.L. (1956) Chemical investigation of the leaves of Anona senegalensis. I. Constituents of the leaf wax. *Journal of the Science of Food and Agriculture*, **7**, 203–209.

22. Kranz, Z., Lamberton, J., Murray, K. and Redcliff, A. (1961) Studies in waxes. XIX. The stem wax of the grass Leptochlona digitata. *Australian Journal of Chemistry*, **14**, 264–271.

23. Tulloch, A. and Weenink, R. (1966) Hydroxy-β-diketones from wheat leaf wax. *Chemical Communications (London)*, **8**, 225–226.

24. Barber, H.N. and Netting, A.G. (1968) Chemical genetics of β-diketone formation in wheat. *Phytochemistry*, **7**, 2089–2093.

25. Wollrab, V. (1969) Secondary alcohols and paraffins in the plant waxes of the family Rosaceae. *Phytochemistry*, **8**, 623–627.

26. Kolattukudy, P.E. and Liu, T.-Y.J. (1970) Direct evidence for biosynthetic relationships among hydrocarbons, secondary alcohols and ketones in Brassica oleracea. *Biochemical and Biophysical Research Communications*, **41**, 1369–1374.

27. Nichols, B. and Wood, B. (1968) New glycolipid specific to nitrogen-fixing blue-green algae. *Nature*, **217**, 767–768.

28. Deswarte, F.E.I., Clark, J.H., Hardy, J.J.E. and Rose, P.M. (2006) The fractionation of valuable wax products from wheat straw using CO2. *Green Chemistry*, **8**, 39–42.

29. Hunt, A.J., Sin, E.H.K., Marriott, R. and Clark, J.H. (2010) Generation, capture, and utilization of industrial carbon dioxide. *ChemSusChem*, **3**, 306–322.

30. Subramaniam, B., Rajewski, R.A. and Snavely, K. (1997) Pharmaceutical processing with supercritical carbon dioxide. *Journal of Pharmaceutical Sciences*, **86**, 885–890.

31. McHugh, M.A. and Krukonis, V.J. (1994) *Supercritical Fluid Extraction: Principles and Practice*, Butterworth-Heinemann, Oxford.

32. Choi, Y., Kim, J., Noh, M. *et al.* (1996) Extraction of epicuticular wax and nonacosan-10-ol from Ephedra herb utilizing supercritical carbon dioxide. *Korean Journal of Chemical Engineering*, **13**, 216–219.

33. Arshadi, M., Hunt, A.J. and Clark, J.H. (2012) Supercritical fluid extraction (SFE) as an effective tool in reducing auto-oxidation of dried pine sawdust for power generation. *RSC Advances*, **2**, 1806–1809.

34. Backlund, I., Arshadi, M., Hunt, A.J. *et al.* (2014) Extractive profiles of different lodgepole pine (Pinus contorta) fractions grown under a direct seeding-based silvicultural regime. *Industrial Crops and Products*, **58**, 220–229.

35. Sin, E.H.K., Marriott, R., Hunt, A.J. and Clark, J.H. (2014) Identification, quantification and Chrastil modelling of wheat straw wax extraction using supercritical carbon dioxide. *Comptes Rendus Chimie*, **17**, 293–300.

36. Lang, Q. and Wai, C.M. (2001) Supercritical fluid extraction in herbal and natural product studies: a practical review. *Talanta*, **53**, 771–782.

37. Özcan, A. and Özcan, A.S. (2004) Comparison of supercritical fluid and Soxhlet extractions for the quantification of hydrocarbons from Euphorbia macroclada. *Talanta*, **64**, 491–495.

38. Zougagh, M., Valcárcel, M. and Ríos, A. (2004) Supercritical fluid extraction: a critical review of its analytical usefulness. *TrAC Trends in Analytical Chemistry*, **23**, 399–405.

39. Vilegas, J.H.Y., de Marchi, E. and Lancas, F.M. (1997) Extraction of low-polarity compounds (with emphasis on coumarin and kaurenoic acid) from Mikania glomerata ('guaco') leaves. *Phytochemical Analysis*, **8**, 266–270.

40. Bott, T. (1982) *Chemistry and Industry*, Society of Chemical Industry, London.

41. Wang, L. and Weller, C.L. (2006) Recent advances in extraction of nutraceuticals from plants. *Trends in Food Science & Technology*, **17**, 300–312.

42. Sjöström, E. (1991) Carbohydrate degradation products from alkaline treatment of biomass. *Biomass and Bioenergy*, **1**, 61–64.

43. Gill, I. and Valivety, R. (1997) Polyunsaturated fatty acids, part 1: Occurrence, biological activities and applications. *Trends in Biotechnology*, **15**, 401–409.

44. Schiestl, F.P., Ayasse, M., Paulus, H.F. *et al.* (1999) Orchid pollination by sexual swindle. *Nature*, **399**, 421–421.

45. Hill, K. (2000) Fats and oils as oleochemical raw materials. *Pure and Applied Chemistry*, **72**, 1255–1264.

46. Gunawan, E.R., Basri, M., Rahman, M.B.A. *et al.* (2005) Study on response surface methodology (RSM) of lipase-catalyzed synthesis of palm-based wax ester. *Enzyme and Microbial Technology*, **37**, 739–744.

47. Bradford, P.G. and Awad, A.B. (2007) Phytosterols as anticancer compounds. *Molecular Nutrition and Food Research*, **51**, 161–170.

48. Majeed, M. Gangadharan, G. K. and Prakash, S. (2007) Compositions and methods containing high purity fatty alcohol C24 to C36 for cosmetic applications. Google Patents, US 20070196507 A1.

49. Marinangeli, C.P.F., Jones, P.J.H., Kassis, A.N. and Eskin, M.N.A. (2010) Policosanols as nutraceuticals: fact or fiction. *Critical Reviews in Food Science and Nutrition*, **50**, 259–267.

50. Ahrens Jun, E., Insull Jun, W., Blomstrand, R. *et al.* (1957) The influence of dietary fats on serum-lipid levels in man. *The Lancet*, **269**, 943–953.

51. Horrobin, D.F. and Huang, Y.S. (1987) The role of linoleic acid and its metabolites in the lowering of plasma cholesterol and the prevention of cardiovascular. *International Journal of Cardiology*, **17**, 241–255.

52. Shepherd, J., Packard, C.J., Patsch, J.R. *et al.* (1978) Effects of dietary polyunsaturated and saturated fat on the properties of high density lipoproteins and the metabolism of apolipoprotein AI. *Journal of Clinical Investigation*, **61**, 1582–1592.

53. Chan, J.K., Bruce, V.M. and McDonald, B.E. (1991) Dietary alpha-linolenic acid is as effective as oleic acid and linoleic acid in lowering blood cholesterol in normolipidemic men. *The American Journal of Clinical Nutrition*, **53**, 1230–1234.

54. Varady, K.A., Wang, Y. and Jones, P.J.H. (2003) Role of policosanols in the prevention and treatment of cardiovascular disease. *Nutrition Reviews*, **61**, 376–383.

55. Montane, D., Farriol, X., Salvadó, J. *et al.* (1998) Application of steam explosion to the fractionation and rapid vapor-phase alkaline pulping of wheat straw. *Biomass and Bioenergy*, **14**, 261–276.

56. Sun, X.F., Sun, R.C., Fowler, P. and Baird, M.S. (2004) Isolation and characterisation of cellulose obtained by a two-stage treatment with organosolv and cyanamide activated hydrogen peroxide from wheat straw. *Carbohydrate Polymers*, **55**, 379–391.

57. Talebnia, F., Karakashev, D. and Angelidaki, I. (2010) Production of bioethanol from wheat straw: An overview on pretreatment, hydrolysis and fermentation. *Bioresource Technology*, **101**, 4744–4753.

58. Deswarte, F.E.I., Clark, J.H., Wilson, A.J. *et al.* (2007) Toward an integrated straw-based biorefinery. *Biofuels, Bioproducts and Biorefining*, **1**, 245–254.
59. Volynets, B. and Dahman, Y. (2011) Assessment of pretreatments and enzymatic hydrolysis of wheat straw as a sugar source for bioprocess industry. *International Journal of Energy and Environment*, **2**, 427–446.
60. Athukorala, Y., Mazza, G. and Oomah, B.D. (2009) Extraction, purification and characterization of wax from flax (Linum usitatissimum) straw. *European Journal of Lipid Science and Technology*, **111**, 705–714.
61. Athukorala, Y. and Mazza, G. (2010) Supercritical carbon dioxide and hexane extraction of wax from triticale straw: Content, composition and thermal properties. *Industrial Crops and Products*, **31**, 550–556.
62. Marcus, Y. and SenGupta, A.K. (2004) *Ion Exchange and Solvent Extraction: A Series of Advances*, Taylor & Francis, Boca Raton.
63. Dodson, J.R., Hunt, A.J., Parker, H.L. *et al.* (2012) Elemental sustainability: Towards the total recovery of scarce metals. *Chemical Engineering and Processing: Process Intensification*, **51**, 69–78.
64. Hunt, A.J., Farmer, T.J. and Clark, J.H. (2013) Elemental sustainability and the importance of scarce element recovery, in *Element Recovery and Sustainability* (ed A.J. Hunt), The Royal Society of Chemistry, Cambridge, pp. 1–28.
65. Fu, W., Chen, Q., Hu, H. *et al.* (2011) Solvent extraction of copper from ammoniacal chloride solutions by sterically hindered β-diketone extractants. *Separation and Purification Technology*, **80**, 52–58.
66. Horn, D., Kranz, Z. and Lamberton, J. (1964) The composition of Eucalyptus and some other leaf waxes. *Australian Journal of Chemistry*, **17**, 464–476.
67. Pechook, S. and Pokroy, B. (2012) Self-assembling, bioinspired wax crystalline surfaces with time-dependent wettability. *Advanced Functional Materials*, **22**, 745–750.
68. Gustavsson, J., Cederberg, C., Sonesson, U. *et al.* (2011) *Global Food Losses and Food Waste*, Food and Agriculture Organization of the United Nations Rome, Italy.
69. Monier, V. (2011) *Preparatory Study on Food Waste Across EU 27*, European Commission.
70. WRAP (2013) *Estimates of Food and Packaging Waste in the UK Grocery Retail and Hospitality Chain*. WRAP. Available at http://www.wrap.org.uk/sites/files/wrap/UK%20 Estimates%20July%2014%20Final.pdf (accessed 12 September 2014).
71. WRAP (2013) Comparing the cost of alternative waste treatment options. *WRAP Gate Fees Report* **2013**. Available at http://www.wrap.org.uk/sites/files/wrap/Gate_Fees_ Report_2013_h%20%282%29.pdf (accessed 12 September 2014).
72. WRAP (2009) Comparing the cost of alternative waste treatment options. *WRAP Gate Fees Report* **2009**. Available at http://www.wrap.org.uk/sites/files/wrap/W504GateFeesWEB2009. b06b2d8d.7613.pdf (accessed 12 September 2014)
73. Dobbs, R. Oppenheim, J. Thompson, F. *et al.* (2011) Resource Revolution: Meeting the World's Energy, Material, Food and Water Needs. McKinsey Global Institute, McKinsey Sustainability & Resource Productivity Practice. Available at http://www.mckinsey.com/ insights/energy_resources_materials/resource_revolution (accessed 26 August 2014).
74. Tuck, C.O., Pérez, E., Horváth, I.T. *et al.* (2012) Valorization of biomass: deriving more value from waste. *Science*, **337**, 695–699.
75. Luque, R. and Clark, J.H. (2013) Valorisation of food residues: waste to wealth using green chemical technologies. *Sustainable Chemical Processes*, **1**, 10.

76. Kazmi, A. and Shuttleworth, P. (eds) (2013) *The Economic Utilisation of Food Co-Products*, RSC Green Chemistry Series, The Royal Society of Chemistry, Cambridge.

77. Lin, C.S.K., Pfaltzgraff, L.A., Herrero-Davila, L. *et al.* (2013) Food waste as a valuable resource for the production of chemicals, materials and fuels. Current situation and global perspective. *Energy and Environmental Science*, **6**, 426–464.

78. Clark, J.H., Pfaltzgraff, L.A., Budarin, V.L. *et al.* (2013) From waste to wealth using green chemistry. *Pure and Applied Chemistry*, **85** (8), 1625–1631.

79. Pfaltzgraff, L.A., De bruyn, M., Cooper, E.C. *et al.* (2013) Food waste biomass: a resource for high-value chemicals. *Green Chemistry*, **15**, 307–314.

80. Balu, A.M., Budarin, V., Shuttleworth, P.S. *et al.* (2012) Valorisation of orange peel residues: waste to bio-chemicals and nanoporous materials. *ChemSusChem*, **5**, 1694–1697.

81. Arancon, R.A.D., Lin, C.S.K., Chan, K.M. *et al.* (2013) Advances on waste valorization strategies: news horizons for a more sustainable society. *Energy Science and Engineering*, **1**, 53–71.

82. Zhang, A.Y., Sun, Z., Leung, C.C.J. *et al.* (2013) Valorisation of bakery waste for succinic acid production. *Green Chemistry*, **15**, 690–695.

83. Atsumi, S., Hanai, T. and Liao, J.C. (2008) Non-fermentative pathways for synthesis of branched-chain higher alcohols as biofuels. *Nature*, **451**, 86–89.

84. Jenkins, R.W., Stageman, N.E., Fortune, C.M. and Chuck, C.J. (2014) Effect of the type of bean, processing, and geographical location on the biodiesel produced from waste coffee grounds. *Energy Fuels*, **28**, 1166–1174.

85. Mukherjee, S. and Litz, R. (1997) *The Mango: Botany, Production and Uses*, CAB International, Wallingford.

86. Evans, E.A. (2008) *EDIS Document FE718, Florida Cooperative Extension Service, Institute of Food and Agricultural Sciences*, University of Florida, Gainesville, FL.

87. Heuzé, V., Tran, G., Bastianelli, D., *et al.* (2013) Mango (Mangifera indica) Fruit and by-Products. Available at http://www.feedipedia.org/node/516 (accessed 12 September 2014).

88. FAOSTAT (2012) Mango. Available at http://www.unctad.info/en/Infocomm/AACP-Products/COMMODITY-PROFILE---Mango/ (accessed 26 August 2014).

89. Berardini, N., Knödler, M., Schieber, A. and Carle, R. (2005) Utilization of mango peels as a source of pectin and polyphenolics. *Innovative Food Science and Emerging Technologies*, **6**, 442–452.

90. Larrauri, J.A., Rupérez, P., Borroto, B. and Saura-Calixto, F. (1996) Mango peels as a new tropical fibre: preparation and characterization. *Food Science and Technology International*, **29**, 729–733.

91. Koubala, B.B., Kansci, G., Mbome, L.I. *et al.* (2008) Effect of extraction conditions on some physicochemical characteristics of pectins from "Améliorée" and "Mango" peels. *Food Hydrocolloids*, **22**, 1345–1351.

92. Berardini, N., Fezer, R., Conrad, J. *et al.* (2005) Screening of mango (Mangifera indica L.) cultivars for their contents of flavonol O- and xanthone C-glycosides, anthocyanins, and pectin. *Journal of Agricultural and Food Chemistry*, **53**, 1563–1570.

93. Ajila, C., Naidu, K., Bhat, S. and Rao, U. (2007) Bioactive compounds and antioxidant potential of mango peel extract. *Food Chemistry*, **105**, 982–988.

94. Solís-Fuentes, J.A. and Durán-de-Bazúa, M.C. (2004) Mango seed uses: thermal behaviour of mango seed almond fat and its mixtures with cocoa butter. *Bioresource Technology*, **92**, 71–78.

95. Masibo, M. and He, Q. (2008) Major mango polyphenols and their potential significance to human health. *Comprehensive Reviews in Food Science and Food Safety*, **7**, 309–319.

96. Rymbai, H., Srivastav, M., Sharma, R., Patel, C. and Singh, A. (2013) Bio-active compounds in mango (Mangifera indica L.) and their roles in human health and plant defence – a review. *Horticultural Science & Biotechnology*, **88**, 369–379.

97. Narasimha Char, B.L., Reddy, B.R. and ThirumalaRao, S.D. (1977) Processing mango stones for fat. *Journal of the American Oil Chemists' Society*, **54**, 494–495.

98. Lakshminarayana, G., ChandrasekharaRao, T. and Ramalingaswamy, P.A. (1983) Varietal variation in content, characteristics and composition of mango seeds and fat. *Journal of the American Oil Chemists' Society*, **60**, 88–89.

99. Schieber, A., Stintzing, F. and Carle, R. (2001) By-products of plant food processing as a source of functional compounds – recent developments. *Trends in Food Science & Technology*, **12**, 401–413.

100. Sirisakulwat, S., Sruamsiri, P., Carle, R. and Neidhart, S. (2010) Resistance of industrial mango peel waste to pectin degradation prior to by-product drying. *International Journal of Food Science and Technology*, **45**, 1647–1658.

101. Maisuthisakul, P. and Gordon, M.H. (2009) Antioxidant and tyrosinase inhibitory activity of mango seed kernel by-product. *Food Chemistry*, **117**, 332–341.

102. Schieber, A., Berardini, N. and Carle, R. (2003) Identification of flavonol and xanthone glycosides from mango (Mangifera indica L. Cv. "Tommy Atkins") peels by high-performance liquid chromatography-electrospray ionization mass spectrometry. *Journal of Agricultural and Food Chemistry*, **51**, 5006–5011.

103. Berardini, N., Carle, R. and Schieber, A. (2004) Characterization of gallotannins and benzophenone derivatives from mango (Mangifera indica L. cv. 'Tommy Atkins') peels, pulp and kernels by high-performance liquid chromatography/electrospray ionization mass spectrometry. *Rapid Communications in Mass Spectrometry*, **18**, 2208–2216.

104. Engels, C., Knödler, M., Zhao, Y.-Y. *et al.* (2009) Antimicrobial activity of gallotannins isolated from mango (Mangifera indica L.) kernels. *Journal of Agricultural and Food Chemistry*, **57**, 7712–7718.

105. Engels, C., Gänzle, M.G. and Schieber, A. (2009) Fractionation of gallotannins from mango (Mangifera indica L.) kernels by high-speed counter-current chromatography and determination of their antibacterial activity. *Journal of Agricultural and Food Chemistry*, **58**, 775–780.

# 3

# Pretreatment and Thermochemical and Biological Processing of Biomass

## Wan Chi Lam[1], Tsz Him Kwan[1], Vitaliy L. Budarin[2], Egid B. Mubofu[3], Jiajun Fan[2] and Carol Sze Ki Lin[1]

[1]*School of Energy and Environment, City University of Hong Kong, Hong Kong*
[2]*Department of Chemistry, Green Chemistry Centre of Excellence, University of York, UK*
[3]*Department of Chemistry, University of Dar es Salaam, Tanzania*

## 3.1 Introduction

Biomass is produced by green plants through photosynthesis. During photosynthesis, solar energy is converted into biomass materials and is stored in chemical bonds. When the chemical bonds are broken, chemical energy is released. Through different conversion technologies, the chemical energy can be transformed into various chemicals, materials, biofuels, or forms of energy. Pretreated biomass can efficiently convert to desirable products or energy through two main platforms: thermochemical or biological. Thermochemical conversion technologies involve thermo-decomposition of biomass into products. The technologies mainly include combustion, torrefaction, pyrolysis, liquefaction, gasification and hydrothermal biomass activation. Biological conversion mainly involves conversion of biomass to sugars for fermentative chemicals production with the help of microorganisms such as bacteria or fungi. Figure 3.1

*Introduction to Chemicals from Biomass*, Second Edition. Edited by James Clark and Fabien Deswarte.
© 2015 John Wiley & Sons, Ltd. Published 2015 by John Wiley & Sons, Ltd.

depicts the various biomass conversion routes for the production of chemicals, materials, and fuels. In this chapter, the concepts of various pretreatment processes including mechanical, physical, chemical, and biological methods are discussed. Different thermochemical and biological conversion processes are also introduced.

## 3.2    Biomass Pretreatments

Direct conversion of biomass into products is not efficient in most cases due to the presence of hemicellulose (20–40%), lignin (20–30%), and extractives in large quantities [1, 2], which serve as a protective layer (Figure 3.2). For this reason, it is necessary to pretreat the biomass before any processing. Pretreatment helps alter the physical properties and chemical composition of biomass, and makes it more suitable for conversion. It can be a biological, chemical, thermal, or mechanical process such as ammonia fiber explosion (AFEX), torrefaction, and steam explosion. These pretreatment processes help alter the amorphous and crystalline regions of the biomass and bring significant changes to structural and chemical compositions.

### 3.2.1    Mechanical Pretreatment of Biomass

#### 3.2.1.1    Chipping, Grinding, and Milling

The objective of the mechanical pretreatment is a reduction of particle size and crystallinity of lignocellulosic materials in order to increase the specific surface area and reduce the degree of polymerization. This can be produced by a combination of chipping, grinding, or milling depending on the desired particle size of the final material (10–30 mm after chipping and 0.2–2 mm after milling or grinding) [3, 4]. Different milling processes (ball milling, two-roll milling, hammer milling, colloid milling, and vibro-energy milling) can be used to improve the enzymatic hydrolysis of lignocelullosic materials [5]. The power requirement of this pretreatment is relatively high depending on the final particle size and the biomass characteristics.

#### 3.2.1.2    Densification

According to the European Commission Green Paper, European energy policy is built on three core objectives [6]:

- *sustainability*: to combat climate change actively by promoting renewable energy sources and energy efficiency;
- *competitiveness*: to improve the efficiency of the European energy grid by creating a truly competitive internal energy market; and
- *security of supply*: to better coordinate the EU's supply of and demand for energy within an international context.

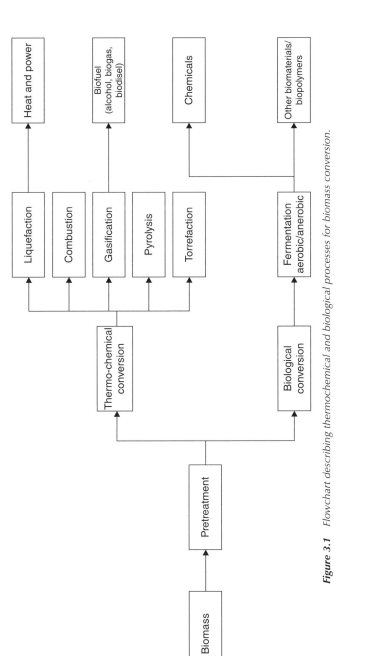

***Figure 3.1*** *Flowchart describing thermochemical and biological processes for biomass conversion.*

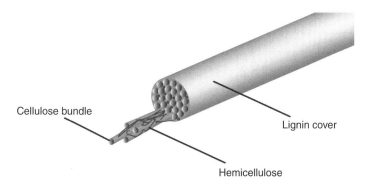

**Figure 3.2**   *Structural arrangement of lignocellulose.*

In addressing those challenges, the EU is increasingly shifting towards policies favoring use of renewable energy sources. Currently biomass delivers around 4% of the EU's primary energy (Eurostat), and in order to reach the future targets set by the EU significant amounts of biomass will be required in the majority of European countries [7]. Since the resources of biomass are situated large distances from energy users, in such countries there will be additional logistic costs and material losses. However, it has been shown that densification prior to international transportation of biomass is crucial, as converting biomass into a densified intermediate can save transport and handling costs. Consequently, pretreatment is a key step in the total supply chain [8].

The different types of biomass densification are bailing, briquetting, extrusion, and pelletization. With respect to the production of high-density solid energy carriers from biomass, briquetting and pelletization are the two most widely used methods. Both processes take place at high pressure and are very closely related techniques. The basic processing difference is the use of a piston or screw press for briquetting, while biomass pellets are produced in a pellet mill. The size of the final products is different; pellets have a cylindrical shape and are only about 6–25 mm in diameter and 3–50 mm in length [9].

The European and North America pellet markets have been studied in great detail by Sikkema *et al.* [10] and Spelter and Toth [11]. Wood pellet production in the EU was estimated to be about 10 million tons and 6.2 million tons for North America in 2009. Recent studies suggest a strong growth for both the European and North American pellet market, and the globally installed pellet production capacity for 2011 was estimated to be about 30 million tons [9].

At present, wood pellet production is limited to sawdust and cutter shavings (dry and wet), predominantly from spruce and pine tree species, as a raw material. As forest-based by-products, feedstocks for conventional wood pellets are finite resources and, due to the existing supply chain, pellet production has a strong dependence on the productivity of the wood-processing sector. To satisfy demand for pellet fuels, agricultural residues and industrial food by-products are being pelletized for fuel, although on a much smaller scale [12]. Pelletizing new forms of biomass is challenging, as non-woody biomass generally has more hemicellulose, less cellulose and lignin than wood, giving it less tensile and compressive strength. One of the ways to overcome this problem is through specific combination of different types of biomass in a proportion which best suits pellet production [13].

The main issue with the future of the biomass pellet sector is the sustainable and price-competitive production of pellets for a growing industrial and domestic pellet market, which place high demands on pellet quality and combustion properties. Current trends are to combine torrefaction (thermal treatment of biomass) and pelletization. Moreover, introducing a broader base of raw materials (i.e. agricultural residues, energy grass, and mixed biomass resources) would help tackle this issue [9].

### 3.2.2 Physical Pretreatment of Biomass

#### 3.2.2.1 ScCO$_2$ and Natural Solvent Extraction

Liquid or supercritical $CO_2$ supercritical carbon dioxide ($ScCO_2$) extractions have been proposed as alternative techniques [14] for plant oil extraction. $ScCO_2$ extraction is based on the solvating properties of supercritical $CO_2$, which can be obtained by employing pressure and temperature above the critical point of a compound, mixture, or element. Extraction by $ScCO_2$ depends on the intrinsic tunable nature of supercritical fluids such as temperature and pressure, and some extrinsic features such as the characteristics of the sample matrix, interaction with targeted analytes, and many environmental factors [15]. To optimize the process of $ScCO_2$ extraction, a large number of variables such as pressure, temperature, flow rate, and modifier used need to be investigated. Proper control of these parameters allows the extractability of $ScCO_2$ to be modified, enabling the process to be applied in different industrial fields from food to pesticide manufacturing [16]. Moreover, a broader range of controllable parameters makes the $ScCO_2$ extraction process more unique, sensitive, and specific compared to conventional extraction methods [17]. Due to its unique properties, effective isolation and fractionation of valuable components with low environmental impact can be achieved. Furthermore, $ScCO_2$ has another advantage in that compounds extracted in this way can be classified as 'natural' under EU regulations in contrast to those extracted with conventional organic solvents. In addition, because $CO_2$ is a major product of fermentation, integrating its use in a biorefinery is advantageous. Moreover, the $ScCO_2$ extraction method leaves no solvent residues in the product(s), making the product suitable for use in food, personal care, or pharmaceutical applications. Products resulting from this extraction method therefore have major commercial advantages over traditional extraction methods. Industrial-scale use of $ScCO_2$ should be applicable within the biorefinery as it has already been employed commercially for hop extraction, decaffeination of coffee and dry cleaning. The utilization of $ScCO_2$ technology within the biorefinery also enables the recycling of the internally generated $CO_2$ from the thermal treatment of the biomass residues.

Thorough comparison of $ScCO_2$ extraction under varying temperatures and pressures with hexane extraction (a traditional solvent of similar polarity) demonstrates that, although the chemical composition of hexane and supercritical extracts do not vary greatly, the proportions of unwanted co-extracted components such as pigments, polar lipids, and free sugars is well known to be far greater in hexane extracts than from $ScCO_2$ [18].

### 3.2.2.2   Ammonia Fiber Explosion

The major processing impediment to the production of economically viable commercial ethanol or other chemicals from biomass through biological processing is the inherent resistance of lignocellulosic materials to conversion to fermentable sugars. In order to improve the efficiency of enzymatic hydrolysis, a pretreatment step is necessary to make the structural carbohydrate fraction accessible to cellulases. Probably one of the most effective biomass pretreatments is ammonia fiber explosion (AFEX), which is a combination of chemical and physical digestion [19]. In AFEX pretreatment, the biomass is treated with liquid anhydrous ammonia at moderate temperatures (60–100°C) and high pressure (250–300 psi) for 5 min followed by a rapid pressure release. In this process, the combined chemical and physical effects of lignin solubilization, hemicellulose hydrolysis, cellulose decrystallization, and increased surface area enables significant enzymatic conversion of cellulose and hemicellulose to fermentable sugars. The AFEX treatment has some unique features that distinguish it from other biomass treatments, as described below:

- Nearly all of the ammonia can be recovered and reused, while the remaining serves as nitrogen source for microbes in downstream processes.
- There is no wash stream in the process. Dry matter recovery following the AFEX treatment is essentially 100%. Treated biomass is stable for long periods and can be fed at very high solids loadings in enzymatic hydrolysis or fermentation processes.
- Cellulose and hemicellulose are well preserved in the AFEX process, with little or no degradation.
- There is no need for neutralization prior to the enzymatic hydrolysis of AFEX-treated biomass.
- Enzymatic hydrolysis of AFEX-treated biomass produces clean sugar streams for subsequent fermentation processes.

The effectiveness of AFEX treatment has been clearly demonstrated with corn stover. The rate and the extent of both glucose and xylose produced were substantially greater than the untreated sample. Clearly all of the pretreatment variables (temperature, moisture content, ammonia loading, and treatment time) influence the reactivity of the treated biomass. AFEX treatment is a dry process, which permits much higher solid loadings in the fermentation process than liquid phase treatments. This implies that much higher product concentrations such as ethanol are potentially achievable in the final fermentation broth. Preliminary experimental data from fed-batch solid-state fermentation (SSF) confirm this assumption [20].

### 3.2.2.3    *Supercritical Water Treatment*

Water near or above its critical point (374°C, 218 atm) is attracting increased attention as a medium for chemical processes. Most of this new attention is driven by the search for more 'green' or environmentally benign chemical processes. Water near its critical point possesses very different properties from those of ambient liquid water. The dielectric constant of sub- and supercritical water (SCW) is much lower, and the number and persistence of hydrogen bonds are both diminished. As a result, high-temperature water behaves like many organic solvents in that organic compounds enjoy high solubilities in near-critical water and complete miscibility with SCW. Gases are also miscible in SCW, so employing a SCW reaction environment that provides an opportunity to conduct chemistry in a single fluid phase that would otherwise occur in a multiphase system under more conventional conditions. The advantages of a single supercritical phase reaction medium is that a higher concentration of reactants can often be attained and there is no interphase mass transport processes which hinder reaction rates [21].

Hydrothermal processing of biomass offers a number of potential advantages over other biomass processing methods, including high throughputs, high energy and separation efficiency, the ability to use mixed feedstocks such as wastes and lignocellulose, the production of direct replacements for existing fuels, and no need to maintain specialized microbial culture or enzymes. In addition, because of the high temperature involved, biofuels produced would be free of biologically active microorganisms or compounds including bacteria, viruses, and even prion proteins [22].

Applications of water as a process media of biomass range from subcritical water extraction to supercritical water oxidation (SCWO). SCWO is a rapidly developing technology for the destruction of organic wastes. In 1994, the world's first commercial SCWO facility for treating industrial waste water became operational. Ongoing catalytic SCWO studies have demonstrated the benefit of utilizing heterogeneous catalysts for reducing energy and processing costs. The low dielectric constant and low polarity of SCW are closer to the properties of non-polar solvents; SCW is therefore capable of dissolving the majority of organic compounds and gases. This characteristic provides an advantage over traditional biomass wet oxidation processes, where oxidation rates are most likely mass-transfer limited due to the low solubility of oxygen and organic compounds in the liquid phase of water [23].

One of the biggest disadvantages of high-temperature water treatment is corrosion, which is a problem for all subcritical and supercritical water systems. Special materials for vessel linings and tubing are needed to resist the highly reactive chemical species generated during the process. These challenges demand superior engineering and design expertise for all system components. The water and process streams must both be pumped to high initial pressures under exact flow and pressure control. The heat exchangers are subjected to high heat transfer rates at high temperatures, but must maintain precise temperature control. The reaction vessel requires accurate temperature, pressure, and flow control. The vessel must seal reliably and be leak free each time it is used.

### 3.2.3    Chemical Pretreatment of Biomass

#### 3.2.3.1    *Acid Hydrolysis*

Acid pretreatment of lignocellulosic biomass can generally be performed using two approaches: either in concentrated acids at relatively lower temperatures or in diluted acids at relatively higher temperatures. The overall goal for acid pretreatment is to solubilize hemicellulose (primarily in the form of xylose), break the lignin seal and decrease the crystallinity of cellulose so that the treated biomass is more susceptible to subsequent enzymatic hydrolysis.

For concentrated acid pretreatment, concentrated sulfuric acid ($H_2SO_4$) of 72%, hydrochloric acid (HCl) of 41%, or trifluoroacetic acid (TFA) of 100% can be used [24]. During the treatment, the milled raw materials (e.g. corncob, switch grass, or bagasse) are treated with pre-heated concentrated acid in a bioreactor under 100°C. Since concentrated acids can efficiently remove lignin and hemicellulose as well as hydrolyzing some cellulose into glucose; a variety of biomass such as wood can be used as feedstock. However, the process also requires a massive use of concentrated acids which is highly corrosive. Acid recovery and the use of special bioreactors are therefore necessary to make the process economically feasible [25]. On the other hand, dilute acid (<4 wt%) hydrolysis is regarded as an efficient and inexpensive pretreatment method [26, 27]. Lu *et al.* [28] reported the pretreatment of corn stover with 2% $H_2SO_4$ at 120°C for 43 min giving a maximum xylose yield of 77%. The study indicated that the majority of hemicellulose is degraded in the pretreatment. In addition to $H_2SO_4$, the use of other dilute acids, including nitric, hydrochloric and phosphoric acids have also been reported [29].

Figure 3.3 depicts a pilot-scale reactor for acid hydrolysis of corn stover [30]. First, biomass is fed to the hopper and it is ground into small particles. Dilute acid is then sprayed on the ground biomass. Finally, it is sent to the pretreatment reactor for incubation at 150–200°C for 3–20 min in order to solubilize the hemicellulose into sugars, while cellulose and lignin remains unaffected during the process. At the end, the acid-pretreated biomass is recovered by filtration.

The major drawback of acid pretreatment is that the by-products formed during the treatment may inhibit the downstream fermentation process, potentially causing a reduction in product yield. These inhibitors can be categorized as organic acids, furan derivatives, and phenolic compounds, which are derived from degradation of hemicellulose, cellulose, and lignin (Figure 3.4) [31, 32]. If the concentration of these by-products is too high, detoxification procedures for the removal of these by-products might be required.

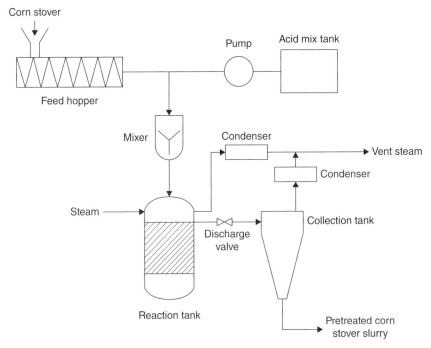

**Figure 3.3** *Simplified diagram of the process of dilute acid hydrolysis of corn stover in a pilot-scale reactor. Reproduced from [30], with kind permission from Springer Science and Business Media.*

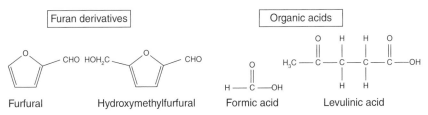

**Figure 3.4** *Chemical structures of some common fermentation inhibitors (furan derivatives and organic acids) generated from acid hydrolysis of lignocellulose.*

### 3.2.3.2 Alkaline Hydrolysis

Alkaline hydrolysis is one of the most commonly used chemical pretreatment processes, which can remove primarily lignin and some hemicellulose from lignocellulosic biomass. Pretreatment of biomass with alkalis causes saponification of intermolecular ester bonds that crosslink lignin and xylem. This subsequently leads to delignification [33, 34]. Sodium hydroxide, calcium hydroxide (lime), potassium hydroxide, aqueous ammonia, and ammonia hydroxide are the most commonly used alkalis for this process. Of those, sodium hydroxide is the most popular due to its low cost and efficiency. Alkaline pretreatment is more effective towards lignocellulosic biomass with lower lignin content such as hardwood, herbaceous crops, and agricultural residues in comparison to those with higher lignin content such as softwood [35].

The process is relatively mild compared to other pretreatment methods and it can be carried out in a batch mode. This pretreatment process involves spraying alkali onto biomass and soaking it for periods from hours to days at ambient temperature. The reaction time can be greatly reduced at elevated temperature [24]. Xu and co-workers reported the use of dilute sodium hydroxide (1%) for pretreatment of switchgrass to efficiently reduce the lignin content by 85.8% at 121°C in 1 h, 77.8% at 50°C, and 62.9% at 21°C for 48 h [36]. At the end of the process, the pretreated biomass was recovered by filtration and neutralized before further processing.

Compared to acid hydrolysis, alkaline hydrolysis is a milder process and thus sugar degradation to furfural, 5-hydroxymethyl-2-furfural (HMF) and organic acids is reduced. Moreover, caustic salts such as calcium carbonate can be recovered from the aqueous liquid/solution generated by the system as insoluble calcium carbonate by neutralizing it with inexpensive carbon dioxide [24]. The energy requirement of the process is low and it can be conducted at ambient temperature and pressure. However, the disadvantage of this method is the use of corrosive chemicals and their associated operating and environmental issues.

### 3.2.3.3  *Ozonolysis*

Treatment of lignocellulosic biomass with ozone gas is referred to ozonolysis. The pretreatment process can effectively degrade lignin and part of the hemicellulose of the biomass [4, 37], while cellulose remains unaffected. Ozone is a gas which is readily soluble in water and is highly reactive towards compounds incorporating conjugated double bonds and functional groups with high electron densities. Since lignin contains high carbon–carbon double bond content compared to hemicelluloses and cellulose, lignin is selectively oxidized in this process, generating low molecular weight compounds such as formic and acetic acids. Ozone treatment for delignification can be applied to a variety of lignocellulosic biomass including wheat straw, bagasse, green hay, peanut, pine, cotton straw, and sawdust [24]. In a recent study, it has been reported that as much as 49% lignin degradation was achieved when corn stover was treated by ozonolysis [34].

Figure 3.5 depicts laboratory-scale ozonolysis apparatus. To enhance the ozonolysis, the biomass is initially hydrated with water in order to adjust the moisture to around 25–35% before feeding the bioreactor [38]. The grinded

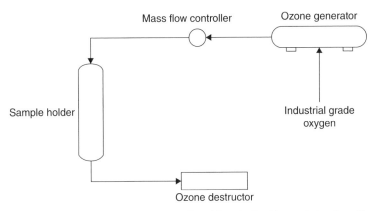

**Figure 3.5**    *Ozonolysis reactor set-up. Reproduced from [38], with permission from Elsevier.*

biomass in the bioreactor is then exposed to ozone gas stream which is generated by an ozone generator, and allows the reaction to proceed. The retention time of the treatment varies with different types of biomass. The ozone concentration used in most cases is around 2–6% (w/w) [38]. The unreacted ozone from the reactor outlet is finally removed by passing through potassium iodide solution.

The major advantage of using ozonolysis as pretreatment process is the low energy requirement, since ozonolysis can take place at ambient temperature and normal pressure [24]. Moreover, inhibitory compounds to the subsequent hydrolysis or fermentation are not produced [39, 40]. However, this process is relatively expensive as a significant amount of ozone gas is required, limiting its application on an industrial scale [4].

### 3.2.4    Microwave-Assisted Hydrothermal Biomass Treatment

The hydrothermal treatment of biomass with the addition of either acid or alkaline catalyst provides an effective pretreatment pathway for the conversion of lignocellulosic biomass to sugars. However, their presence within the aqueous phase subsequent to the process mean that further treatments are required to render the products pH neutral.

When these methods are applied to cellulose, it is often found that the yield of hexose sugars is minimal with little selectivity towards fermentable sugars, for instance glucose. This is attributed to the limited accessibility of cellulose chains resulting from their high degree of crystallinity. However, it has been found that subjecting cellulose to microwave radiation as a source of heating results in the opening of the crystalline structure (from crystalline to amorphous structure) at temperatures lower than that required under conventional thermal means (180°C in comparison to 250°C). This reduces the need for the addition of aggressive chemical agents with the additional benefit of preferential depolymerization to glucose.

This has been reported by Fan *et al.* [[41], whereby it is believed that the method of heating through microwave interaction (through dipolar polarization) with celluloses $CH_2O(6)H$ mean that, above 180°C, these $CH_2OH$ groups could be involved in a localized rotation in the presence of microwaves. The latter act in similarly to 'molecular radiators', allowing the transfer of microwave energy to the surrounding environment.

With the limited presence of water inside the rigid cellulose framework, this is likely to involve collisions between the $CH_2OH$ groups and the anomeric C1 of the same glucose ring, thus forming levoglucosan. The latter can easily hydrolyze to glucose, which is thought to be directly related to microwave activation of the $CH_2OH$ pendant groups. This is summarized as the proposed decomposition mechanism in Figure 3.6.

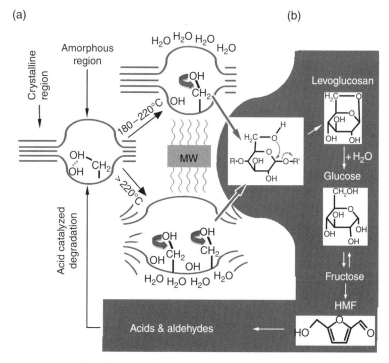

**Figure 3.6**  *Proposed mechanism for the microwave decomposition of cellulose. Representation of the cellulose–microwave interaction as a function of temperature: (a) mechanism of $CH_2OH$ group activation; (b) scheme of cellulose degradation toward acids and aldehydes. Reproduced with permission from [41]. Copyright © 2013, American Chemical Society.*

A result of this unique interaction is that, at 220°C under microwave treatment, it was found that glucose yield was nearly 50 times higher than that under similar conventional hydrolysis conditions. According to the proposed model, the yield of glucose obtained below 220°C is only related to the depolymerization of amorphous cellulose and not the crystalline content that becomes active above 220°C.

It has also been found that glucose yield under microwave conditions is power density dependent, suggesting that the microwave activation/decomposition of cellulose has a strong kinetic dimension. These are thought to result from two competitive processes determined by the speed of the $CH_2OH$ group rotation: (1) acceleration by interaction with microwave photons; and (2) deceleration through interaction (e.g., collision, electromagnetic) with neighboring groups. The dominance of either process depends on the degree of freedom of the $CH_2OH$ groups. For depolymerization of cellulose to occur, the $CH_2OH$ groups need to acquire the activation energy necessary to provoke the proposed $SN_2$ reaction described above. This can already be achieved at high microwave power densities, while more elevated temperatures are required to liberate the $CH_2OH$ groups when using lower microwave densities.

### 3.2.5    Biological Pretreatment

#### 3.2.5.1    *Fungal Digestion*

Although thermal chemical methods provide efficient means for biomass pretreatment to reduce the recalcitrance of the lignocellulosic biomass, these methods require expensive equipment, the use of corrosive chemicals, and intensive energy consumption, which limits their industrial applications and may contribute to environmental pollution. In contrast, biological pretreatments that utilize wood-degrading microorganisms for lignin removal can offer a safer, cheaper, and more environmentally friendly biomass pretreatment process. White-rot fungi are regarded as the most effective microorganisms for delignification of many different lignocullulosic biomass including wood chips, wheat straw, grass and softwood [42]. This is due to its remarkable ability to secrete a series of extracellular lignin-degrading enzymes such as lignin peroxidases, manganese peroxidases, and laccases to degrade lignin and hemicellulose [43, 44]. These can degrade lignin completely into carbon dioxide and water. *Phanerochaetechrysosporium*, *Ceriporiosissubvermispora*, *Cyathusstercoreus*, *Dichomitussqualens*, *Phlebia radiate*, *Pleurotllsostreatus*, and *Trametesversicolor* are some widely studied white-rot fungi that show high delignification efficiency [45, 46].

Solid-state fermentation is a preferred process to perform biological delignification over submerged fermentation because it leads to lower costs for product recovery, drying and lower risk of contamination [47]. Delignification of lignocellulosic biomass with white-rot fungi can be conducted at solid state in a bioreactor under controlled conditions. Initial moisture content, temperature, and aeration rate of the biomass are the three crucial factors that directly affect the fungal growth and its ability to produce lignin-degrading enzymes. These factors play an important role in delignification of biomass. For example, it has been reported that when the moisture content of the biomass is too high, it may promote fungal mycelium formation and inhibit the delignification process [48]. In contrast, low moisture content may retard fungal growth and lead to poor delignification. The optimal temperature for delignification depends greatly on the selection of the fungal strains, which ranges from 20 to 30°C [49]. Good aeration should also be ensured throughout the process.

This approach is however time consuming. Normally, several weeks are required for significant delignification and thus use of white-rot fungi for delignification in industries has been limited to date [50].

### 3.2.6  Summary

Pretreatment is a critical step to increase biomass conversion efficiency. In general, pretreatment can be divided into four major groups: mechanical, chemical, physicochemical, and biological methods. Each pretreatment method has its own advantages and limitations depending on the type of biomass used. There is no single method that can be regarded as the best option in biomass pretreatment. In general, an efficient pretreatment method should be able to eliminate the unwanted biomass fraction selectively in a cost-effective, environmentally friendly, and time-efficient manner. The process conditions, major effects, advantages, and disadvantages of the pretreatment methods reported in this chapter are summarized in Table 3.1.

## 3.3  Thermochemical Processing of Biomass

### 3.3.1  Direct Liquefaction

Direct liquefaction is a high-pressure thermal decomposition and hydrogenation process which converts biomass into hydrocarbon oil as fuel in solvent under elevated pressure (up to 250 bars) and temperature (up to 500°C). Through this process, biomass is depolymerized into unstable monomers and then re-polymerized to a broad range of hydrocarbon compounds under an oxygen-deficient or depleted environment [51]. Water and carbon dioxide are formed to remove the oxygen in the biomass. Basically, direct liquefaction has a high feed flexibility but only a single type of feedstock can be processed in a single batch under the specific condition with a suitable catalyst in order to obtain consistent products and yields. Since direct liquefaction is performed in solvent, no pre-drying is required. The process can therefore be applied to wet biomass with up to 78% moisture content [52], while other processing technologies such as gasification and combustion are only suitable for biomass with low moisture content [53].

**Table 3.1** Summary of process conditions, major effects, advantages and disadvantages of biomass pretreatment methods.

| Pretreatment methods | Major process conditions | Major effects | Advantages | Disadvantages |
|---|---|---|---|---|
| Mechanical processes | | | | |
| Chipping, grinding, and milling | Final particle size of the material (10–30mm after chipping and 0.2–2 mm after milling or grinding) | Reduction of particle size and the crystallinity of lignocellulosic materials | Control of the final particle size | High energy requirements |
| Densification (pelletization) | High pressure | Reduction of particle size; conversion of biomass into high-density solid energy carriers | Increased biomass energy value | Limited choice of raw materials |
| Physical processes | | | | |
| ScCO₂ and natural solvent | Pressure and temperature above the critical point of the compounds | Extraction of bio-oil and extractives | Specific compounds extraction; low environmental impact; no remaining solvent residues | High energy requirements |
| AFEX | Treated with liquid anhydrous ammonia; temperature: 60–100°C; pressure: 250–300 psi | Lignin solubilization; hemicellulose hydrolysis; cellulose decrystallization; increased surface area | No wash stream; no need for neutralization; free of contaminants | High energy requirements |
| Sub- and supercritical water treatment | Temperature: near or above 374°C; pressure: near or above 218 atm | N/A | Allow high concentrations of reactants; free of biologically active organism or compounds; high solubility of various organic compounds and gases | High energy requirements; corrosive; high operating costs |

(Continued)

**Table 3.1** (Continued)

| Pretreatment methods | Major process conditions | Major effects | Advantages | Disadvantages |
| --- | --- | --- | --- | --- |
| Chemical processes<br>Acid hydrolysis | Dilute acids;<br>HCl/$H_2SO_4$/$HPO_3$;<br><4% acid; <160°C;<br>concentrated acid;<br>HCl/$H_2SO_4$/TFA;<br>>40% for concentrated acid;<br>160–220°C; treatment period<br>for hours or less | Dilute acid; hemicellulose<br>removal; concentrated acid;<br>cellulose and hemicellulose<br>hydrolyzed | Low cost; effective for wide<br>range of biomass | Formation of<br>fermentation<br>inhibitors;<br>corrosive |
| Alkaline hydrolysis | NaOH/KOH/$CaOH_2$ of<br>1–10%;<br>treatment time of hours– days<br>at ambient temperature | Lignin and hemicellulose removal | Low cost; low energy<br>requirement; low level of<br>inhibitors; low energy<br>requirement | Only suitable for<br>biomass with low<br>lignin content;<br>incorporation of<br>high amount of<br>salts |
| Ozonolysis | 2–3% (w/w) ozone with flow<br>rate of 60 L hr$^{-1}$; treatment<br>time for 1–2 hr at ambient<br>temperature and pressure | Lignin removal; partial<br>hemicellulose removal | | Expensive |
| Biological processes<br>Fungal digestion | Fermentation temperature at<br>20–30°C; treatment period<br>from weeks to months;<br>initial biomass moisture<br>content of 60–85% | Lignin removal | Very low cost; no harsh<br>chemicals; environmentally<br>friendly | Very time<br>consuming |

Some chemical and physical parameters are critical to the hydrocarbon yields [51]. These include temperature, pressure, residence time, selection of solvent and catalyst, as well as their concentrations. Intensive research has been carried out to optimize these parameters to increase hydrocarbon oils yield. Influence of the temperature, residence time, and concentrations of catalysts on the oil yield have been found to follow a volcano-type pattern in which the hydrocarbon oil yield reaches a maximum in the intermediate range and drops after any further increase, while the pressure shows a positive correlation with the hydrocarbon oil yield [53]. On the other hand, the addition of solvent, catalyst, and reducing gas have been demonstrated to improve the yield of hydrocarbon oil and prevent the formation of solid char condensed by the monomers formed in the depolymerization [51]. Catalysts such as NaOH and $H_2SO_4$ are employed to decrease the required reaction temperature and enhance reaction kinetics. A solvent is usually added to prevent the monomer units from re-polymerizing or condensing into undesirable solid chars. Alcohols with a short carbon chain such as methanol and ethanol are recognized as favorable solvents due to the lowest formation of solid char. Furthermore, reducing gases such as $H_2$ and CO are used to increase the ratio of hydrogen to carbon in order to facilitate an efficient degradation of the oxygen-containing functional groups.

Direct liquefaction of biomass has been tested at a pilot scale in several countries such as Denmark, Germany, and the USA [54]. The CatLiq® technology developed by the Danish company SCF Technologies has been successfully scaled up in a pilot facility in Copenhagen, which is currently one of the largest direct liquefaction facilities in the world. It processes $20\,L\,hr^{-1}$ of dried distillers grains with solubles (DDGS), which are the crude protein, crude fiber, and crude fat generated from first-generation ethanol production. The schematic flow of the CatLiq process is shown in Figure 3.7. First, the feed from the feed tanks is pressurized by the feed pump and preheated by the feed heater. On the other hand, the circulation pump provides a high flow rate for a uniform distribution of heat and instantaneously heats up newly added feed. The feed then passes through a trim heater and enters a fixed-bed reactor filled with a zirconia-based catalyst at subcritical condition (280–370°C and 250 bars) for liquefaction. Water is used as

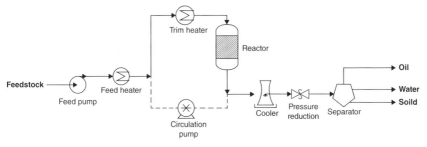

**Figure 3.7**   *Flow diagram representing the processes in a direct liquefaction plant. Reproduced from [55], with permission from Elsevier.*

solvent in this process. The product stream is then cooled, depressurized, and centrifuged to separate the hydrocarbon oil from the water. The oil produced is used onsite or transported for further processes into fuels for power generation [53, 55].

Nowadays, direct liquefaction is not a common biomass valorization technology. The core problem is the high oxygen level of hydrocarbon oil, which prevents it from substituting gasoline or diesel fuel. Research on direct liquefaction of biomass mainly focuses on considerably reducing the oxygen level of the oil in order to facilitate more diverse applications such as transportation fuel [51].

### 3.3.2   Direct Combustion

Direct combustion is one of the oldest methods of energy production. It is the burning of biomass to release the stored chemical energy in the form of heat or combined heat and power (CHP) in excess oxygen. The combustion process consists of very complicated reactions that can generally be categorized into different stages, namely devolatilization and char combustion. The chemical reactions involved are shown in Figure 3.8.

In the devolatilization stage, a mixture of volatile fuel gases such as volatile hydrocarbon, carbon monoxide, and hydrogen and liquid are generated as the biomass is dried in the heating process. The fuel gases generated are then further oxidized in the combustion process to provide more heat energy in the presence of oxygen. With the use of various devices such as boilers, burners, turbines, or internal combustion engine generators, the energy released can also be transformed into electricity. A wide range of biomass including forestry waste, barks, sawdust, and animal droppings could be used for combustion. Although pollutants such as nitrogen oxides, sulfur dioxide, dioxin, and carbon dioxide are also produced during biomass combustion, the process is regarded as carbon neutral.

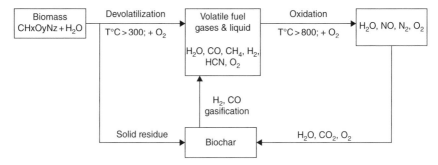

**Figure 3.8**   *Flowchart describing the direct combustion process and the products generated. Reproduced with permission from [56]. Copyright © 2003, American Chemical Society.*

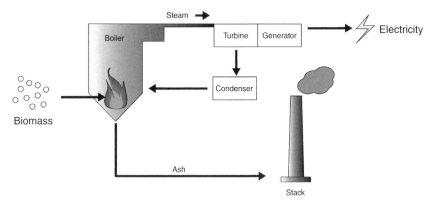

**Figure 3.9**  *Direct combustion process of biomass for electricity generation through a steam-turbine generator system.*

Figure 3.9 depicts the process of electricity generation from biomass combustion through a steam turbine-generator system. Initially, biomass is milled and burnt in a boiler to produce high-pressure steam. The steam passes through a series of turbines which are connected to electric generators.

The direct combustion of biomass for energy generation faces several problems. Biomass is diverse in nature, and generally has the detrimental combination of low physical and low energy density as a fuel. The calorific value (heat of combustion) of biomass is typically around 50% of coal. Poorer physical characteristics such as higher water content and poorer grind ability generally mean that collection, transportation, and use is not only challenging in terms of logistics and processing, but also questionable in terms of overall environmental benefits. In addition, the direct combustion of biomass also results in a lost opportunity with regards to the co-production of chemicals and energy.

Several options are available for the valorization of biomass by condensing the energy within. In terms of thermally based treatments, the most common are as follows.

1. *Gasification*: a high-temperature process (between 750 and 1800°C) of degradation of biomass to produce energy-rich gases.
2. *Pyrolysis*: a medium- to high-temperature process (400–750°C) whereby the biomass is degraded to yield mainly liquid products for energy generation.
3. *Torrefaction*: a low-temperature treatment (below 300°C) of biomass to produce a higher-energy-density solid fuel.

Commercial uptake of these technologies faces a number of barriers. The set-up and processing costs are typically high, the quality of products in comparison to fossil fuels is typically poorer, and the enhancement in the energy/calorific value is low (e.g., on average only about 10% in the case of torrefaction). Furthermore, the image associated with pyrolysis as a technology is often negative [51]. It is clear that new, clean, and efficient technologies for the conversion of biomass to fuels and other valuable products are necessary to help us meet the challenges of a growing population, diminishing resources, and climate change.

### 3.3.3    Gasification

Gasification is a thermochemical process converting biomass into syngas, which primarily consists of CO and $H_2$, by partial oxidation of biomass in the presence of an oxidizing agent. The process conditions of gasification are similar to those of pyrolysis, in which the biomass is converted into liquid (hydrocarbon oil), gaseous (fuel gas), and solid (charcoal) fractions in the absence of $O_2$. Gasification takes place at higher temperatures ranging from 700 to 1000°C leading to relatively higher gas yields (up to 85%) compared to pyrolysis, which is up to 35% [57, 58]. The syngas obtained can be used for heat and power generation and productions of chemicals such as methanol, dimethyl ether, alcohol, organic acids, and polyesters [59]. Gasification is able to process various types of biomass, including agricultural residues, forestry residues, food waste, and organic municipal wastes [59]. In order to facilitate biomass gasification, ground biomass with a moisture content lower than 10% is recommended for this process [58]. Upon heating to 350°C in the absence of air or $O_2$, the biomass is thermally decomposed and reduced to solid carbonized biomass by devolatilization. At 700–1000°C, oxidizing agent (air, steam, $CO_2$, $O_2$, or a mixture of these) is added to oxidize the solid carbonized biomass. Reduction reaction also takes place at high temperature (800–1000°C) when the oxidizing agent is consumed. Table 3.2 summarizes the reactions taking place during the gasification process using steam as oxidizing agent. As a result, a gas mixture consisting of CO, $CO_2$, $CH_4$, $H_2$ and $H_2O$ is produced together with impurities including tars, char, alkali, sulfur compounds, and nitrogen compounds. As a result, the gas mixture needs to undergo a series of downstream processing steps such as barrier filtering, hot gas cleaning, and wet scrubbing to remove the impurities before it can be used as syngas.

**Table 3.2**  *Chemical reactions which occur during process of biomass gasification [58, 59].*

| | | |
|---|---|---|
| Oxidation | Partial oxidation reaction | $2C + O_2 \rightarrow 2CO$ |
| | Complete oxidation reaction | $C + O_2 \rightarrow CO_2$ |
| Reduction | Water-gas reaction | $C + H_2O \rightarrow CO + H_2$ |
| | Bounded reaction | $C + CO_2 \leftrightarrow 2CO$ |
| | Shift reaction | $CO_2 + H_2 \leftrightarrow CO + H_2O$ |
| | Methane reaction | $C + 2H_2 \leftrightarrow CH_4$ |
| Overall | $CH_xO_y$ (biomass) + $O_2$ (21% of air) + $H_2O$ (steam) → | |
| | $CH_4 + CO + CO_2 + H_2 + H_2O$ (unreacted steam) + C (char) + tar | |

Apart from syngas, significant amount of pollutants such as tar and polycyclic aromatic hydrocarbons (PAHs) are also generated from the gasification process. This is one of the most significant technical barriers to its commercialization [60].

### 3.3.4 Pyrolysis

Pyrolysis is a thermal treatment that decomposes organic biomass into low-molecular-weight liquid, solid, and gaseous products in the absence of air or oxygen. The process is conducted in the absence of air to prevent the combustion of biomass into carbon dioxide. The liquid fraction obtained at the end of pyrolysis refers to pyrolysis oil or bio-oil. It is a mixture of hydrophilic organics, water, and tars [61]. Bio-oil can be used directly as fuel to generate heat or electricity through a turbine or a boiler. However, due to its poor purity, low heating value, incompatibility with conventional fuels, high viscosity, incomplete volatility, and chemical instability [62, 63], its direct application as engine fuel is restricted. With further upgrading processes to purify, condition, and deoxygenate, bio-oil could potentially be used as transportation fuel, generating much interest from industry. Figure 3.10 shows the potential applications of bio-oil as raw materials for fuels and chemicals production. The remaining solid of pyrolysis refers to as bio-char or charcoal consists of around 85% carbon [65]. Bio-char is usually returned to the ground for soil remediation. The gas fraction consists mainly of $CO$, $CO_2$, $H_2$ and $CH_4$. Forestry residues and agricultural by-products such as wood residue, plant stalks, rice shells, or corn stovers are the commonly used renewable feedstock in pyrolysis [66].

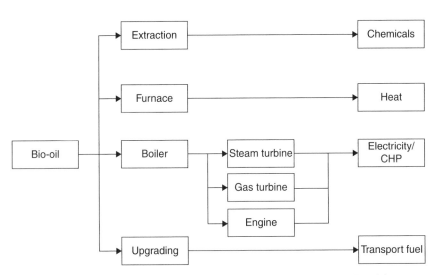

**Figure 3.10** *Potential applications of bio-oil after further upgrading. Reproduced from [64], with permission from Elsevier.*

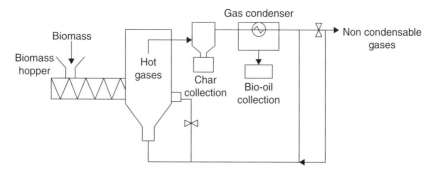

**Figure 3.11**   *Pyrolysis process with a fluidized bed pyrolysis reactor for bio-oil production.*

Pyrolysis can generally be divided into two main categories – fast or slow pyrolysis – depending on the heating regime which ranges from seconds to minutes. By varying the heating rate, the proportion of the ultimate products also varies. In fast pyrolysis, bio-oil is the main product. The heating rate of the process can be as high as $1000$–$10,000°C\,s^{-1}$ with a peak temperature of about $500°C$. At the end of the process, approximately $50$–$75\%$ of biomass (on a wet basis) is converted to bio-oil [61]. To maximize the bio-oil production, a high heating rate, strictly controlled reaction temperature range, a short residence time ($<3\,s$) of vapour in the reactor, and rapid quenching of the product gas should be ensured [64]. In contrast, the heating rate and final temperature of slow pyrolysis is lower, while residence times of solid and vapour are longer. Bio-char is produced as the primary product through carbonization in slow pyrolysis. The heating rate of the process is $0.1$–$2°C\,s^{-1}$ with a peak temperature of around $400°C$.

Figure 3.11 illustrates a fluidized bed pyrolysis reactor for bio-oil production from biomass. Biomass is fed into a pyrolysis chamber (a hot sand bed) through the hopper and the biomass is heated to the required pyrolysis temperature. The biomass starts to decompose in this chamber and both the condensable and non-condensable vapours released from the biomass leave the chamber. The bio-char formed remains partly in the chamber and partly in the gas. When the condensable vapour passes through the condenser, it condenses as bio-oil which is collected in the chamber. While part of the non-condensable vapour leaves the chamber as gas products, another part of it is returned to the system to provide heat.

### 3.3.5   Torrefaction

The primary goal of torrefaction is to refine raw biomass to an upgraded solid fuel with better handling qualities and enhanced combustible properties similar to those of fossil coal, leading to decreased costs (see Table 3.3). The torrefaction process reduces oxygen content and thus increases the energy content (high heating value) from typically $10$–$17\,MJ\,kg^{-1}$ to as much as $19$–$22\,MJ\,kg^{-1}$ (typically

**Table 3.3** *Comparison of biomass properties before and after torrefaction.*

| Original biomass | Torrefied biomass |
|---|---|
| Bulky | Densified |
| Wet | Dry and hydrophobic |
| Expensive to grind | Easily crushed |
| Low energy density | High energy density |
| Biodegradable | Non-biodegradable |
| Expensive to transport | Less expensive to transport |

lower for agricultural residues) with energy densification ratios of 1.3. Torrefaction also yields a material, which can be stored over prolonged periods of time and feed more economic, year-round operations. Biomass is bulky and vulnerable to degradation over time, making it difficult to store.

The main torrefaction product is the solid phase which, as for pyrolysis, is usually referred to as the charred residue (or char). In the field of torrefaction the solid product is also frequently called torrefied wood or torrefied biomass. By mass, important reaction products other than char are carbon dioxide, carbon monoxide, water, acetic acid, and methanol [67]. All these non-solid reaction products contain relatively more oxygen compared to the untreated biomass. The O/C ratio of torrefied biomass is therefore lower than untreated biomass, resulting in an increase of the calorific value of the solid product. After the biomass has been torrefied it can be densified, usually into briquettes or pellets using conventional densification equipment, to further increase the density of the material and hence its energy content, as described in Section 3.2.1.2. In addition, the biomass exchanges its hydrophilic properties to hydrophobicity, allowing for effortless storage that goes hand-in-hand with a greater resistance against biological degradation, self-ignition, and physical decomposition.

The results of torrefaction are different for different types of biomass and very much dependent on the cell structure of the biomass. The different structural components (cellulose, hemi-cellulose, and lignin) of biomass are affected by heat very differently. Whereas the hemicellulose decomposes within the torrefaction temperature range (about 290°C), the cellulose decomposition occurs at a much higher temperature (about 360°C). In contrast to polysaccharides (cellulose and hemicellulose), phenolic-based lignin is pyrolyzed at a broad range of temperatures between 250 and 400°C (see Figure 3.11). It can be seen from Figure 3.12 that sample mass loss under torrefaction conditions is lower than 20%, comprising: (1) 10% mass loss which corresponds to water removal; and (2) 10% of actual organic matter decomposition. Hemicellulose is the main component which is affected by torrefaction.

### 3.3.5.1 *Microwave Torrefaction*

Researchers have recently investigated the microwave-assisted torrefaction of biomass and demonstrated microwave activation effects [41, 68–71]. Surprisingly, microwave-mediated low-temperature torrefaction yields a variety of products

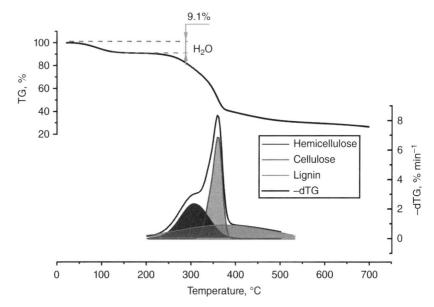

**Figure 3.12** *Comparison of thermal behavior of biomass components within the torrefaction temperature range.*

typical of high-temperature conventional pyrolysis ranging from gases, liquids, and solids. Furthermore, this low-temperature microwave process is scalable, energy efficient, and generates five major types of products from lignocellulosic biomass:

1. a high-quality char with properties superior to those achieved by most conventional methods and with an enhanced energy value, feedability, and grindability, making it suitable as a coal replacement;
2. bio-oil, suitable for upgrading to liquid fuel;
3. an aqueous solution of organic acids and aldehydes;
4. an aqueous solution of sugars; and
5. a gas fraction containing combustible organic compounds, which could be used for energy production.

Commercial development of torrefaction is currently in its early phase. Several technology companies and their industrial partners are moving towards commercial market introduction. The current demand for torrefied biomass of utilities alone greatly exceeds the production capacity that can be realized in the coming years. An overview of reactor technologies that are applied for torrefaction is provided in Table 3.4.

Most torrefaction technology developers listed in Table 3.4 are companies with extensive backgrounds in biomass processing and conversion technologies such as carbonization and drying. The reactors being developed are, in most cases,

**Table 3.4** *Overview of reactor technologies and associated suppliers. Reproduced from [72], with permission from Elsevier.*

| Reactor technologies | Torrefaction equipment suppliers |
|---|---|
| Rotary drum reactor | CDS (UK), Torr-Coal (NL), Bio3D(FR), EBES AG (AT),4 Energy Invest (BE), Bioendev/ETPC (SWE), Atmosclear,S.A.(CH) |
| Screw conveyor reactor | BTG (NL), Biolake (NL), Foxcoal (NL), Agri-tech producers (US) |
| Multiple Hearth Furnace (MHF)/TurboDryer | CMI-NESA (BE), Wyssmont (US) |
| TORBED reactor | Topell (NL) |
| Microwave reactor | Rotavawe (UK); Sairem (France) |
| Compact moving bed | ECN (NL), Thermya (FR), Buhler (US) |
| Oscillating belt conveyor | Stramproy Green investment (NL), New Earth Eco Technology (US) |

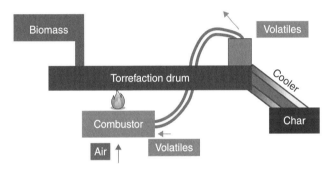

**Figure 3.13** *Industrial-scale torrefaction processor.*

established technologies that companies are familiar with and have been optimizing for torrefaction applications. Currently, no single technique is fundamentally superior to others, each technology having its advantages and drawbacks. A standard processor for biomass torrefaction is shown in Figure 3.13. Four sections can be distinguished: (1) the biomass feeding section; (2) the reactor section, where biomass is converted into torrefied material and a combustible gas; (3) the cooling section (the torrefaction product is highly flammable, hence the need for a cooling system); and (4) the combustor section, where the produced gases and vapours are burned in an excess of oxygen and the heat generated is used to heat the process. Proper selection of a reactor for given biomass properties and applications is important, as each reactor has unique characteristics and is designed to handle specific types of biomass.

Despite global efforts to develop torrefaction technology, there are still several technical and economic challenges that need to be overcome before the technology is fully commercialized. The torrefaction process itself is not fully understood scientifically, and the effects of reaction conditions are still being investigated [73, 74]. However, significant progress has been made within a relatively short period; we may see some systems in commercial operation in the near future [72].

For full commercialization, torrefaction reactors still require to be optimized to economically meet end-user requirements and achieve market standardization of the product [12, 44]. Certain characteristics need to be proven or scaled-up in order to meet commercial expectations [3, 44, 75]. A major operation challenge is achieving optimum torrefaction process control of the biomass feed, process temperature and residence time to ensure optimal thermal efficiency and consistent torrefied product [12, 51, 70, 76].

## 3.4    Biological Processing

### 3.4.1    Fermentation

Fermentation involves the use of microorganisms such as bacteria and fungi to transform sugars into products. Bioconversion of biomass into products through fermentation is a very flexible process which could lead to a wide range of products including biofuels, biochemicals, or biomaterials. Traditionally, sugar or starch crops such as sugarcane and corn have been used for bioethanol production [1]. More recently, agricultural by-products from cereals [77–80] or municipal waste such as food waste [81–84] have been used as feedstock for bioconversion to further lower the production cost.

An overview of bioconversion processes, from biomass through fermentation, is summarized in Figure 3.14. Biomass is pretreated before undergoing the fermentation step. For lignocellulosic biomass, the pretreatment step is aimed at removing lignin and hemicellulose in order to increase the subsequent hydrolysis step for glucose production, thereby increasing the product fermentation yield. Chemical pretreatment methods such as acid/alkaline hydrolysis are commonly used. Subsequently, the remaining cellulosic-rich biomass is digested by cellulase in order to produce glucose. The hydrolysate formed is a nutrient-rich medium, which serves as a generic feedstock in the subsequent microbial fermentation.

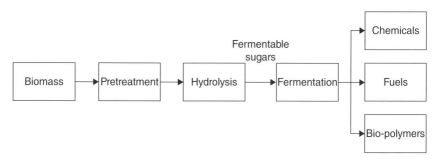

**Figure 3.14**    *Biomass conversion routes of bioproduct formation through fermentation.*

### 3.4.2   Anaerobic Digestion

Anaerobic digestion (AD) involves a mixture of bacteria including syntrophic bacteria, fermentative bacteria, acetogenic bacteria and methanogenic bacteria to decompose biomass under anaerobic conditions in order to produce biogas (methane and hydrogen) as fuel. The process is divided into four stages: hydrolysis, acidogenesis, acetogenesis, and methanogenesis. In the hydrolysis stage, insoluble organic compounds are broken down into water-soluble monomers by hydrolases secreted by a consortium of bacteria. The barely decomposable polymer, which is not decomposed by hydrolases, remains solid and thus limits the efficiency of AD. The products of hydrolysis are then converted into short-chain organic acids, alcohols, aldehydes, and carbon dioxide in the acidogenesis stage and transformed into acetates, carbon dioxide, and hydrogen in the acetogenesis stage. Finally, methanogenic bacteria utilize acetic acid, hydrogen, and carbon dioxide to produce methane. About 70% of the methane is converted from acetic acid while the remaining 30% results from carbon dioxide reduction. Biogas and digestate, which is the remaining solids with the mineralized nutrient, are the final products [85].

The biogas composition varies with substrates due to the composition difference. Lipids give the highest methane yield (1,014 m$^3$ ton$^{-1}$) while the methane yields of carbohydrates and proteins are much lower (415 and 496 m$^3$ ton$^{-1}$, respectively). Figure 3.15 depicts the yield of methane of different substrates [86].

Figure 3.16 shows the process flow diagram of the Hong Kong Organic Waste Treatment Facilities Phase II, which is an AD plant [87]. This facility can process

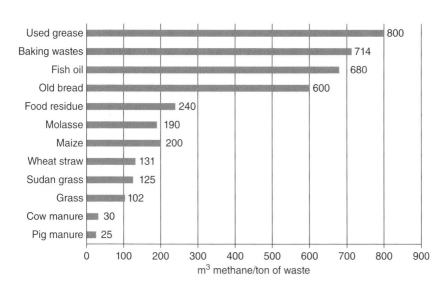

**Figure 3.15**   *Methane yield from different substrates [86].*

300 tons/day of source-separated organic waste in which food waste is the major component. Soil fertilizer and biogas are produced from the process. The process can be divided into pretreatment, AD, energy recovery, post-treatment of digestate, and air and waste water treatment. In the pretreatment process, organic waste undergoes manual sorting, trammel screen, metal separation, and shredding to separate inorganic and oversized materials, reduce particle size, and adjust the moisture content in order to facilitate an efficient AD. The inert materials that are not suitable for bio-conversion are disposed to landfills. The substrate is then heated to 70°C for 4 hr for pasteurization before entering the digesters. During the AD, the biogas generated is collected and delivered to the gas-cleaning system for the removal of impurities such as hydrogen sulfide, water, and particulates, followed by the CHP generator for onsite energy recovery or exported to the gas network. After digestion, the digestate passes through a separator for dewatering. The solid fraction is conveyed to the composting system while the liquid fraction is pumped to the waste water treatment system. The composting system comprises enclosed composting tunnels and a maturation area. Air is injected into the floor of the tunnels and passes through the composting mass to facilitate aerobic microbial digestion for 8 days. The composting mass undergoes maturation for 14 days before being used as soil fertilizer. On the other hand, the waste water generated from dewatering of the digestate and the leachate from composting and biogas drying are diverted to an onsite waste water treatment facility. It is expected that $0.89\,m^3$ of waste water is generated per 1 ton of waste. For the air treatment system, the air is processed by a scrubber for pH adjustment and ozone/UV for odor treatment.

Today, almost all of the biogas produced worldwide is used for heat and electricity production. It is also deemed a versatile renewable energy that can be utilized for many applications, including vehicle fuel and town gas. However, improved biogas quality is needed to meet the specific requirements of these applications. This mainly involves increasing the methane content in biogas by removing the impurities such as $CO_2$, $H_2S$, $NH_3$, and solid particles. In Sweden,

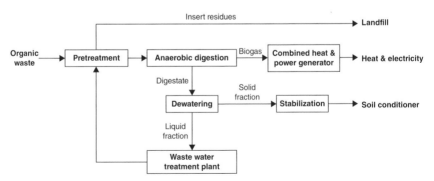

**Figure 3.16**    *The proposed process flow in the Hong Kong Organic Waste Treatment Facility [87].*

**Table 3.5** Summary of the major thermochemical and biological biomass conversion processess.

| Conversion method | Process conditions | Major products | Advantages | Disadvantages |
|---|---|---|---|---|
| Direct liquefaction | Biomass moisture content <10%; temperature: 700–1000°C; presence of solvent | Hydrocarbon oil | Effective for wide range of wet biomass; high product flexibility | Require external heat and power supply; require downstream processes to upgrade hydrocarbon oil |
| Direct combustion | Excess oxygen; biomass moisture content <50% | CHP | Short conversion time | Energy intensive; low technology |
| Gasification | Oxygen-deficient/depleted environment; biomass moisture content can be >50%; pressure: up to 240 bar; temperature: up to 500°C; presence of oxidizing agent | Syngas | Effective for wide range of biomass; high product flexibility | Require external heat and power supply; require downstream processing for purification |
| Pyrolysis | Absence of air or oxygen; Biomass moisture content <50% **Fast pyrolysis** Heating rate: 1000–10,000°C s$^{-1}$; peak temperature: 500°C **Slow pyrolysis** Heating rate: 0.1–2°C s$^{-1}$; peak temperature: 400°C | Bio-oil, biochar, combustible gases (e.g., H$_2$ and CH$_4$), CHP | Short conversion time; various products are obtained | Energy intensive |

(Continued)

**Table 3.5** (Continued)

| Conversion method | Process conditions | Major products | Advantages | Disadvantages |
|---|---|---|---|---|
| Torrefaction | Inert atmosphere at temperatures between 200 and 300°C<br>**Microwave torrefaction**<br>Temperature: >350°C; in the frequency band of 300 MHz to 300 GHz | Torrefied biomass with increased energy density<br>**Microwave torrefaction**<br>Bio-oil, gas, and charcoal | Mild condition | Limited availability of microwave apparatus and at extremely high cost; the microwaves can be reflected, absorbed and/or transmitted by the material before reaching the biomass |
| Fermentation | Fermentation temperature at 20–30°C; treatment period for about a weeks; medium with around 30 g L$^{-1}$ glucose and 1.5 g L$^{-1}$ nitrogen source for example yeast extract | Chemicals, biofuels, biopolymers | Low energy requirement; wide range of chemicals can be produced | Long conversion time; labor intensive |
| AD | Oxygen depleted environment; C/N ratio of feedstock: 15–30 | Biogas | Low energy requirement; capable of processing a wide range of biomass; power generation from biogas combustion | Methane yield depends on substrate; require downstream processes to upgrade biogas |

biogas has already been used as vehicle fuel. Similarly, biogas is also injected into the natural gas grid for public use in several European countries, for example Germany and Switzerland [88]. In view of the advantages provided by AD, which include organic waste valorization, the production of energy and soil conditioner, and reduction of GHG emissions, AD is expected to become more common as a biomass treatment.

## 3.5   Summary

With the global increase in energy demand and the depletion of fossil fuel sources, it is necessary to search for alternative sources of chemicals and energy. Due to its worldwide abundance, biomass represents an important source of renewable chemicals and energy. Through various conversion methods, biomass can be used to produce a wide range of biofuels, heat, electricity, and chemicals. In this chapter, we have introduced the major thermochemical and biological conversion processes available. The major products, process conditions, advantages, and disadvantages of these conversion processes are summarized in Table 3.5.

## References

1.  McKendry, P. (2002) Energy production from biomass (part 1): conversion technologies. *Bioresource Technology*, **83**, 37–46.
2.  Chandra, R.P., Bura, R., Mabee, W.E., *et al.* (eds) ( 2007) *Biofuels (Advances in Biochemical Engineering/Biotechnology).* Springer, Berlin, Heidelberg.
3.  Amarasekara, A.S. (2014) *Handbook of Cellulosic Ethanol*, John Wiley & Sons, Hoboken, NJ.
4.  Sun, Y. and Cheng, J. (2002) Hydrolysis of lignocellulosic materials for ethanol production: a review. *Bioresource Technology*, **83**, 1–11.
5.  Taherzadeh, M.J. and Karimi, K. (2008) Pretreatment of lignocellulosic wastes to improve ethanol and biogas production: a review. *International Journal of Molecular Sciences*, **9**, 1621–1651.
6.  COTE COMMUNITIES (2006) Commission of the European Communities, Green Paper: A European Strategy for Sustainable, Competitive and Secure Energy, COM(2006) 105, 8 March 2006. Available at http://europa.eu/documents/comm/green_papers/pdf/com2006_105_en.pdf (accessed 4 September 2014).
7.  Uslu, A., Faaij, A.P.C. and Bergman, P.C.A. (2008) Pre-treatment technologies, and their effect on international bioenergy supply chain logistics. Techno-economic evaluation of torrefaction, fast pyrolysis and pelletisation. *Energy*, **33**, 1206–1223.
8.  Hamelinck, C.N., Suurs, R.A.A. and Faaij, A.P.C. (2005) International bioenergy transport costs and energy balance. *Biomass & Bioenergy*, **29**, 114–134.
9.  Stelte, W., Sanadi, A.R., Shang, L. *et al.* (2012) Recent developments in biomass pelletization – a review. *BioResources*, **7**, 4451–4490.
10.  Sikkema, R., Steiner, M., Junginger, M. *et al.* (2011) The European wood pellet markets: current status and prospects for 2020. *Biofuels, Bioproducts & Biorefining*, **5**, 250–278.

11. Spelter, H. and Toth, D. (2009) North America's Wood Pellet Sector, United States Department of Agriculture Forest Service. Forest Products Laboratory Research Paper FPL-RP-656.

12. Agar, D. and Wihersaari, M. (2012) Torrefaction technology for solid fuel production. *Global Change Biology Bioenergy*, **4**, 475–478.

13. Christiansen, R.C. (2006) The art of biomass pelletizing, in *Biomass Magazine*, vol. **3**, BBI International, USA, pp. 25–29.

14. Chen, C.C. and Ho, C.T. (1988) Gas chromatographic analysis of volatile components of ginger oil (Zingiber officinale Roscoe) extracted with liquid carbon dioxide. *Journal of Agricultural and Food Chemistry*, **36**, 322–328.

15. Pereira, C.G. and Meireles, M.A.A. (2010) Supercritical fluid extraction of bioactive compounds: fundamentals, applications and economic perspectives. *Food and Bioprocess Technology*, **3**, 340–372.

16. Azmir, J., Zaidul, I.S.M., Rahman, M.M. *et al.* (2013) Techniques for extraction of bioactive compounds from plant materials: a review. *Journal of Food Engineering*, **117**, 426–436.

17. Sharif, K.M., Rahman, M.M., Azmir, J. *et al.* (2014) Experimental design of supercritical fluid extraction - a review. *Journal of Food Engineering*, **124**, 105–116.

18. Budarin, V.L., Shuttleworth, P.S., Dodson, J.R. *et al.* (2011) Use of green chemical technologies in an integrated biorefinery. *Energy & Environmental Science*, **4**, 471–479.

19. Alizadeh, H., Teymouri, F., Gilbert, T.I. and Dale, B.E. (2005) Pretreatment of switchgrass by ammonia fiber explosion (AFEX). *Applied Biochemistry and Biotechnology*, **121**, 1133–1141.

20. Teymouri, F., Laureano-Perez, L., Alizadeh, H. and Dale, B.E. (2005) Optimization of the ammonia fiber explosion (AFEX) treatment parameters for enzymatic hydrolysis of corn stover. *Bioresource Technology*, **96**, 2014–2018.

21. Savage, P.E. (1999) Organic chemical reactions in supercritical water. *Chemical Reviews*, **99**, 603–621.

22. Peterson, A.A., Vogel, F., Lachance, R.P. *et al.* (2008) Thermochemical biofuel production in hydrothermal media: a review of sub- and supercritical water technologies. *Energy & Environmental Science*, **1**, 32–65.

23. Ding, Z.Y., Frisch, M.A., Li, L.X. and Gloyna, E.F. (1996) Catalytic oxidation in supercritical water. *Industrial & Engineering Chemistry Research*, **35**, 3257–3279.

24. Kumar, P., Barrett, D.M., Delwiche, M.J. and Stroeve, P. (2009) Methods for pretreatment of lignocellulosic biomass for efficient hydrolysis and biofuel production. *Industrial & Engineering Chemistry Research*, **48**, 3713–3729.

25. von Sivers, M. and Zacchi, G. (1995) A techno-economical comparison of three processes for the production of ethanol from pine. *Bioresource Technology*, **51**, 43–52.

26. Torget, R., Himmel, M. and Grohmann, K. (1992) Dilute-acid pretreatment of two short-rotation herbaceous crops. *Applied Biochemistry and Biotechnology*, **34–35**, 115–123.

27. Esteghlalian, A., Hashimoto, A.G., Fenske, J.J. and Penner, M.H. (1997) Modeling and optimization of the dilute-sulfuric-acid pretreatment of corn stover, poplar and switchgrass. *Bioresource Technology*, **59**, 129–136.

28. Lu, X.B., Zhang, Y.M., Yang, J. and Liang, Y. (2007) Enzymatic hydrolysis of corn stover after pretreatment with dilute sulfuric acid. *Chemical Engineering & Technology*, **30**, 938–944.

29. Zheng, Y., Pan, Z. and Zhang, R. (2009) Overview of biomass pretreatment for cellulosic ethanol production. *International Journal of Agricultural and Biological Engineering*, **2**, 51–68.

30. Schell, D.J., Farmer, J., Newman, M. and McMillan, J.D. (2003) Dilute-sulfuric acid pretreatment of corn stover in pilot-scale reactor: investigation of yields, kinetics, and enzymatic digestibilities of solids. *Applied Biochemistry and Biotechnology*, **105–108**, 69–85.

31. Galbe, M. and Zacchi, G. (2012) Pretreatment: the key to efficient utilization of lignocellulosic materials. *Biomass and Bioenergy*, **46**, 70–78.

32. Chandel, A.K., da Silva, S.S. and Singh, O.V. (2011) Detoxification of lignocellulosic hydrolysates for improved bioethanol production, in *Biofuel Production-Recent Developments and Prospects*, vol. **10** (ed M.A.d.S. Bernardes), InTech, p. 225.

33. Brodeur, G., Yau, E., Badal, K. *et al.* (2011) Chemical and physicochemical pretreatment of lignocellulosic biomass: a review. *Enzyme Research*, **2011**, 1–17.

34. Haghighi Mood, S., Hossein Golfeshan, A., Tabatabaei, M. *et al.* (2013) Lignocellulosic biomass to bioethanol, a comprehensive review with a focus on pretreatment. *Renewable and Sustainable Energy Reviews*, **27**, 77–93.

35. Chen, Y., Stevens, M.A., Zhu, Y. *et al.* (2013) Understanding of alkaline pretreatment parameters for corn stover enzymatic saccharification. *Biotechnology for Biofuels*, **6**, 1–10.

36. Xu, J., Cheng, J.J., Sharma-Shivappa, R.R. and Burns, J.C. (2010) Sodium hydroxide pretreatment of switchgrass for ethanol production. *Energy and Fuels*, **24**, 2113–2119.

37. Quesada, J., Rubio, M. and Gómez, D. (1999) Ozonation of lignin rich solid fractions from corn stalks. *Journal of Wood Chemistry and Technology*, **19**, 115–137.

38. Panneerselvam, A., Sharma-Shivappa, R.R., Kolar, P. *et al.* (2013) Potential of ozonolysis as a pretreatment for energy grasses. *Bioresource Technology*, **148**, 242–248.

39. Vidal, P.F. and Molinier, J. (1988) Ozonolysis of lignin — improvement of in vitro digestibility of poplar sawdust. *Biomass*, **16**, 1–17.

40. Neely, W.C. (1984) Factors affecting the pretreatment of biomass with gaseous ozone. *Biotechnology and Bioengineering*, **26**, 59–65.

41. Fan, J., De Bruyn, M., Budarin, V.L. *et al.* (2013) Direct microwave-assisted hydrothermal depolymerization of cellulose. *Journal of the American Chemical Society*, **135**, 11728–11731.

42. Akin, D.E., Rigsby, L.L., Sethuraman, A. *et al.* (1995) Alterations in structure, chemistry, and biodegradability of grass lignocellulose treated with the white rot fungi Ceriporiopsis subvermispora and Cyathus stercoreus. *Applied and Environmental Microbiology*, **61**, 1591–1598.

43. Lundell, T.K., Mäkelä, M.R. and Hildén, K. (2010) Lignin-modifying enzymes in filamentous basidiomycetes – ecological, functional and phylogenetic review. *Journal of Basic Microbiology*, **50**, 5–20.

44. Alvira, P., Tomás-Pejó, E., Ballesteros, M. and Negro, M.J. (2010) Pretreatment technologies for an efficient bioethanol production process based on enzymatic hydrolysis: a review. *Bioresource Technology*, **101**, 4851–4862.

45. Kumar, R. and Wyman, C.E. (2009) Effects of cellulase and xylanase enzymes on the deconstruction of solids from pretreatment of poplar by leading technologies. *Biotechnology Progress*, **25**, 302–314.

46. Shi, J., Chinn, M.S. and Sharma-Shivappa, R.R. (2008) Microbial pretreatment of cotton stalks by solid state cultivation of Phanerochaete chrysosporium. *Bioresource Technology*, **99**, 6556–6564.

47. Karpl, S.G., Woiciechowski, A.L., Socco, V.T. and Socco, C.R. (2013) Pretreatment strategies for delignification of sugarcane bagasse: a review. *Brazilian Archives of Biology and Technology*, **56**, 679–689.

48. Zadražil, F. and Brunnert, H. (1981) Investigation of physical parameters important for the solid state fermentation of straw by white rot fungi. *European Journal of Applied Microbiology and Biotechnology*, **11**, 183–188.

49. Reid, I.D. (1989) Solid-state fermentations for biological delignification. *Enzyme and Microbial Technology*, **11**, 786–803.

50. Wang, W., Yuan, T., Cui, B. and Dai, Y. (2013) Investigating lignin and hemicellulose in white rot fungus-pretreated wood that affect enzymatic hydrolysis. *Bioresource Technology*, **134**, 381–385.

51. Behrendt, F., Neubauer, Y., Oevermann, M. *et al.* (2008) Direct liquefaction of biomass. *Chemical Engineering & Technology*, **31**, 667–677.

52. Minowa, T., Yokoyama, S.-y., Kishimoto, M. and Okakura, T. (1995) Oil production from algal cells of Dunaliella tertiolecta by direct thermochemical liquefaction. *Fuel*, **74**, 1735–1738.

53. SCF Technologies A/S (2010)SCF Technologies:Green oil from waste. Available at www.stylepit.com/investor/file.asp?id=502 (accessed 4 September 2014).

54. Toor, S.S., Rosendahl, L. and Rudolf, A. (2011) Hydrothermal liquefaction of biomass: a review of subcritical water technologies. *Energy*, **36**, 2328–2342.

55. Toor, S.S., Rosendahl, L., Nielsen, M.P. *et al.* (2012) Continuous production of bio-oil by catalytic liquefaction from wet distiller's grain with solubles (WDGS) from bio-ethanol production. *Biomass and Bioenergy*, **36**, 327–332.

56. Nussbaumer, T. (2003) Combustion and co-combustion of biomass: fundamentals, technologies, and primary measures for emission reduction. *Energy & Fuels*, **17**, 1510–1521.

57. Bridgwater, A.V. (2012) Review of fast pyrolysis of biomass and product upgrading. *Biomass and Bioenergy*, **38**, 68–94.

58. Puig-Arnavat, M., Bruno, J.C. and Coronas, A. (2010) Review and analysis of biomass gasification models. *Renewable and Sustainable Energy Reviews*, **14**, 2841–2851.

59. Kumar, A., Jones, D. and Hanna, M. (2009) Thermochemical biomass gasification: a review of the current status of the technology. *Energies*, **2**, 556–581.

60. Kuo, J.H., Lian, Y.H., Rau, J.Y., *et al.* (2010) Effect of Operating Conditions on Emission Concentration of PAHs During Fluidized Bed Air Gasification of Biomass. 2010 International Conference on Chemistry and Chemical Engineering (ICCCE), August 1–3, Kyoto, p. 76.

61. Hossain, A.K. and Davies, P.A. (2013) Pyrolysis liquids and gases as alternative fuels in internal combustion engines – a review. *Renewable and Sustainable Energy Reviews*, **21**, 165–189.

62. Ahmad, M.M., Nordin, M.F.R. and Azizan, M.T. (2013) Upgrading of bio-oil into high-value hydrocarbons via hydrodeoxygenation. *American Journal of Applied Sciences*, **7**, 746–755.

63. Mortensen, P.M., Grunwaldt, J.D., Jensen, P.A. *et al.* (2011) A review of catalytic upgrading of bio-oil to engine fuels. *Applied Catalysis A: General*, **407**, 1–19.

64. Bridgwater, A.V., Toft, A.J. and Brammer, J.G. (2002) A techno-economic comparison of power production by biomass fast pyrolysis with gasification and combustion. *Renewable and Sustainable Energy Reviews*, **6**, 181–246.

65. Basu, P. (2013) Chapter 5: Pyrolysis. In: *Biomass Gasification, Pyrolysis and Torrefaction*, 2nd edition (ed. P. Basu). Academic Press, Boston, p. 147.

66. Yaman, S. (2004) Pyrolysis of biomass to produce fuels and chemical feedstocks. *Energy Conversion and Management*, **45**, 651–671.

67. Bourgois, J. and Guyonnet, R. (1988) Characterization and analysis of torrefied wood. *Wood Science and Technology*, **22**, 143–155.

68. Budarin, V.L., Clark, J.H., Lanigan, B.A. *et al.* (2009) The preparation of high-grade bio-oils through the controlled, low temperature microwave activation of wheat straw. *Bioresource Technology*, **100**, 6064–6068.

69. Budarin, V.L., Clark, J.H., Lanigan, B.A. *et al.* (2010) Microwave-assisted decomposition of cellulose: a new thermochemical route for biomass exploitation. *Bioresource Technology*, **101**, 3776–3779.

70. Budarin, V.L., Zhao, Y.Z., Gronnow, M.J. *et al.* (2011) Microwave-mediated pyrolysis of macro-algae. *Green Chemistry*, **13**, 2330–2333.

71. Masek, O., Budarin, V., Gronnow, M. *et al.* (2013) Microwave and slow pyrolysis biochar-Comparison of physical and functional properties. *Journal of Analytical and Applied Pyrolysis*, **100**, 41–48.

72. Batidzirai, B. (2013) Design of sustainable biomass value chains – optimising the supply logistics and use of biomass over time. PhD thesis, Department of Innovation, Environmental and Energy Sciences, Utrecht University, Utrecht.

73. Peng, J.H., Bi, H.T., Sokhansanj, S. and Lim, J.C. (2012) A study of particle size effect on biomass torrefaction and densification. *Energy & Fuels*, **26**, 3826–3839.

74. Tapasvi, D., Khalil, R., Skreiberg, O. *et al.* (2012) Torrefaction of Norwegian birch and spruce: an experimental study using macro-TGA. *Energy & Fuels*, **26**, 5232–5240.

75. Azargohar, R. and Dalai, A.K. (2008) Steam and KOH activation of biochar: experimental and modeling studies. *Microporous and Mesoporous Materials*, **110**, 413–421.

76. Allan, G.G., Krieger, B.B. and Work, D.W. (1980) Dielectric loss microwave degradation of polymers: cellulose. *Journal of Applied Polymer Science*, **25**, 1839–1859.

77. Dorado, M.P., Lin, S.K.C., Koutinas, A. *et al.* (2009) Cereal-based biorefinery development: Utilisation of wheat milling by-products for the production of succinic acid. *Journal of Biotechnology*, **143**, 51–59.

78. Du, C., Lin, S.K.C., Koutinas, A. *et al.* (2008) A wheat biorefining strategy based on solid-state fermentation for fermentative production of succinic acid. *Bioresource Technology*, **99**, 8310–8315.

79. Lin, C.S.K., Luque, R., Clark, J.H. *et al.* (2011) A seawater-based biorefining strategy for fermentative production and chemical transformations of succinic acid. *Energy & Environmental Science*, **4**, 1471–1479.

80. Anto, H., Trivedi, U.B. and Patel, K.C. (2006) Glucoamylase production by solid-state fermentation using rice flake manufacturing waste products as substrate. *Bioresource Technology*, **97**, 1161–1166.

81. Leung, C.C.J., Cheung, A.S.Y., Zhang, A.Y.-Z. *et al.* (2012) Utilisation of waste bread for fermentative succinic acid production. *Biochemical Engineering Journal*, **65**, 10–15.

82. Zhang, A.Y.Z., Sun, Z., Leung, C.C.J. *et al.* (2013) Valorisation of bakery waste for succinic acid production. *Green Chemistry*, **15**, 690–695.

83. Pleissner, D., Lam, W.C., Sun, Z. and Lin, C.S.K. (2013) Food waste as nutrient source in heterotrophic microalgae cultivation. *Bioresource Technology*, **137**, 139–146.

84. Shahid, M., Shahid ul, I. and Mohammad, F. (2013) Recent advancements in natural dye applications: a review. *Journal of Cleaner Production*, **53**, 310–331.

85. Weiland, P. (2010) Biogas production: current state and perspectives. *Applied Microbiology and Biotechnology*, **85**, 849–860.

86. Weiland, P. (2006) Biomass digestion in agriculture: a successful pathway for the energy production and waste treatment in Germany. *Engineering in Life Sciences*, **6**, 302–309.

87. Environmental Protection Department, Hong Kong (2013) Development of Organic Waste Treatment Faiclities. Available at http://www.epd.gov.hk/eia/register/profile/latest/esb226/esb226.pdf (accessed 26 August 2014).

88. Petersson, A. and Wellinger, A. (2009) Biogas upgrading technologies – developments and innovations. IEA Bioenergy. Available at http://www.en.esbjerg.aau.dk/digitalAssets/80/80449_iea-biogas-upgrading-report-2009.pdf (accessed 4 September 2014).

# 4

# Platform Molecules

## Thomas J. Farmer[1] and Mark Mascal[2]

[1] Department of Chemistry, Green Chemistry Centre of Excellence, University of York, UK
[2] Department of Chemistry, University of California Davis, USA

## 4.1   Introduction

This book highlights the potential for biomass to be used as feedstock for our chemical industry, supporting and eventually supplanting the plethora of chemicals and materials we use in everyday life that are currently derived from non-renewable fossil resources. Other chapters within this book demonstrate the potential for biomass to be grown, harvested, processed and utilised as an effective replacement for fossil resources for the production of polymers, materials, fuel and energy. However, to ultimately supplant fossil resources the bio-economy needs a means by which to produce the array of chemical products familiar and important to us, such as plastics, synthetic textiles, dyes and pigments, surfactants, pharmaceuticals, agrichemicals and home and personal care products. The modern petrochemical industry produces the vast majority of these products and materials from a small set of simple, cheap base chemicals. Research over the last 20 years has endeavoured to demonstrate that the bio-economy can deliver the same, via a set of simple, cheap biomass-derived building block chemicals, also referred to as platform molecules.

*Introduction to Chemicals from Biomass*, Second Edition. Edited by James Clark and Fabien Deswarte.
© 2015 John Wiley & Sons, Ltd. Published 2015 by John Wiley & Sons, Ltd.

Within this chapter the status of the current fossil-based chemical industry is briefly described with regards to the use of building-block chemicals, and an analogy to a bio-based equivalent discussed. Current rationale around the concept of platform molecules is discussed, leading to a working definition for these bio-derived chemicals. The constituent parts of the biomass feedstock are reviewed, and some possible chemicals derivable from each constituent are highlighted, showing the importance of the composition of the biomass with regards to the molecules obtainable from it. The processing technologies used to produce platform molecules from biomass are described and a list of current platform molecules given, with four examined in further detail (one from each of the identified processing technologies).

The concept of bio-based platform molecules stems from the late twentieth century where it was envisaged that simple, small molecules derivable from biomass could be utilised as building blocks for higher-value chemicals and materials [1, 2]. However, it was not until the US Department of Energy's (US DOE) report on Top Value Added Chemicals from Biomass in 2004 that such focus and intensive research was directed towards this type of biomass-derived product [3]. It has been widely purported that the utilisation of biomass as a replacement feedstock to fossil fuels requires the successful implementation of integrated biorefineries, and that these biorefineries will need to simultaneously produce fuels, energy and chemical/material products to operate competitively against petroleum refineries. To assist in the progression towards a bio-based economy and to accelerate the technical development of biorefineries, the US DOE set out to critically assess the large range of biomass-derivable chemicals to generate a succinct list of the most promising candidates, thus focusing global research efforts. It was anticipated that the chemicals and materials produced from a biorefinery would be the major source of profit in such operations, much in the same way that petrochemicals, although representing less than 10% of annually extracted fossil resources, enjoy a much higher profit margin than fossil fuels and energy.

Interestingly, the US DOE 2004 report does not make reference to the term 'platform molecule', though it does discuss the analogy to fossil-derived base chemicals. The 2004 report also focused on molecules derivable from sugars, syngas and triglycerides, omitting chemicals derivable from lignin which were addressed in a later report [4]. The building-block chemicals identified by the US DOE were not all fully analogous to base chemicals; some (such as 2,5-furandi-carboxylic acid, FDCA) are themselves produced from a building-block molecule (i.e. 5-hydroxymethylfurfural, HMF). The US DOE report was instead aiming to list molecules that could, with further development, be produced on a significant scale and have a range of options for conversion to higher-value chemicals and materials. In addition, some biomass-derivable chemicals were omitted as they were already established on a commodity scale, such as bioethanol, and did not require extensive research and development. Since publication of the US DOE report, commercial development of the various molecules listed has been diverse [5]. The four-carbon diacids (malic, fumaric and succinic acid) have received considerable attention with Myriant, BioAmber, BASF-Purac (Succinity joint venture) and DSM-Roquette (Reverdia joint venture) all pursuing commercial-scale production of one or more of these. Plant capacities for these diacids vary between 10,000 and 100,000 tonnes per annum, most likely for use in polymer applications. 2,5-FDCA has also enjoyed strong commercial development, with Avantium's YXY technology leading towards a 100% bio-based replacement for polyethylene terephthalate (PET). 3-Hydroxypropionic acid (3HP) is also under advanced commercialisation. OPXBio is producing 3HP via fermentation and, in collaboration with Dow, converting it to bio-derived acrylic acid [6]. Metabolix and a consortium of Cargill, Novozymes and BASF are also targeting bio-based acrylic acid via 3HP [7, 8]. Levulinic acid will be produced commercially by Biofine Technologies, due online in 2015 [9]. Finally, Itaconix was established in 2008 as a company demonstrating the potential to produce useful polymers from itaconic acid, the major current product being detergent-builder Dispersant DSP 2K [10]. In contrast, other US DOE Top 12 molecules such as 3-hydroxybutrolactone, glutamic acid, sorbitol and xylitol have not seen any major developments since 2004 [5].

A later report from the US DOE addressed the potential to produce building-block molecules from lignin, though no distinct list of molecules with the greatest potential was agreed [4]. It was concluded that in the shorter to medium term lignin would be better utilised as a carbon source for power and fuels (via combustion, pyrolysis or gasification) or to produce macromolecules. However, a clear statement from the US DOE was that an eventual move away from simple combustion of lignin for power generation would be a key component in the successful implementation of biorefineries. Production of distinct building-block molecules from the depolymerisation of lignin was seen as a long-term goal, which would require advances in the technology. The widespread availability of lignin globally means it should not be overlooked when seeking the production of platform chemicals from biomass, and the effective valorisation of lignin will support progression towards a sustainable bio-based economy. Nevertheless, some compounds derivable from lignin were highlighted by the US DOE; these were in agreement with opinions elsewhere regarding the most viable small aromatic molecule targets from lignin.

## 4.2   Fossil-Derived Base Chemicals

The overarching aim in the progression towards a sustainable bio-based chemical industry is to supplant the unsustainable chemical industry currently in existence. Typically, publications relating to the generation of bio-derived platform molecules begin by making a direct comparison to fossil-derived base chemicals, the fundamental building blocks for almost all of the current chemical industry.

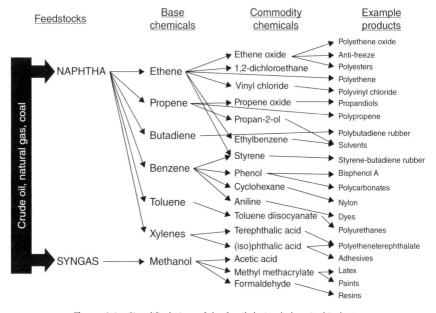

**Figure 4.1**    *Simplified view of the fossil-derived chemical industry.*

**Table 4.1**    *Global annual production (2010) [11] of the seven fossil-derived base chemicals and their global production from biomass (2012).*

| Base/bulk chemical | Predominant feedstock | Annual production from fossil sources (t a$^{-1}$) [11, 12] | Annual production from biomass (t a$^{-1}$) |
|---|---|---|---|
| Ethene | Oil, gas | 123 300 000 | 200 000 [13, 14] |
| Propene | Oil, gas | 74 900 000 | Pilot scale [15] |
| Butadiene | Oil, gas | 10 200 000 | Pilot scale [16] |
| Benzene | Oil | 40 200 000 | >100 [17] |
| Toluene | Oil | 19 800 000 | >100 [17] |
| Xylenes (o-, m-, p-) | Oil | 42 500 000 | >100 [17] |
| Methanol | syngas | 49 100 000 | 340 000 [18] |

The majority of all chemical and material products created from the chemical industry are derived from seven simple building-block molecules (Figure 4.1), the 'base chemicals' (Table 4.1). The base chemicals are derived from fossil resources (predominately crude oil and natural gas) via distillation, cracking, reforming and fractionation. All seven base chemicals are produced on a global annual scale in excess of 10 million tonnes, with ethene produced in excess of 100 million tonnes [11]. All seven are also available at low cost and therefore economically upgradeable to an array of other chemicals and materials (Figure 4.1). With the exception of methanol (MeOH), all are devoid of heteroatoms with only double bonds or aromatic rings as functional groups. As a result, further chemical modifications, such as oxidation (from air), are required to introduce chemical complexity and allow the production of a broader set of commodity chemicals. Total production of these base chemicals exceeded 360 million tonnes in 2010; it is estimated that around 3.6 billion tonnes of crude oil was produced in the same year, showing that around 10% of global crude oil is directed to base chemical production [11, 12]. Sulphur and ammonia are sometimes also considered as base chemicals due to their large-scale production (77 million and 134 million tonnes in 2010, respectively) and link to fossil resources (hydrogen gas $H_2$ for $NH_3$ production from syngas and sulphur from desulphurisation of fossil fuels), but are not discussed here in order to simplify the comparison to biomass-derived platform molecules. The sulphur content of biomass is less than that of typical fossil fuels and may therefore bring about a supply chain issue in a future chemical industry more reliant on biomass as a feedstock. Conversion of biomass into syngas is possible and, as such, bio-derived ammonia is a mid- to long-term possibility for a sustainable bio-based economy assuming the volumes required can be supplied by the available biomass.

A successful bio-based chemical industry needs to be able to generate an equivalent quantity and diversity of starting materials and final products as those seen from the petrochemical industry. It is for this reason that routes to simple building-block chemicals from biomass, to replace the base chemicals, is currently under extensive investigation by numerous research groups and chemical companies across the globe. Regional, national and international funding bodies are also now significantly supporting research efforts in the field of bio-based and sustainable chemicals. We are clearly in the midst of a bio-based revolution, one where platform molecules will play a fundamental role.

## 4.3   Definition of a Platform Molecule

The majority of research articles in the field of platform molecules make reference to the US DOE reports described above. There has been a lack of consensus regarding the definition of a platform molecule however, and the US DOE provided no such definition. Several research articles, reports, patents and books have been published using a range of different terms when referring to building-block chemicals produced from biomass, which we refer to here as bio-based platform molecules. Some examples of the terms that have been used are listed below:

- bio-platform molecules [19];
- platform molecules [20];
- biomass-based platform chemicals [21];
- biomass-derived platform molecules [22];
- base chemicals from biomass [23];
- building-block chemicals from biomass [3]; and
- platforms (e.g. syngas platform, triglyceride platform).

Some authors make reference to 'platforms' rather than platform molecules, and this is used where no one distinct molecule forms the building block (i.e. triglycerides with a varied fatty acid composition) or where the composition of the different molecules can vary (i.e. varying ratios of $H_2$, carbon monoxide CO and $CO_2$ in synthesis gas). A platform therefore includes several building-block chemicals grouped together, resulting in a broad range of downstream chemical products. Two example platforms, syngas and triglycerides, are discussed in greater detail later in this chapter (see Sections 4.8.1 and 4.8.4).

The first issue to highlight in some of the terms used to describe these molecules is the application of the 'bio-' prefix (i.e. bio-platform molecules). According to the Oxford English Dictionary the bio-prefix is used as a direct relationship to 'life', and although it can be argued that its use in this instance is correct, it does lead to confusion when considering other uses of the bio-prefix elsewhere. Some would argue that the bio-prefix is only valid for molecules where the entirety of their production route is via biological processes or that it is a molecule produced and utilised by nature, and is therefore not applicable to molecules that have required thermal or chemical treatment in their production. Issues also arise as the bio-prefix is applied to products, chemicals or materials that are biodegradable, such as in the field of bio-lubricants. A more suitable alternative would perhaps be 'bio-based' or 'bio-derived' platform molecules, as these imply that the molecules' constituent elements have originated from biological processes, but that chemical or thermal treatments could have been applied during their production. Recent European and international projects and governmental, academic or industrial bodies have taken a similar approach when establishing terms for products derived from biomass, or bio-based products. For instance, the US 2002 Farm Bill defined bio-based products as:

> commercial or industrial products (other than food or feed) that are composed in whole, or in significant part, of biological products, renewable agricultural materials (including plant, animal, and marine materials), or forestry materials.

As interest in the bio-based economy continued to grow in the early twenty-first century, the European Commission also published a definition for a bio-based product in 2007 [24]:

> A bio-based product is a non-food product derived from biomass (plants, algae, crops, trees, marine organisms and biological waste from households, animals and food production). Bio-based products may range from high value-added fine chemicals such as pharmaceuticals, cosmetics, food additives, etc., to high volume materials such as general bio-polymers or chemical feedstocks. The concept excludes traditional bio-based products, such as pulp and paper, and wood products, and bio-mass as an energy source.

The European Commission definition makes no reference to the extent to which the article or product needs to be bio-based, while the US Farm Bill highlighted that the product should be composed wholly or to a significant extent from bio-based materials. A recent update in the definition of biomass by the European Committee on Standardisation (CEN) clarifies the term further, considering the fact that fossil fuels themselves are originally sourced from biological material and have undergone natural processes in their transformation:

> Biomass is material of biological origin excluding material embedded in geological formations and/or fossilised.

These above definitions set a useful marker for the application of 'bio-based' with respect to the term bio-based platform molecule. It is the opinion of the authors that interchange between bio-based and bio-derived is allowable in this context, both essentially referring to the fundamental source of an article being from biomass. Considering the above terms, a succinct definition for a bio-based platform molecule would be:

> A bio-based (or bio-derived) platform molecule is a chemical compound whose constituent elements originate wholly from biomass (material of biological origin, excluding fossil carbon sources), and that can be utilised as a building block for the production of other chemicals.

However, if a direct comparison of a platform molecule to a fossil-derived base chemical is required then the quantity of the chemical produced is also relevant. As the bio-economy matures it will become apparent which of the bio-based building-block chemicals are most like the fossil-derived base chemical by the volumes they are produced and used in.

There are some instances where several compounds in a sequential reaction pathway could each be viewed as platform molecules, further complicating correct assignment of the true platform molecule. For example, fructose (feed-stock) can be converted to HMF (platform molecule if isolated), but it can also be converted directly to levulinic acid [25, 26] and γ-valerolactone [27] (both via HMF but without isolation of the HMF). These two compounds could therefore be described as bio-derived platform molecules, depending on the production route, as they can be produced directly from biomass (Figure 4.2).

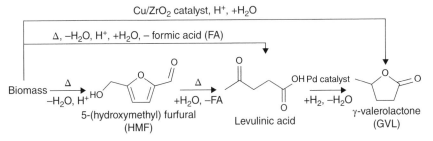

**Figure 4.2** *Formation of levulinic acid and γ-valerolactone (GVL) either directly from biomass or via 5-HMF.*

Process economics will eventually determine which are formally platform molecules and which are simply bio-based chemicals (i.e. derived from platform molecules or produced in too small a quantity to be relevant as a comparison to base chemicals).

There have also been difficulties in deciding if some chemicals are platform molecules or biomass feedstocks. For example, mono/disaccharides can be viewed as biomass feedstocks as they can be isolated from biomass in its crude form (extracted from sugar beet, fruits and sugar cane), but can also be derived from polysaccharides via biological and/or chemical processes; in this sense, they are more like a platform molecule. For the purposes of this chapter, saccharides are always viewed as biomass feedstocks.

The term 'sugar platform' is also used in some articles relating to biorefineries, and this term is applied when a differentiation to other platforms (e.g. syngas) is desirable. The sugar platform covers a very wide range of chemical products and includes all the platform molecules derived via the biological (fermentation) or chemical-catalytic processing of sugars.

It is important that platform molecules, and eventual bio-based products, endeavour to avoid the use of fossil-derived elements and fuels in their production. For example, for fatty acid methyl esters (FAMEs) to be wholly bio-derived platform molecules they must use bio-based MeOH for the transesterification, ensuring all carbons in the molecule are bio-based. It is also possible that some of the bulk or commodity chemicals currently derived from petroleum could potentially be sourced from biomass. These are known colloquially as 'drop-in replacements', since they can be integrated into current petrochemical processes, production chains and consumer markets without infrastructural changes. For example, ethene is currently commercially available from bio-ethanol [28].

## 4.4    Where Platform Molecules Come From

Bio-based platform molecules must be derived from biomass and, as such, biomass represents for the biorefinery what fossil resources (coal, gas and oil) represent for the fossil-based refinery; biomass is the feedstock for the biorefinery. With regard to platform molecules, biomass can therefore be defined as:

> A biologically produced material (non-fossil) that occurs in sufficient abundance to serve as a feedstock for the production of commodity products or platform molecules.

Although simple in definition, plant biomass is complex and diverse in composition [29]. Biorefinery feedstocks, and therefore platform chemical feedstocks, are typically further differentiated into polysaccharides, lignin, protein and extractives (e.g. triglycerides and terpenes), as all are found as constituent parts within typical biomass in varying quantities. Mono- and disaccharides, such as glucose, fructose and sucrose, should additionally be treated as biomass feedstock as they are also found within some types of biomass or readily formed from polysaccharides. As discussed, simple saccharides can be viewed either as platform chemicals or biomass, but for this chapter are viewed only as biomass [30, 31]. For the sake of simplifying further discussions on platform molecules, the following are treated as biomass feedstocks for the biorefinery and therefore feedstocks of platform chemical production:

- polysaccharides (starch, cellulose, hemicellulose, chitin, inulin, etc.);
- lignin;
- protein;
- mono/disaccharides (glucose, fructose, sucrose, xylose, mixture of sugars);
- extractives (triglycerides, terpenes, pigments, waxes, etc.); and
- any combination of the above constituting raw, native biomass.

Biorefineries will normally process biomass in crude form containing mixtures of these constituents, but the division in a theoretical sense is useful as each constituent typically leads to a different set of molecules (with the exception of gasification routes). Platform molecules are therefore typically categorised based on the constituents of the biomass they are derived from. For example, furans are derived solely from the carbohydrates and not from lignin or protein, while vanillin will be derived from lignin alone. These various biomass constituents are discussed in greater detail in the following sections, with some examples given of the types of platform molecules derivable from each. Considering both economics and avoidance of competition with food production, biomass for platform chemical production should ideally be waste from other industries (e.g. lignocellulosic agricultural or forestry residues) and not specifically grown for chemical or energy production (i.e. second-generation) [32]. Waste biomass will predominately be in mixed lignocellulose form, although some examples exist where one constituent is the dominant component of a waste stream, such as lignin from the pulp and paper industry. It is also possible that the components may be separated prior to processing.

Along with the constituents described above, biomass additionally contains ash, a variable mixture of inorganics. Although not specifically relevant to bio-based platform molecules, the ash content of biomass is nevertheless important when considering the production of chemicals from biomass as it could represent a significant waste stream. On the other hand, ash minerals may also represent an opportunity for biorefineries, for example offering routes to bio-based aluminosilicate catalysts that could be used for the production of chemicals from biomass [33].

### 4.4.1   Saccharides

Saccharides as a whole generally represent the most dominant component of biomass. Edible carbohydrates such as starch, sucrose, glucose and fructose have been used extensively for the formation of platform molecules (Figure 4.3). While mono- and disaccharides can be used directly in fermentative, thermal and

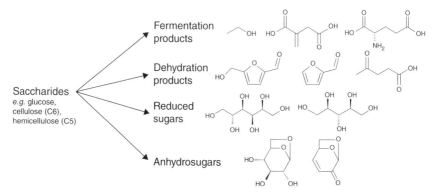

**Figure 4.3**  *Example platform molecules derived from saccharides.*

chemical treatment to produce platform molecules, the polysaccharides (e.g. starch and cellulose) may require prior hydrolysis to their monomer sugars (e.g. glucose, fructose and xylose). Saccharification of polysaccharides (i.e. formation of sugars from polysaccharides) is particularly important when targeting platform molecules derived via fermentation, such as succinic and itaconic acid, since the bacterium or yeast used in these fermentations requires carbohydrates in the form of simple sugars [34]. On an industrial scale, glucose is currently produced by the enzymatic hydrolysis of starch from corn, wheat, potato and tapioca [35]. In the short to medium term, starch hydrolysis will continue to be the major industrial route to glucose for use in platform molecule production.

Utilisation of cellulose and hemicellulose represents a greater challenge for platform molecule production, but the successful implementation of a biorefinery converting cellulose or hemicellulose into platform molecules also offers potentially greater rewards. There is variation in the amounts of cellulose and hemicellulose found within plant matter, and this variation is linked to species diversity, specific sections of a plant and seasonal changes. Nevertheless, cellulose and hemicellulose combined typically represent 70 wt% of lignocellulosic biomass. Cellulose and hemicellulose represent a far greater proportion of global biomass composition compared to starch; these polysaccharides are therefore likely to become the major constituent of the biomass feedstock for platform molecules in the long term. Additionally, neither cellulose nor hemicellulose derived from lignocellulose would be in competition for direct or land use in food, as is the case for starch and sugars. Utilisation of the polysaccharides in lignocellulose for ethanol production ('cellulosic ethanol') has been extensively investigated and shows promise [36, 37]. Significant developments are still required and diversity in the nature of the lignocellulose continues to cause challenges when developing suitable pretreatment methods [38]. These pretreatments are vital as the recalcitrant nature of plant biomass, borne out from the structural integrity of the lignocellulose matrix, means the efficacy of any hydrolysing enzyme is reduced until the structure is made more accessible [39]. However, as the processes involved for the production of ethanol from lignocellulose are by the hydrolysis of cellulose to the free monosaccharide sugars, technological advancements here could benefit any platform molecules produced via biological, chemical or thermal treatment of monosaccharides.

Hemicellulose is more hydrolytically sensitive than cellulose and is therefore typically hydrolysed first when lignocellulose is being processed. The residual cellulose is treated under harsher conditions once the hemicellulose sugars are removed, or recovered and used such as in the pulp and paper industry [40, 41]. Several catalysed hydrolytic methodologies are applied to hemicellulose including enzymatic, mineral acids and bases [42], and supported acids and bases [43, 44] and metal catalysts. Both enzymatic and mineral acid and base catalysed hydrolysis suffers from uneconomical catalyst recovery, and residual mineral acid or base may inhibit subsequent fermentation [45]. Supported catalysts offer benefits in the ease of recovery (via filtration), but reduced rates of molecular diffusion of the polymeric constituents of biomass (e.g. cellulose and lignin) into the pores of the catalyst can reduce their suitability [46]. Hemicellulose is a branched polymer and may contain several different sugar monomers, both pentoses and hexoses, though xylose is usually present in the largest quantity. This diversity of sugars within hemicellulose results in several enzymes being required for hydrolysis to free monosaccharide monomers [38]. Hexoses recovered from the hemicellulose can be used in the same manner as those isolated from starch and cellulose hydrolysis, while the pentoses are ideally applied to the formation of furfural [47, 48], as shown in Figure 4.3. Indeed, furfural is currently the platform molecule with the largest global production from lignocellulose, annually in excess of 200,000 tonnes, and has the potential to lead to a diverse set of higher-value chemicals such as 2-methylTHF, ethyl levulinate and furoic acid [47, 49]. Ethanol and butanol have also been produced by fermentation of monosaccharides derived from hemicellulose, and other higher-value platform molecules can also be readily derived from the sugars including xylitol, levulinic acid and itaconic acid [40, 50, 51].

The conditions for the hydrolysis of cellulose to glucose are more severe than those required for the hydrolysis of starch and hemicellulose. Such conditions often lead to unwanted by-products, including isomeric sugars (fructose, mannose), HMF and levoglucosan [52]. Some of these by-products, especially furan derivatives such as HMF and furfural, are known inhibitors in some fermentations; improved selectivity and/or post-hydrolysis detoxification are therefore needed. As for hemicellulose, enzymes, mineral acid and base-supported acids and metals have also been applied to the hydrolysis of cellulose [53, 54]. The milder conditions involved in enzyme-catalysed hydrolysis avoid the issues of by-product formation and are typically high yielding, but cellulose containing biomass requires pretreatment to improve accessibility of the molecular structure of this robust polysaccharide [55]. Enzymatic hydrolysis of the cellulose also typically leads mainly to oligomers rather than monomer sugar production, though 'cocktails' of enzymes can be used to overcome this issue [56]. Ionic liquids [57], supercritical water [58], plasmas [59] and microwaves [60, 61] have also been utilised in the pretreatment or selective hydrolysis of cellulose, though these protocols require further development to be applied on an industrially significant scale. Achieving selective, economically viable and reproducible methods for the conversion of cellulose to fermentable sugars is a vital requirement for the development of biorefineries that produce platform molecules by fermentation, especially if competition with food continues to remain high on the bio-economy's agenda. Despite the difficulties of cellulose utilisation, it still represents an ideal feedstock for the formation of platform molecules from biomass, being cheap, widely abundant and often the major waste product of food production.

An alternative to the saccharification of polysaccharides is the chemical and/or thermal conversion from their polymeric form direct to platform chemicals [62]. Examples of platform molecules obtained directly from the chemical and/or thermal treatment of polysaccharides include HMF, levulinic acid, furfural and sorbitol (Figure 4.3). The production of furans (furfural, HMF and (chloromethyl) furfural or CMF) and their derivatives (e.g. levulinic acid) requires conditions to promote both hydrolysis and dehydration, and this is most commonly achieved using soluble or solid acid catalysts in water or the application of ionic liquids [63, 64]. Formation of the sugar alcohols requires hydrolytic hydrogenation conditions (i.e. hydrolysis in the presence of $H_2$ or a source of $H_2$) [52], cellulose being converted to sorbitol and mannitol [65] and hemicellulose predominately forming xylitol and arabitol (Figure 4.4) [66]. These hydrolytic hydrogenation conditions can further lead to dehydration products such as sorbitan, 1,4-dianhydroxylitol and isosorbide [67]. Even more extreme conditions for hydrolytic hydrogenation

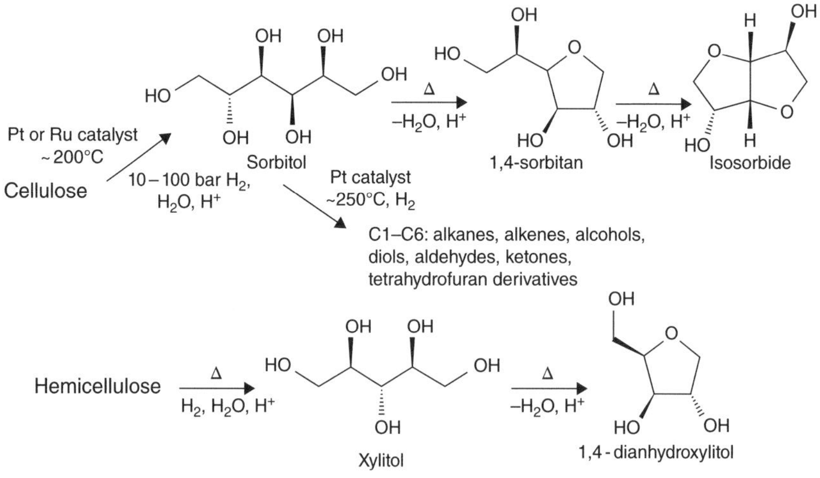

**Figure 4.4**  *Isosorbide and 1,4-anhydroxylitol from cellulose and hemicellulose, respectively.*

(>250°C, 50 bar $H_2$) have been used for the conversion of sorbitol to alkanes and various oxygenates (alcohols, aldehydes and ketones), but poor selectivity, high energy demands and the use of platinum group catalysts are required [68].

Although cellulose and hemicellulose are separated from the lignocellulose matrix prior to use in most of the examples above, cellulosic biomass can be converted into platform molecules directly. This approach can substantially reduce costs by removing the need for saccharification or solubilisation pretreatments [69]. The term lignocellulose is often used to describe biomass left as a waste from agricultural or industrial processes; examples of common lignocellulosic biomass include corn stover, wheat straw, olive pomace, sugarcane bagasse and sawdust. Since the majority of waste biomass is in the form of lignocellulose, the development of technologies for the conversion of lignocellulose into platform molecules is a major objective of the bio-based economy. The Biofine Process is one such example where lignocellulose can be processed directly, converting the constituent cellulose and hemicellulose into levulinic acid, furfural and formic acid (Figure 4.5). Using dilute sulphuric acid in a two-stage reaction at temperatures from 190 to 220°C with short reaction times (seconds to minutes), cellulose is converted into HMF and this in turn converts into levulinic acid with formic acid produced as by-product. The furfural product is a result of the dehydration of the xylose units of hemicellulose. Residual degraded lignin along with humic material is left as a solid char that is currently burnt for energy recovery. The levulinic acid, furfural and formic acid are all useful platform molecules, with a diverse range of chemical and material derivatives possible from each. In nearly all instances

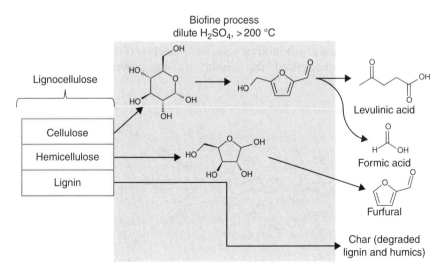

**Figure 4.5**  *The Biofine Process for the conversion of lignocellulose to levulinic acid, formic acid, furfural and lignin-derived char.*

where lignocellulose is utilised as a feedstock for platform molecule production, the lignin is left as a waste and often burnt for energy recovery. There are examples where the use of ionic liquids for the processing lignocellulose has produced aromatic compounds (e.g. guaiacol, eugenol and vanillin) from lignin, where it is found that different biomasses lead to different aromatic products [70]. However, commercialisation of an ionic liquid process could prove difficult due to the cost, toxicity and difficult product recovery when using this class of solvent. Alternatively, technologies for the specific depolymerisation of lignin can be applied to lignin residues, some examples of which are described in the following section.

Various other polysaccharides are also available from waste biomass in appreciable quantities and can also be applied to the production of platform molecules. Inulin for example is a fructan found in various plants that is historically isolated from chicory, though Jerusalem artichoke tubers also contain inulin in excess of 70% dry weight and are commercially cultivated [71]. Although inulin can be hydrolysed to fructose prior to conversion to platform molecules [72], it is also often found that the polysaccharide itself can be converted directly into these same molecules (HMF, CMF, etc.) [19, 73]. A greater issue in its use as feedstock for platform chemicals is its overall abundance, which is much lower than for the lignocellulose components (lignin, cellulose and hemicellulose). There are some natural polysaccharides that contain heteroatoms other than oxygen and, as such, these could present alternative possibilities in the future of platform molecule production (Figures 4.6 and 4.7). Chitin for instance is a nitrogen-containing polysaccharide that is abundant and widely available as a by-product of the seafood industry, forming the major constituent of the exoskeletons of crustaceans. Chitin, which can be de-acetylated to chitosan, is predominately used in commercial products in its polymeric form (e.g. thickener in foods and binder in textiles) but can be hydrolysed to its glucosamine monomer and, with further processing, this could also be utilised a platform molecule (Figure 4.6) [74]. It has recently been demonstrated that chitin can be converted directly into

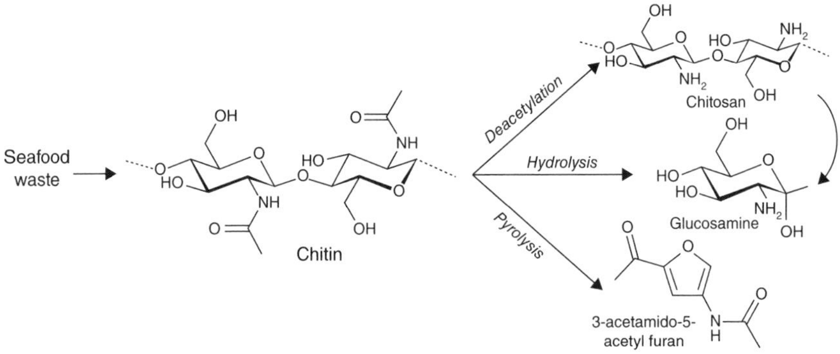

**Figure 4.6**  *Potential for platform molecules derived from chitin.*

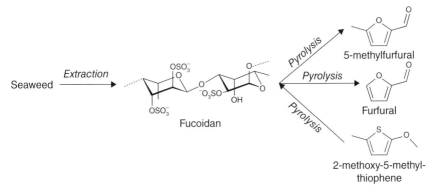

***Figure 4.7*** *Potential for platform molecules from seaweed-derived sulphur-containing polysaccharides.*

3-acetamido-5-acetyl furan, a nitrogen-containing furan, and although yields were low this still represents an interesting application of this widely available polysaccharide [75]. Sulphur-containing polysaccharides can also be found naturally, examples including fucoidan and carrageenan, which are naturally present in seaweeds. Pyrolysis of these polysaccharides has led to the formation of typical platform molecules produced from cellulose and hemicellulose, such as furfural and 5-methylfurfural, but also to some uncommon heteroatom derivatives such as 2-methoxy-5-methyl-thiophene (Figure 4.7) [76, 77].

## 4.4.2   Lignin

Lignin represents a significant proportion of lignocellulosic material and is considered the largest potential source of bio-based aromatics. However, the complex character and physical and chemical robustness of the lignin structure has resulted in little success in the effective valorisation of lignin beyond burning (energy recovery), gasification and use in niche applications as a dispersant, binder, adhesive and as a precursor to carbon fibre [78–80].

Lignin is highly recalcitrant relative to the other major constituents of lignocel-lulose, and has proven resistant to most biological and chemical degradation treatments [81]. Some technologies are however available, or under development, for the conversion of polymeric lignin into smaller molecules that are useful as building blocks for the biorefinery. Research efforts to date focus on the produc-tion of platform chemicals from lignin (Figure 4.8) either by pyrolysis or chemical depolymerisation (including hydrothermal hydrolysis and hydrogenolysis).

Pyrolysis is performed rapidly at high temperatures (>500°C) and pressures, producing a mixture of gases, char and bio-oil (a highly complex mixture of organic compounds) [82]. It is unlikely that pyrolysis will be used to produce specific platform molecules from lignin due to the complexity of the bio-oil

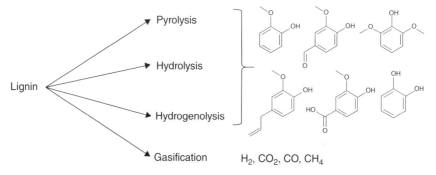

**Figure 4.8**   *Example platform molecules derived from lignin.*

collected, though gasification followed by Fischer–Tropsch processing is one avenue to produce small building blocks from lignin. Chemical depolymerisation covers a broader range of conditions where various temperatures, catalysts (acid, base, metals and ionic liquids) and reagents (e.g. water and hydrogen) are applied [83]. Hydrothermal hydrolysis is performed at lower temperatures (<400°C) than pyrolysis, though higher pressures are often used to increase efficiency of the process. Hydrothermal treatments can be tailored towards gasification or liquefaction depending upon temperature, with higher temperatures favouring gas formation. Hydrogenolysis is performed at similar temperatures and pressures to hydrothermal hydrolysis but steam is replaced with $H_2$, or another source of hydrogen such as formic acid, and often requires the addition of catalysts. The milder conditions for hydrolysis and hydrogenolysis hold greater promise for the production of platform molecules from lignin compared to pyrolysis. However, all the technologies currently available for the formation of platform chemicals from lignin are energy intensive and require costly purification of complex mixtures. Investigations into the controlled enzymatic depolymerisation of lignin have so far proven ineffective for the formation of platform chemicals. The microbes used to degrade lignin do so by the enzymatic formation of hydrogen peroxide, and the products from the initial peroxide-mediated depolymerisation are themselves further degraded before effective isolation can take place [81].

Several platform molecules derivable from lignin have been highlighted in the various reviews discussed above, though the molecules with greatest potential are guaiacol, catechol, vanillin, cresol, syringol, phenol, vanillic acid and syringic acid (Figure 4.8) [84, 85]. There are a couple of additional interesting candidates observed occasionally in lignin degradation, including 4-vinylphenol, 4-vinyl-guaiacol (both via hydrothermal depolymerisation in the presence of ethanol) and β-ketoadipic acid (by bacterial degradation of phenolics), but more research is required in all three cases [4, 86, 87]. Of all the molecules mentioned above, only vanillin production from the lignosulphonate generated during the sulphite pulping process is currently commercialised. This route is however low yielding

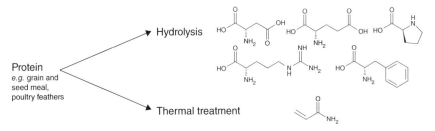

**Figure 4.9**    *Example platform molecules derived from protein.*

(<8%) and fails to compete economically with current synthetic routes to vanillin from petrochemicals; as such, it is only viable if consumers are happy to pay a premium for bio-based over fossil-based [32].

### 4.4.3    Protein

Protein fractions of biomass are essential in human and animal nutrition and are generally of greater economic value than edible carbohydrate fractions (sugars, starch) [88]. Nevertheless, interest is increasing in the potential utilisation of proteins for the production of platform chemicals [32]. The simplest route to platform molecules from protein is via hydrolysis of the peptide bonds, which yields a mixture of the various amino acids (Figure 4.9). Although complicated by the difficult separation of the amino acids from one another, the hydrolysis route results in a diverse set of potentially very useful molecules (Table 4.2). Options for use of amino acids are extensive due to their diverse functionality and, with the exception of glucosamine, amino acids represent the only nitrogen-containing platform molecules under recent investigation. Some amino acids, such as proline, aspartic acid and glutamic acid, have been researched extensively for conversion to other chemicals or in catalysis for directing enantioselective reactions [89–91]. Others such as arginine, phenylalanine and lysine are of interest due to the chemicals that can be produced from them, although their abundance is a matter of concern. Arginine is present in the cyanobacterial polymer cyanophycin in an equimolar ratio to aspartic acid in the peptide backbone, and it is proposed that the arginine could be converted into 1,4-diaminobutane via sequential hydrolysis (to ornithine) and decarboxylation [92–94]. Growth of cyanobacteria is slow and therefore unlikely to prove economically feasible on a large scale, though modified *Escherichia coli* is being developed to increase rates of cyanophycin production [95]. Phenylalanine can be converted to styrene via a deamination (to cinnamic acid) and decarboxylation and, in a newer iteration, can simultaneously produce acrylates with styrene via ethenolysis while lysine can be used to produce caprolactam, a Nylon 6 precursor [94, 96, 97].

An important issue in the utilisation of amino acids as platform molecules is the availability of the individual amino acids. Both glutamic and aspartic acid have

***Table 4.2***    *List of amino acids as potential platform molecules.*

| Amino acid | Structure | Typical wt% (dry) of amino acid in agricultural waste |
|---|---|---|
| Alanine (Ala) | | 0.2–3.4% (sorghum DDGS) |
| Arginine (Arg) | | 0.1–5.7% (jatropha seed meal) |
| Asparagine (Asn) | | Included in value for aspartic acid |
| Aspartic acid (Asp) | | 0.3–6.4% (jatropha seed meal) |
| Cysteine (Cys) | | 0–1.1% (rapeseed meal, 5.2% available in poultry feather meal) |
| Glutamic acid (Glu) | | 0.4–18.4% (sugarbeet vinasse) |
| Glutamine (Gln) | | Included in value for glutamic acid |

**Table 4.2**   *(Continued)*

| Amino acid | Structure | Typical wt% (dry) of amino acid in agricultural waste |
|---|---|---|
| Glycine (Gly) | 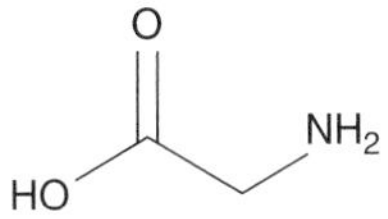 | 0.2–2.5% (cassava leaves, 7.1% available in poultry feather meal) |
| Histidine (His) | 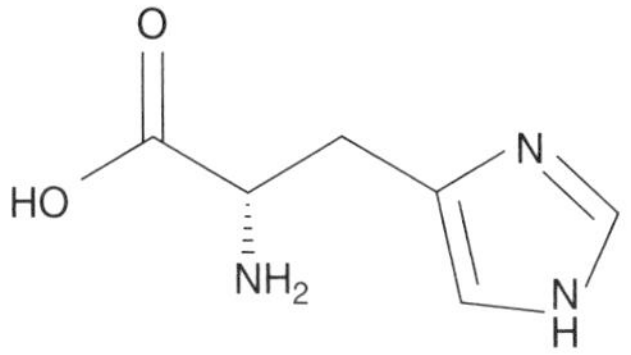 | 0.1–1.4% (jatropha seed meal) |
| Isoleucine (Ile) | 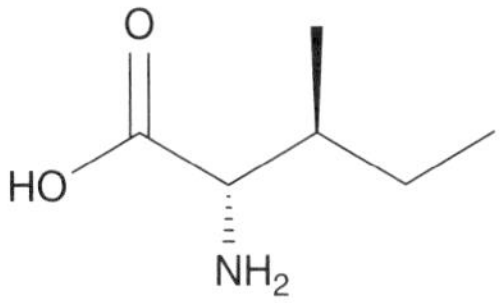 | 0.1–2.7% (soybean meal) |
| Leucine (Leu) | 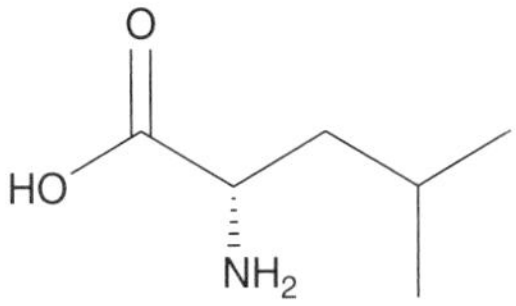 | 0.2–5.0% (sorghum DDGS) |
| Lysine (Lys) | 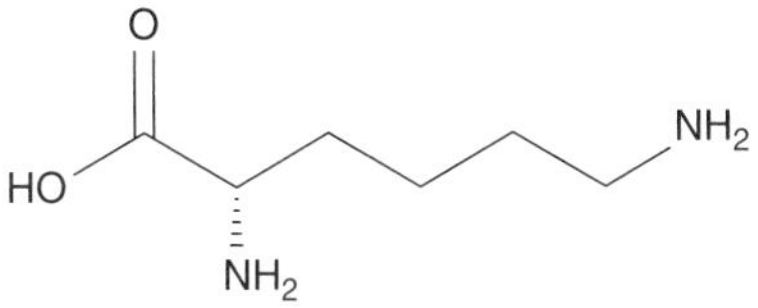 | 0.2–3.4% (soybean meal) |
| Methionine (Met) | 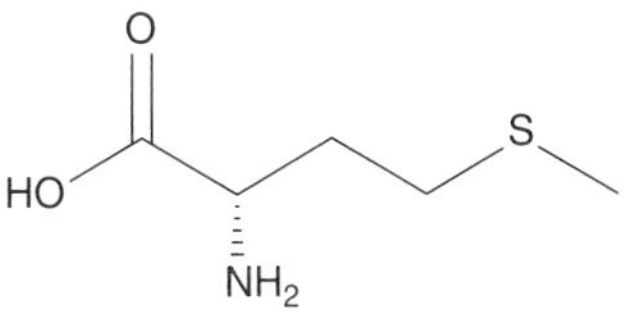 | 0.1–1.4% (sugarbeet vinasse) |
| Phenylalanine (Phe) | 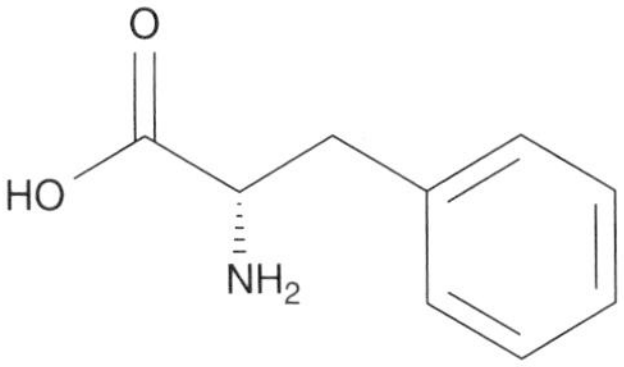 | 0.1–2.6% (cassava leaves and soy bean meal) |
| Proline (Pro) | 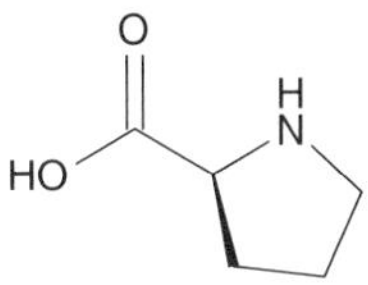 | 0–3.7% (wheat DDGS) |

*(Continued)*

**Table 4.2**    (Continued)

| Amino acid | Structure | Typical wt% (dry) of amino acid in agricultural waste |
| --- | --- | --- |
| Serine (Ser) | | 0.1–2.7% (jatropha seed meal) |
| Threonine (Thr) | | 0–1.8% (rapeseed meal) |
| Tryptophan (Trp) | | 0–1.0% (cassava leaves) |
| Tyrosine (Tyr) | | 0–2.1% (cassava leaves) |
| Valine (Val) | | 0–2.7% (soybean meal) |

Molecules are ordered alphabetically. Stereoisomer shown is the predominant natural isomer. Typical wt% of amino acid in agricultural waste is based on data from Lammens *et al.* [98] where wt% (dry) of protein and wt% of each amino acid in the protein was given for a range of agricultural wastes. The biomass in brackets after the higher wt% value is the biomass from the study that contained this highest quantity for each specific amino acid.

good recovery and general high abundance over all biomass types. Others may be limited in their availability and this will be dictated by the feedstock processed in the biorefinery. A comprehensive review by Lammens *et al.* in 2012 investigated the availability of protein-derived amino acids [98]. A range of biomasses suitable as a feedstock for biorefineries were studied for their protein content, residue composition and global protein potential. Protein containing biomass under investigation for amino acid recovery includes agricultural residues such as maize and wheat, dried distiller's grains with solubles (DDGS), sugar beet and sugar cane vinasse and leaf, wheat straw, rapeseed, soybean, sunflower seed and jatropha seed meal, as well as micro- and macroalgae and poultry feather meal. Algae is seen as a particularly promising source of protein for chemical production; the large quantities of algae likely to be grown will produce significant volumes of protein-rich waste that is currently viewed as unpalatable and therefore only of value for animal feed [99]. By combination of the wt% of protein in the biomass and the amino residue composition of that protein, a representative wt% for each amino acid could be calculated [98]. The range of wt% (dry) values for each amino acid over the selection of investigated agricultural residues is included in Table 4.2.

The lower wt% values in the range for each amino acid are typically from wheat straw or sugar beet leaves as each contains less than 5% protein. Higher yields are possible from protein-rich sources such as poultry feather meal (>85% protein content) and microalgae (>60% protein content in *Spirulina*). These other protein sources also have distinctly different compositions of amino acids, with some residues present in significantly higher abundances compared to agricultural wastes. For example proline, serine and glycine are each present in poultry feather meal in quantities greater than 7 wt%, while their highest content in the agricultural residues investigated was 3.7, 2.3 and 2.5%, respectively.

Examples exist which demonstrate the production of small building-block molecules from the thermal treatment (>120°C) of protein. For example, it is known that acrylamide is formed during thermal decomposition of proteins rich in aspartic acid, alanine and methionine residues, such as olive water [100–102]. These studies of the thermal treatment of proteins are focused more on the potential risk of producing harmful chemicals such as acrylamide in the processing of food than the generation of platform molecules; nevertheless, they highlight the potential for this route in the future if yields and selectivity can be increased [103].

### 4.4.4    Extracts

Many bio-based chemicals utilised today fall within the category of extracts, which are chemicals present within biomass that can be removed via an extraction process such as the use of solvents or the pressing of seeds. Most of these chemicals traditionally extracted from biomass are secondary metabolites such as triglycerides, terpenes, phenolics and tannins, carotenoids, sterols and flavonoids. However, the vast majority of secondary metabolites will likely prove too expensive and present in too little abundance to be classed as significant platform molecules. Most extracts are isolated from biomass in yields of less than 1 wt%, typically as impure mixtures, and involve expensive purification processes, resulting in products of low volume but potentially high value. For most extracts, their use in the form as isolated from biomass will prevail, with high-value

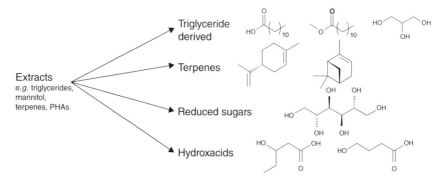

**Figure 4.10**    *Example platform molecules obtained via direct extraction or derived from extracts of biomass.*

applications such as in dietary supplements, food additives, flavours and fragrances, pharmaceuticals, dyes and cosmetics. However, some extracts are available from biomass in larger quantities and via cheaper processing routes; they could therefore be treated as viable platform molecules (Figure 4.10) as part of an integrated biorefinery, the two most promising being the triglycerides (vegetable oils) and terpenes.

Of the platform molecules obtained via extraction from biomass, those derived from triglycerides represent the largest by volume. Indeed, the utilisation of plant triglycerides for the production of chemicals has been practised for centuries, the most obvious example being the formation of soap. Other uses of molecules derived from vegetable oils include lubricants, dicarboxylic acids, resins, alkyds, stabilisers, plasticisers, fatty alcohols (including di- and polyols) and methyl esters, the last being predominately utilised as bio-diesel [104].

The fatty acids chains of the triglycerides can be of varied length and have a varied functionality (Table 4.3) from saturated, one or multiple carbon–carbon double bonds (mono- or polyunsaturated), and even include hydroxy or epoxide groups. Different biomass sources will give different oils with varied fatty acid composition. Palm oil, for instance, is rich in the saturated palmitic acid (>40%), along with high levels of oleic acid (>30%) [105]. Genetically engineered crops have been used to produce high-oleic acid oils where levels of oleate chains in the oils can be in excess of 80%, while olive oil is naturally high in oleate (>70%). Linoleic acid is found in high quantities (>50%) in corn and soybean oil, while linolenic is prevalent in linseed oil (>50%). Ricinoleic acid is the dominant fatty acid (>80%) in castor oil while vernolic acid, with its natural epoxide, is found in vernonia oil (>70%) [106]. Other important unsaturated fatty acids include erucic acid ($C_{22}$), prevalent in rapeseed oil, and eicosapentaenoic acid (EPA, $C_{20}$) and docosahexanenoic acid (DHA, $C_{22}$) from marine oils (e.g. fish and seaweed), though the importance of these fatty acids in food and as essential oils could limit their use as platform chemicals.

**Table 4.3** List of some of the fatty acids found in vegetable that could be used for the production of bio-based chemicals.

| Fatty acid | Structure |
|---|---|
| Lauric ($C_{12}$) | |
| Myristic ($C_{14}$) | |
| Palmitic ($C_{16}$) | |
| Steric ($C_{18}$) | |
| Oleic ($C_{18}$) | |
| Linoleic ($C_{18}$) | |

(Continued)

**Table 4.3** *(Continued)*

| Fatty acid | Structure |
|---|---|
| α-Linolenic (C$_{18}$) | |
| α-Eleostearic acid (C$_{18}$) | |
| Vernolic (C$_{18}$) | |
| Ricinoleic (C$_{18}$) | |

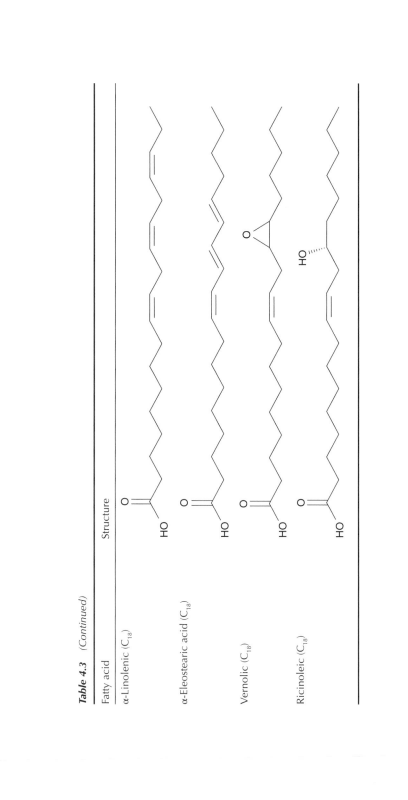

The recent growth in the biodiesel industry has led to a surge in research and development in the valorisation of triglyceride-derived chemicals, more specifically the large quantities of the glycerol co-product. Since being highlighted by the US DOE 2004 report, interest in the use of glycerol to make higher-value chemicals has steadily increased [107]. However, an issue regarding the valorisation of glycerol from biodiesel production is the presence of salts, monoglycerides, MeOH, soap and fatty acids present within the crude glycerol co-product. Products derivable from glycerol are reviewed later in Section 4.8.4 where the triglyceride platform is given as an example of best-in-class for extracts.

Although glycerol is important as a platform molecule, the free fatty acids (FFAs) or fatty acid esters (FAEs) also show strong potential as platform molecules [108]. Conversion of fatty acids to other chemicals for a variety of applications is well established. All fatty acids or esters contain a carboxyl group that can be used for chemical modification. Many of the vegetable-based fatty acids are also unsaturated or polyunsaturated (Table 4.3), and these carbon–carbon double bonds and their allylic sites can also be targeted for chemical modification. Options for chemical modifications of these sites include oxidation, hydration, reduction, cross-metathesis, etherification, isomerisation and oxidative cleavage. Occasionally other functionalities are found naturally in fatty acids, such as the epoxide in vernolic acid from vernonia oil or the hydroxyl in ricinoleic acid from castor oil. Uses for ricinoleic acid include production of sebacic acid for polymer applications and reduction to 12-hydroxysteric acid for use in lubricants. Triglycerides could themselves be viewed as platform chemicals (i.e. converted to other chemicals with the triglyceride moiety intact); as a result of varied fatty acid functionality and composition, they are best referred to as the triglyceride platform when used in this manner.

Terpene isolation from biomass has been extensively investigated and some terpenes represent excellent candidates for platform molecules. Tree toppings following the felling of spruce or pine are known to contain high levels of woody terpenes, especially α-pinene, and turpentine oil (a by-product of the pulping industry) is an abundant source of pinenes [109]. Citrus oil derived from waste peel contains D-limonene in significant quantities.

Terpenes themselves find direct use as solvents, monomers, flavours and fragrances. Dimerisation of terpenes has been used to produce high-density renewable fuels [110, 111]. Additionally, they can be converted to higher-value chemicals via oxidation, isomerisation, reduction and hydration. Of the diterpenes, only pinene and limonene are currently produced on an appreciable scale (~300000 tonnes for pinene and >25000 tonnes for limonene per year) [112]. Notable transformations of terpenes include formation of bioactive compounds such as vitamin A, perillic acid and campholenic aldehyde or derivation to a wide selection of fragrance and flavour compounds, while limonene and pinene have also been used to produce solvents (*para*-cymene) and monomers [109]. Natural rubber is also classified as a terpene and can be used to produce platform molecules via pyrolysis. However, instead of harvesting expensive natural rubber for platform molecule production, it would probably be more economical to convert spent rubber into chemicals such as monomers and limonene ethers [113, 114].

Other notable platform molecules obtained from the extraction from biomass include reduced sugars such as D-mannitol, readily isolated from brown seaweed (Phaeophyceae), and lipids (wax esters, sterols, fatty alcohols, etc.) found on the exterior and within terrestrial plants. Extraction of waxes and sterols from the surfaces of plants is often practised as a pretreatment of lignocellulosic wastes for the removal of the hydrophobic layers, which improves penetration of chemical or biological agents into the cellular structure. However the wax content in the biomass is low and the wax comprises many compounds, however; it is therefore more likely that waxes and sterols will instead continue to be used directly in higher-value applications such as cosmetics.

Hydroxy acids derived from polyhydroxyalkanoates (PHAs) also represent an interesting opportunity for platform molecules. PHAs are produced in bacteria and therefore require the use of carbohydrates (starch, sugars) in a biological process and subsequent recovery of the lipophilic polymer via extraction. Hydroxy acids can be produced from the hydrolysis of the extracted PHAs allowing access to a range of potentially useful platform molecules.

## 4.5   Process Technologies: Biomass to Platform Molecules

Biomass can be converted into platform molecules via several processing technologies, the most common and practical being thermal (sometime referred to as thermochemical), chemical-catalytic (sometimes referred to as chemocatalytic) [55], biological or extraction (see Chapter 3 for more information).

Thermal treatment involves the application of elevated temperatures (>200°C) to biomass, and can be in the absence (e.g. pyrolysis) or presence (e.g. partial combustion) of oxygen [115]. Slow pyrolysis of biomass has been practised by humans for centuries, the goal primarily being formation of charcoal for use as fuel. During the last century a link between the rate of heating and yield of liquid from the pyrolysis process was inferred, and this led to a rise in the interest of fast pyrolysis via the rapid heating of biomass in a controlled manner. Yields of liquid from fast pyrolysis can reach 70 wt%, and it is from this liquid that a range of platform molecules can be derived. Conditions for pyrolysis can be altered to favour higher gas (gasification) or liquid (liquefaction) yields, while keeping solid yields below 10 wt%. From pyrolysis it is the liquid fraction that has the greatest scope for platform molecule isolation, but the complex mixture of up to hundreds of molecules means expensive separation could prevent fast pyrolysis from becoming economical for platform molecule production. Instead, fast pyrolysis liquefaction is seen as a means of producing an intermediate bio-oil platform, and this liquid product, after sufficient upgrading to eliminate phase separation and tar formation, may serve better as a fuel as opposed to a source of chemicals [116].

Formation of syngas (a mixture of CO and $H_2$ in varying proportions) via gasification thermal processing involves the controlled partial combustion of biomass, which however necessitates considerable product upgrading due to the diversity of organic components in the feedstock. The syngas platform is reviewed in greater depth later in Section 4.8.1 as a best-in-class example from the thermal processing of biomass. The generation of hydrogen from biomass is another important thermal process, especially as the reduction of highly oxygenated platform molecules becomes a prevalent conversion route to higher-value chemicals (see Section 4.6). The method of heating in thermal treatment can be varied and can include fast or slow heating via conventional means or the use of emerging technologies such as microwaves (see Chapter 3). In some cases, thermal processing also involves the addition of chemical agents to influence the product selectivity and yields, and therefore some biomass to platform molecule processes sit between chemical-catalytic and thermal (i.e. both high temperatures and chemical agents required). Examples include cases where an acid, base, or metal catalyst has been added, hydrothermal treatments or the use of solvents such as ionic liquids or sulfolane [117, 118]. Catalysts and solvents can dramatically alter the product profiles, altering gas/liquid/char ratios, increasing rates of reactions and lowering the temperatures needed for some reactions, while hydrothermal treatments can increase levels of hydrolysis.

Chemical-catalytic processes are those where a chemical agent(s) has been included in the process. These reactions proceed at moderate temperatures, generally much lower than those required for thermal treatments, though as mentioned above the differentiation between thermal and chemical-catalytic processes can in some instances be unclear. Examples of chemical-catalytic processes include reduction or oxidation of saccharides, acid treatment of saccharides to form HMF or CMF, transesterification or hydrolysis of triglycerides, deacetylation and depolymerisation of chitin and hydrolytic hydrogenation of lignin.

Biological processes are those involving the use of isolated enzymes and microorganisms [119]. Biological processes may form a useful pretreatment prior to other processing methods. For example, developments in the biological degradation of cellulose to glucose could be vital in reducing the cost of producing reduced sugars such as sorbitol and xylitol from biomass. A key consideration of biological processing of biomass is the requirement for various pretreatments, especially of lignocellulose. Recovery of products following fermentation can also add to costs; diacids such as succinates and itaconates require acidification while alcohols are isolated via distillations from the broth. Options for biological pre-treatments and current methodologies are described in greater detail in Chapter 3.

Extraction processes are those that isolate platform molecules or platform molecule precursors direct from biomass via the use of physical operations (e.g. pressing) or solvents (e.g. ethanol, super-heated water and supercritical $CO_2$).

**Table 4.4**   *Comparison of the four main processing technologies used for the production of platform molecules from biomass.*

|  | Thermal | Chemical-catalytic | Biological | Extraction |
|---|---|---|---|---|
| Advantages | Widely applicable to various biomass types | Widely applicable to various biomass types | Mild conditions | Higher-value products |
|  | Can be decentralised (liquefaction at site of biomass production) | Good selectivity | Good selectivity | Natural products |
|  | Very fast | Fast | Natural products |  |
| Disadvantages | Complex mixtures produced | Toxic/corrosive reagents may be needed | Slow | Low quantity of products |
|  | Harsh conditions | May require specialised heterogeneous catalysts | Expensive pretreatment, recovery and purification | Scalability |
|  | Unstable product | Mass transfer issues when using heterogeneous catalysis | Specific feedstock (sugars) required | Limited to a small range of products |

Extraction can also form a useful pretreatment and can be a means by which to obtain higher-value components such as waxes, sterols, pigments, flavours and fragrances. For extraction of platform molecules, the focus should generally be on those compounds that are available in large quantities such as triglycerides and terpenes (pinenes and D-limonene).

Clearly apparent when considering all the processing techniques above is the variety of molecules obtainable from the various biomass types by applying these different technologies. Different biomasses and different desired platform molecules require different processing technologies. Shown in Table 4.4 are some general advantages and disadvantages for each of the four main technologies. No one technology is inherently superior to another, and all are required for the biorefinery to allow diversity in the platform molecules produced.

A biorefinery dealing with varied biomass feedstocks will most likely require an approach to platform molecule production that integrates all four of the processing technologies described above. This integrated approach is needed as the constituent variability in biomass and the desire to access a range of platform molecules warrants all four technologies to be available. Some platform molecules require more than one processing technology for their production such as, for example, FAMEs, which require extraction followed by chemical-catalytic transesterification, or hydroxyl acids that involve first biological routes to PHAs, followed by extraction and finally chemical-catalytic hydrolysis.

## 4.6    Bio-Derived v. Fossil-Derived: Changing Downstream Chemistry

A switch from fossil-derived base chemicals to bio-derived platform molecules will likely require changes to the types of reactions typically used for the production of commodity chemicals (those produced on a large scale direct from base chemicals). Fossil-derived base chemicals, with the exception of MeOH, are totally devoid of heteroatoms and contain only carbon and hydrogen. The result of this is that many of reactions used to produce commodity chemicals in the current petrochemical industry require the introduction of heteroatoms, most often oxygen. Platform chemicals, inherently high in oxygen content, will not typically require the same initial oxidation steps to reach useful commodity chemicals, but instead may need reduction and dehydrations to reduce or tailor functionality. An example of this difference in steps can be seen for the theoretical production of ethyl acrylate from biomass compared with the current fossil-based route (Figure 4.11). The primary base chemicals for the fossil route are propylene (for acrylic acid formation) and ethene (for ethanol formation); both steps of base to commodity chemical require the introduction of oxygen, either from oxidation or

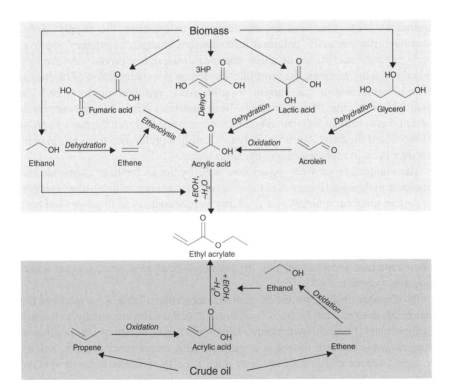

**Figure 4.11**    *Fossil-derived and bio-derived routes to ethyl acrylate.*

hydration. Several routes may be considered for the production of acrylates from biomass and a few of these are shown in Figure 4.11. The ethanol needed for the esterification is simply produced from fermentation of glucose, and this ethanol can also be used to form ethene needed for the metathesis in the fumaric acid route. Glucose is also needed as the feedstock for fermentation and can be used to produce (1) lactic acid and 3HP, giving acrylic acid via dehydration; or (2) fumaric acid giving acrylic acid via ethenolysis. As described later in Section 4.8.4, glycerol can also be used for acrylic acid formation via dehydration to acrolein and oxidation to acrylic acid.

The example of acrylic acid above is one of a bio-based 'drop-in replacement' for a former fossil-derived equivalent, that is, a chemical of the exact same structure but produced from a biomass feedstock as opposed to being petroleum-derived. These drop-in replacements are of particular interest to the chemical industry as the downstream technologies, facilities and markets are already established. An alternative to drop-in replacements would be to move to new materials and products (e.g. poly(ethylenefuranoate) replacing PET). It is likely that some new products, different from those derived from base chemicals, will become preferable in the bio-based chemical industry either as a result of superior properties or more favourable economics. The limitation of this approach is the time and money required to establish the supply chain, manufacturing capacity, potential infrastructure changes, regulatory approval and consumer acceptance of a new chemical, material or product. As the bio-based economy matures, the most likely outcome is a combination of bio-based drop-in replacements for current fossil-derived consumer products and the gradual introduction of new products. Both consumer demand for greener products and government legislation and incentives could increase the rate of uptake of bio-based products, but is unlikely to result in a preference for novel materials over drop-in replacements in the short term.

The comparison of wt% oxygen and wt% hydrogen for base chemicals and platform molecules (Figure 4.12) also highlights another valuable characteristic of certain platform molecules: a similarity in functionality to fossil-derived base chemicals. Both terpenes and fatty acids/esters occupy an area of the plot (Figure 4.12) populated by the fossil base chemicals, and clearly these two classes of platform molecule are of importance if the user seeks applications where a lack of heteroatoms is important (e.g. hydrophobic head of a surfactant or water-repelling polymers).

As demonstrated in the list of platform molecules (Table 4.5), many of the chemicals derived from biomass contain chiral centres and are usually, as a result of their natural origin, enantiopure. This is obviously advantageous in circumstances where the user wishes to target a specific stereoisomer product; indeed, levoglucosenone and proline are both often used in synthesis to target specific isomers of a given product. Examples also exist where several stereoisomers are potentially derivable from bio-based feedstocks, resulting in some interesting

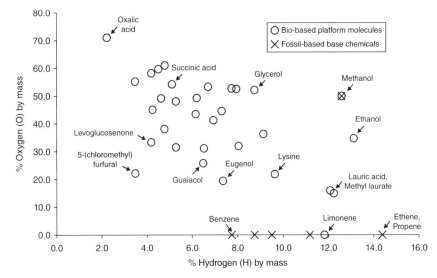

***Figure 4.12***    *% Hydrogen (by mass) versus % oxygen (by mass) for fossil-derived base chemicals (X) and bio-based platform molecules (O).*

opportunities for varied applications. One such example is 1,4:3,6-dianhydrohexitols (isosorbide, isomannide and isoidide) formed from the three $C_6$ reduced sugars sorbitol, mannitol and iditol (Figure 4.13), each possessing distinctly different reactivity in polymerisations (due to the varied hydrogen bonding of the hydroxyls to the ethers) and characteristics of the final polyesters due solely to the initial isomeric form of the free starting sugars [120]. Achieving a similar variability in fossil-derived polyesters would require the use of chiral catalysts or auxiliaries in the monomer synthesis, adding greatly to the cost of their production, while the effect can be obtained from these bio-based monomers simply by the inherent chirality of the natural sugar feedstocks.

## 4.7    List of Platform Molecules

A list of current suggested platform molecules is provided in Table 4.5, grouped together on the basis of the process technology use to produce them from biomass. The list has been compiled using the authors' knowledge of the area and by the application of the definitions described previously. As the field of bio-based platform molecules matures, it is likely that some chemicals from this list will become more prevalent than others (the main drivers being economics and feedstock availability). In a fully matured bio-based chemical industry, the definition of a platform molecule will be refined to those that become available on a large scale and at a low cost. An extensive list has been given here to

**Table 4.5** List of suggested platform molecules.

| Platform molecule | Processing technology | Constituent of biomass molecule is derived from | Structure |
|---|---|---|---|
| Synthesis gas (syngas) platform Methanol | Thermal Thermal | Any carbonaceous material Any carbonaceous materials (via syngas, see above) | $H_2$, CO, $CO_2$ HO— |
| Guaiacol | Thermal | Lignin | (2-methoxyphenol structure) |
| Levoglucos-anone | Thermal | Cellulose or sugars | (levoglucosenone structure) |
| Vanillin | Thermal | Lignin | (vanillin structure) |

| | | |
|---|---|---|
| Levoglucosan | Thermal | Cellulose or sugars |
| Eugenol | Thermal | Lignin |
| Vanillic acid | Thermal | Lignin |

(Continued)

**Table 4.5** *(Continued)*

| Platform molecule | Processing technology | Constituent of biomass molecule is derived from | Structure |
|---|---|---|---|
| Oxalic acid | Chemical-catalytic | Cellulose or sugars | |
| Furfural | Chemical-catalytic | Hemicellulose or xylose | |
| γ-valerolactone | Chemical-catalytic | Cellulose or sugars | |
| Levulinic acid | Chemical-catalytic | Cellulose or sugars | |
| 5-(Hydroxy methyl)furfural | Chemical-catalytic | Cellulose or sugars | |

| Compound | Method | Source | Structure |
|---|---|---|---|
| 5-(Chloromethyl)furfural [121] | Chemical-catalytic | Cellulose or sugars | |
| Xylitol | Chemical-catalytic | Hemicellulose or xylose | |
| Glucosamine | Chemical-catalytic | Chitin | |
| Sorbitol | Chemical-catalytic | Cellulose or sugars | |
| Glucaric acid | Chemical-catalytic | Glucose or cellulose | |

(Continued)

**Table 4.5** *(Continued)*

| Platform molecule | Processing technology | Constituent of biomass molecule is derived from | Structure |
|---|---|---|---|
| Ethanol | Biological | Sugars | |
| Butanols (n-, iso-, sec-) | Biological | Sugars | |
| Acetoin [122, 123] | Biological | Sugars | |
| Lactic acid | Biological | Sugars | |
| 3-Hydroxy propionic acid | Biological | Sugars | |
| Fumaric acid | Biological | Sugars | |

| Succinic acid | Biological | Sugars | |
| Itaconic acid | Biological | Sugars | |
| Amino acids (e.g. aspartic acid and glutamic acid) | Biological | Protein or sugars | |
| Malic acid | Biological | Sugars | |

(Continued)

**Table 4.5** (Continued)

| Platform molecule | Processing technology | Constituent of biomass molecule is derived from | Structure |
|---|---|---|---|
| Muconic acid [124, 125] | Biological | Sugars | |
| Aconitic acid | Biological | Sugars | |
| Citric acid | Biological | Sugars | |
| 3-hydroxy-butanoic acid | Biological + extraction + chemical-catalytic | Sugars (to PHAs) | |

| Compound | Process | Structure |
|---|---|---|
| 3-hydroxy-pentanoic acid | Biological + extraction + chemical-catalytic | Sugars (to PHAs) | |
| Glycerol | Extraction + chemical-catalytic | Triglycerides | |
| Terpenes | Extraction | Direct extraction | |
| Mannitol | Extraction or chemical-catalytic | Direct extraction or Sugars | |
| Fatty acids (e.g. oleic acid) | Extraction and chemical-catalytic | Triglyceride (see below) | |

(Continued)

**Table 4.5** (Continued)

| Platform molecule | Processing technology | Constituent of biomass molecule is derived from | Structure |
|---|---|---|---|
| Fatty acid akyl esters (e.g. methyl oleate) | Extraction and chemical-catalytic | Triglyceride (see below) | |
| Triglycerides platform | Extraction | Oil containing biomass | |

Molecules are group together based on the process technology primarily used to produce them from biomass; some require two or more process technologies. All amino acids are obtainable from biomass (either from fermentation or protein depolymerisation) but only glutamic and aspartic acid are given here as examples.

demonstrate the current diversity in functionality and to ensure the reader has the broadest possible vision when considering bio-based chemicals for any process or product. Some entries in Table 4.5 only contain selected examples from a larger range of molecules. For example, all natural amino acids are derivable from biomass, either from fermentation or protein depolymerisation, but only aspartic and glutamic acid are highlighted in the list below. Lammens *et al.* analysed a range of biomass sources for the amino acid mass fraction following protein hydrolysis and concluded that aspartic and glutamic acid were the most abundant residues and therefore the most likely to become key platform molecules [98]. The structures shown for terpenes (α-pinene and D-limonene) and fatty acids (lauric acid) are also only representative examples of a wider range of possible platform molecules. PHAs, produced via sugar-consuming bacteria, can be hydrolysed yielding various hydroxy acids, with only representative examples shown in Table 4.5.

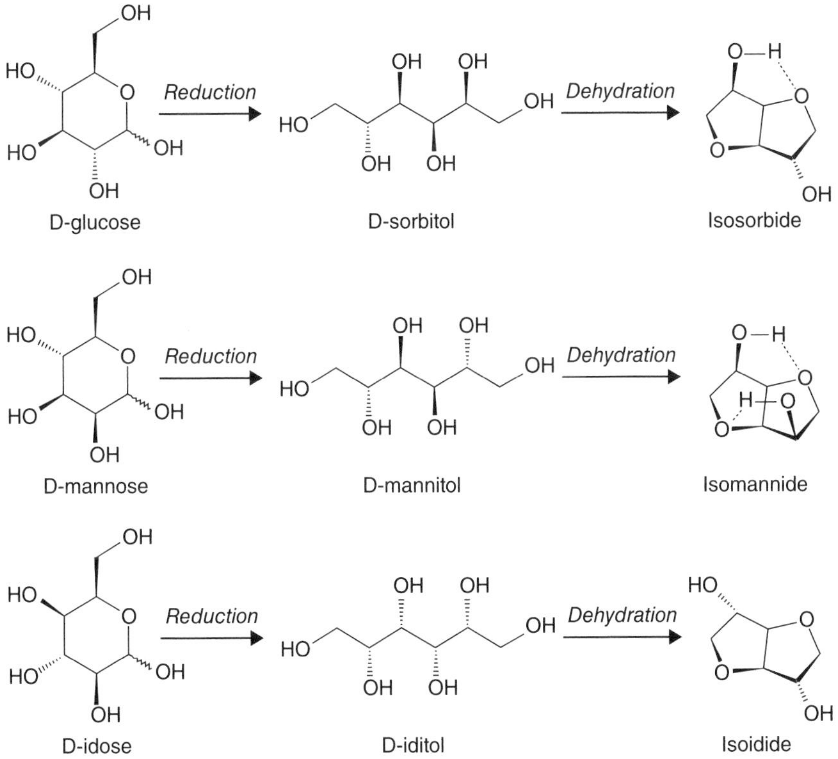

**Figure 4.13** *Three possible isomers of the 1,4:3,6-dianhydrohexitols derivable from platform molecules.*

## 4.8   Example Platform Molecules

For platform molecules to offer value to the bio-based chemical industry they need to be useful in the formation of marketable chemicals. While all of the platform molecules listed in Table 4.5 are actually or potentially useful, in the following we describe one mainstream representative from each of the four biomass-processing routes. In each example, a set of possible derivatives is provided and discussed. The suggested products in each case are not exhaustive; instead, they are representative of the potential of that molecule to be converted into a diverse range of higher-value chemicals and materials, representing the full range of chemical transformations for the different functional groups of each platform molecule. These four examples have been selected as they represent what the authors consider to be the current platforms or platform molecules of exceptional promise for each of the processing technologies, with well-established production routes from biomass and numerous investigated products for each.

### 4.8.1   Synthesis Gas Platform: Thermal Treatment

The thermal/thermochemical treatment of biomass is viewed as one of the primary platforms for the biorefinery [126]. Included within the thermal treatment of biomass is the syngas platform. The syngas platform can be used to produce fuels, chemicals and energy via the intermediate $H_2$, $CO_2$ and CO constituents, while the broader thermal treatment platform additionally includes the generation of bio-oil and bio-char (Figure 4.14). Although possibly utilised as an energy source for the biorefinery, the nutrient-rich bio-char has also shown promise as soil amendment/enhancer, while additionally acting as a means of carbon sequestration [127–130]. Bio-oil can be used directly as a source of energy [131] or for syngas production [132]. It can also be steam reformed to give bio-hydrogen [133] or catalytically upgraded to a transportation fuel [134–136]. Bio-oil contains some potential platform molecules, such as furfural, levoglucosan, alcohols and various phenolics, though the complex mixture of the oil means purification costs are high and yields of individual components are low [137]. Although bio-oil and bio-char are important products from the thermal treatment of biomass, it is therefore likely that the use of synthesis gas will prove to be the favoured route to platform molecules.

Synthesis gas (syngas), primarily a mixture of CO and $H_2$, can be prepared from any carbonaceous material via controlled thermal treatment (>700°C) in the presence of oxygen and/or steam, a process generally referred to as gasification. Both the production and subsequent utilisation of syngas is well established, with commercial facilities using coal feedstocks dating back to the early 1800s [138]. Although initially used as a fuel (town gas) during the First and Second World Wars, technology advanced for the conversion of synthesis gas to other mainstream chemicals such as MeOH, olefins, alkanes, ethers and ammonia. The drive towards a bio-based economy has resulted in renewed interest in the biomass gasification,

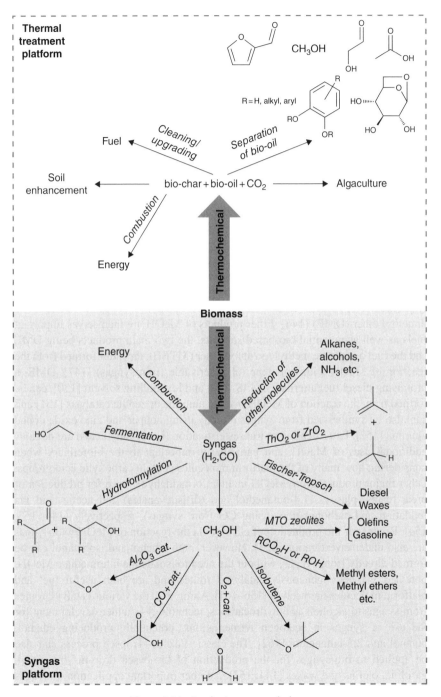

***Figure 4.14*** *Synthesis gas as a platform.*

thus producing bio-based syngas [139–141]. Similar technological advances for traditional fossil-derived syngas to chemicals are directly applicable to the bio-based equivalent as the fundamental constituents of syngas (CO and $H_2$) are the same. However, purification of the crude syngas following gasification is required prior to further processing, removing any particulates, ash, sulphur compounds, nitrogen compounds and tars [142, 143]. A key advantage of the biomass gasification route, and the subsequent syngas platform, is that any carbonaceous material can be used to produce a similar composition of CO and $H_2$, thus allowing lower-value fractions such as lignin to be converted to platform molecules. Of the chemicals produced from syngas (Figure 4.14), MeOH is dominant in terms of quantity and is usually generated by a high-temperature and -pressure exothermic reaction over various copper catalysts [144]. The yields of MeOH from synthesis gas depend on the ratio of $CO/CO_2$ and sulphur content; it is therefore desirable for gasification methodologies to focus on reducing $CO_2$ and sulphur impurities levels if MeOH is the target product [145].

Methanol itself can then be viewed as an important platform molecule when derived from bio-syngas, with a wide array of derivative chemicals accessible from it [144]. Olefins and gasoline can be produced from MeOH over zeolite catalysts (the Mobil process) at temperatures in excess of 400°C, via the intermediate dimethyl ether (DME) [146]. Ether products of MeOH are themselves important fuels as well as potential bio-based solvents, the two main products being DME and the fuel oxygenate methyl *tert*-butyl ether (MTBE), the latter formed from the reaction of MeOH with isobutene (also derivable from syngas) [147]. DME, a promising diesel fuel alternative [148, 149] and low boiling solvent [150], can be formed from the reaction of MeOH over an alumina or zeolite catalysts [151] and can also be synthesised from syngas directly using copper and zinc oxide doped alumina [152]. Esterification or transesterification (i.e. of fatty acids) are obvious additional uses of MeOH, and particularly important to the biorefinery when considering how many of the platform molecules contain carboxylic acid groups. Other major products from MeOH include formaldehyde (used for production of urea, resins, plastics, 1,4-butanediol and (di)isocyanates) and acetic acid via oxidation and carbonylation (using CO from syngas), respectively [144, 153, 154]. Routes for the production of ethanol via the reaction of MeOH with syngas are also under investigation [155]. However, both ethanol and isobutanol can be formed directly from syngas, without the need to isolate the intermediate MeOH, obtained via heterogeneously catalysed routes and are both useful fuel and platform molecules themselves [156–161]. Along with the various catalytic routes from syngas to alcohols and hydrocarbons, technology is under development for the use of syngas in acetogen fermentations, potentially producing ethanol, butanol and 2,3-butanediol [162]. The classic Fischer–Tropsch process can also be applied to bio-syngas for the production of bio-based drop-ins for diesel, gasoline, olefins and waxes [144, 163]. Another important application of syngas within the biorefinery will be for use in hydroformylation and reduction, both very relevant to platform molecules considering the various functionalities present. As a whole, the syngas and broader thermal treatment platform are useful in obtaining a wide range of chemical products, fuel and energy from almost any carbonaceous biomass; it is the fact that any biomass can be used to generate syngas that makes this route to platform molecules so appealing to the bio-based economy.

### 4.8.2  5-(Chloromethyl)furfural: Chemical-Catalytic Treatment

The preparation of 5-CMF was first described in 1901 [164], its bromo analogue having been characterised as early as 1899 [165]. Remarkably, these halomethyl-furfurals were assigned correct structures even before the structure of 5-HMF had been confirmed, based on a comparison with the known 5-(bromomethyl)-2-furoic acid [166]. These early efforts involved the treatment of fructose, sucrose or cellulose with a saturated solution of the hydrogen halide in an organic solvent, giving yields in the range 12–30%. The first high-yielding preparation of CMF was published in a 1977 patent and involved the use of a biphasic aqueous acid/organic solvent reactor, by which a 77% yield of CMF from fructose was obtained [167]. This result was followed up in a 1981 paper by Szmant and Chundury, who performed a detailed optimisation study and found that CMF could be produced from fructose in up to 95% yield, although glucose and starch gave much lower yields [168]. Notable research groups of the twentieth century that involved CMF in their research include those of Henry J.H. Fenton [164, 165], Emil Fischer [169], Norman Haworth [170], and Donald Cram [171].

In 2008, Mascal and Nikitin reported the production of CMF from either glucose, sucrose or cellulose in 71–76% yield, alongside 14–18% yield of a mixture of HMF, levulinic acid and 2-(2-hydroxyacetyl)furan [172]. This was significant in that it described the first synthesis of CMF in good yield from practical biomass feedstocks (glucose, cellulose). This was followed up by a study involving raw biomass, which reported similar findings [173]. In 2009, an optimised procedure was published whereby either glucose, sucrose, cellulose or corn stover was heated in a biphasic aqueous hydrochloric acid/organic solvent reactor at 80–100°C for 3 hr to give CMF in 80–90% isolated yield, depending on the feedstock, alongside 5–8% levulinic acid [174]. The method has recently been adapted to a continuous flow reactor, and to processing under microwave irradiation [175, 19]. CMF has therefore been established as a viable alternative to the highly popular platform chemical HMF, which is presently derived in good yield only from fructose and has serious issues associated with its isolation [176].

The derivative chemistry of CMF is essentially identical to that of HMF, the only differences being that CMF is more reactive towards nucleophilic substitution at the methylene carbon and that it is more lipophilic than HMF.

CMF has basically two derivative manifolds: furanic and levulinic. Under mild conditions, CMF reacts with nucleophiles to give halide substitution products

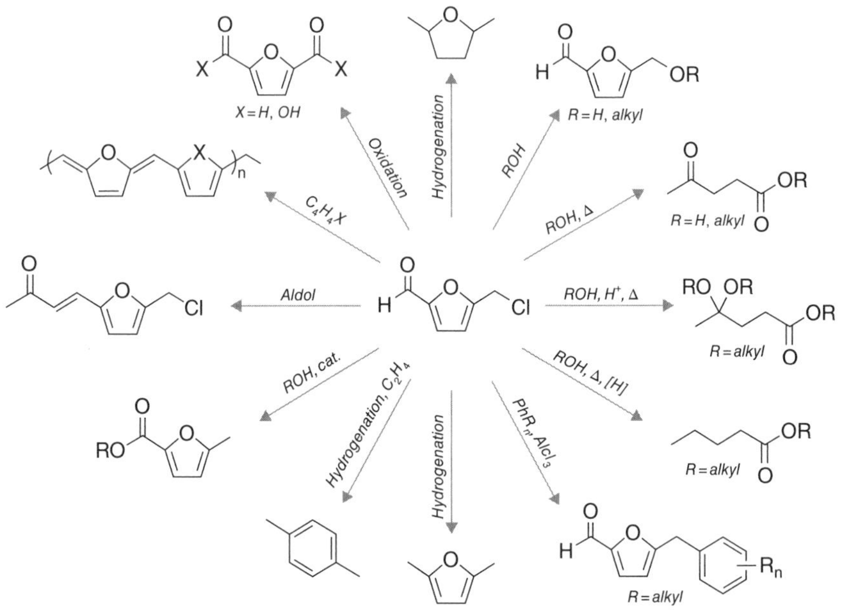

***Figure 4.15*** *5-(Chloromethyl)furfural as a platform molecule.*

and/or aldehyde condensation products. Under more forcing conditions, the furan ring opens with the loss of formate to give levulinate products. Several useful derivatives have been described, some examples of which are shown in Figure 4.15. From the top and proceeding clockwise, complete hydrogenation of CMF gives 2,5-dimethyltetrahydrofuran, a promising fuel oxygenate [177–179]. Reaction with water gives HMF, an icon of the renewable chemistry movement [176]. Likewise, reaction with alcohols at room temperature gives alkoxymethyl furfurals, which have also been considered for use as biofuels [180–182]. Reaction with either water or alcohols at higher temperatures gives levulinic acid or levulinic esters, respectively, plus the corresponding formates [121]. Like HMF, levulinic acid is considered a key biomass-derived platform chemical [26], while levulinate esters have been proposed as diesel additives [183]. Further exposure of levulinate esters to alcohols in the presence of acid (either added acid or auto-catalytically via liberated HCl) gives levulinate ester acetals which have applications as novel monomers, plasticisers and solvents [184]. Finally, the hydrogenation of levulinate esters gives valeric esters, which have been shown to possess outstanding fuel properties [185].

The Friedel–Crafts reaction was one of the first derivatisations performed on the CMF molecule [186], and yields aryl derivatives that may be useful as biofuel precursors [187]. Gentle hydrogenation of CMF gives 2,5-dimethylfuran, an outstanding biofuel in its own right [188] but also highly valuable as the precursor to *para*-xylene by cycloaddition with ethylene [189–192]. Treatment of CMF with alcohols in the presence of an N-heterocyclic carbene catalyst gives furoate esters, which have also been shown to have excellent fuel properties [193]. Longer carbon chain lengths are required for the engineering of diesel-like hydrocarbons, which can be achieved by aldol condensation of CMF or CMF-derived HMF with ketones [194]. Conjugated, electro-active polymers are produced when CMF reacts with electron-rich aromatics such as pyrrole, furan or thiophene [195]. Finally, oxidation of CMF with nitric acid gives either 2,5-diformylfuran or 2,5-FDCA. Both of these are interesting monomers, in particular the latter which is considered as a renewable replacement for petroleum-derived terephthalic acid [175, 186, 196]. The equivalent polymer generated from FDCA, called polyethylene furanoate (PEF), has been shown to be competitive with PET in terms of performance.

### 4.8.3    n-Butanol (Biobutanol): Biological Treatment

Classically, biobutanol has been derived from the anaerobic ABE (acetone-butanol-ethanol) fermentation of sugars by *Clostridium* sp., a process that was first described by Pasteur in 1862 and first industrialised in the UK in the early part of the twentieth century [197]. But even using the organism of choice for this process (*Clostridium acetobutylicum*), the production rate is modest and the solvent titre peaks at around a 2% solution which involves high water usage and product isolation costs; this is the main reason that biobutanol has not until recent times attracted stronger market interest. Significant recent advances in the selection or engineering of strains of *Clostridium* for higher butanol selectivity and stress tolerance, as well as the insertion of the butanol pathway into heterologous organisms such as *E. coli* and *Saccharomyces cerevisiae*, are however leading towards the production of butanol in industrially competitive yields [198]. In fact, this platform molecule has attracted sufficient attention to merit the publication of 25 review articles on biobutanol over the past 4 years. Parallel developments in feedstock selection, reactor management and downstream processing are also encouraging further interest in biobutanol commercialisation. The current movement away from the use of food crops in favour of lignocellulosic feedstocks is therefore advantageous for *Clostridium*, which can utilise both 5- and 6-carbon sugars, thereby increasing the useable biomass input into the process over those that can only ferment hexoses. Production limitations may also be alleviated by transitioning away from batch fermentation towards continuous processes by immobilisation of organisms on a support. The application of a two-stage, dual-path procedure that relies on different species of *Clostridium* has been shown to mitigate selectivity issues by decoupling carboxylic acid formation from its reduction to alcohols [199, 200]. Finally, *in situ* butanol removal during fermentation can improve productivity and diminish product inhibition. For example, gas stripping is a comparatively

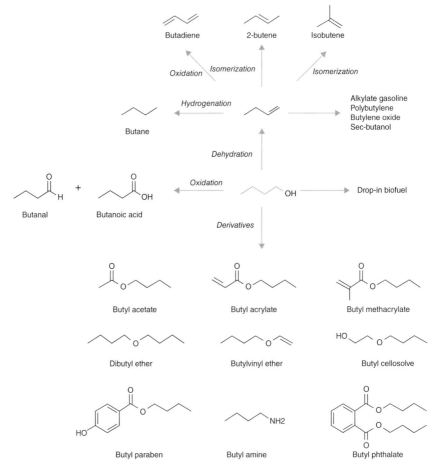

**Figure 4.16**   *n-Butanol as a platform molecule.*

simple approach to reducing butanol concentration without affecting culture, intermediates or other components of the medium [201].

*n*-Butanol has fuel properties that are superior to those of ethanol: it is less volatile, has a higher volumetric energy content and, unlike ethanol, gasoline–butanol blends do not separate in the presence of water. The octane number of *n*-butanol is 87, which is equivalent to that of commercial gasoline; spark-ignition engines can utilise gasoline-butanol blends in any proportion up to B100, without modification. Additionally, existing ethanol capacity can be cost-effectively retrofitted to biobutanol production. Butanol is therefore poised to become the 'new ethanol'.

The advantages of butanol as a biofuel are mirrored in its versatility as a chemical intermediate [202]. As shown in Figure 4.16, butanol can be derivatised in multiple ways to provide mainstream industrial chemicals. Esterification provides butyl acetate, a major industrial solvent, and butyl acrylate monomers, which are the largest-volume commercial derivatives of butanol. Butyl phthalate is a common plasticiser, while butyl paraben is widely used in personal care formulations. Ethers of butanol also find sundry markets as solvents and monomers. Dibutyl ether in particular is also a diesel additive of considerable interest [203]. Finally, functional derivatives of butanol, such as amines, are used in the production of agrochemicals and various other commodities.

Butanol can be catalytically oxidised to either butanal or butanoic acid, both of which branch out towards their own sets of derivative markets. Alternatively, dehydration gives 1-butene or, under some conditions, mixtures of 1- and 2-butenes. Butenes are a highly versatile platform in their own right, serving as a feedstock for alkylate gasoline, a high-octane $C_7$–$C_9$ product that is widely used in commercial fuels. An alternative, cationic oligomerisation of butanes, leads to a mixture of $C_8$ and $C_{12}$ alkane products called polymer gasoline. The largest non-fuel market for butenes is in polymer formulations, that is, in polybutene, polybutylene oxide and as a co-monomer with ethylene in polyethylene-type polymers. Sec-butanol, available by hydration of *n*-butanol, is useful solvent alcohol with its own range of applications. Simple hydrogenation of butene gives butane, which is an important component of winter blend gasolines.

The dehydrogenation of butene to 1,3-butadiene is a mature technology, and the vast market of butadiene is chiefly applied to the production of synthetic elastomers. Isobutene is available from *n*-butenes by isomerisation, with multiple additional applications to elastomers, polymers and industrial chemicals.

The virtually unlimited market reach of butanol in transportation fuels, materials and chemicals provides a strong impetus for the biotechnological approach to biomass valorisation to transition away from ethanol towards butanol. The full realisation of this goal however awaits further refinements in biobutanol production and the commercialisation of these technologies.

### 4.8.4   Triglyceride Platform: Extraction

Triglycerides are the main component of plant and animal lipids and they, along with their constituent glycerol and fatty acids, have been utilised as building blocks for chemical production since antiquity. Triglycerides are composed of esters of long-chain fatty acids bonded to a central glycerol (1,2,3-trihydroxy-propane) unit, therefore containing a total of three esters. Each fatty acid chain can be different to the others of the same triglyceride and therefore vegetable oils are not typically one distinct molecule but a range of different molecules. It is for this reason that vegetable oils are typically viewed as a 'platform' (i.e. triglyceride platform) rather than as individual platform molecules. Instead, the constituent fragments of triglycerides are seen as individual platform molecules (e.g. ricinoleic acid, oleic acid and glycerol), some of which are highlighted

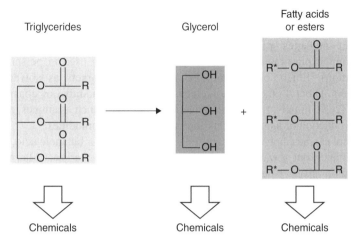

**Figure 4.17**   *Routes to utilise triglycerides and building blocks for chemical production: the triglyceride platform.*

below. Global annual vegetable oil production in 2009–2010 was estimated to be in excess of 140 million tonnes with the majority used for food and feed, though roughly 20% was in non-food applications: 7.7 million tonnes as fatty acids (primarily for soaps) and 2 million as fatty alcohol oleochemicals [116, 204, 205]. Triglycerides and the chemicals derived from them clearly represent an already well-established bio-based sector of the chemical industry. Even during the last century when the petrochemical industry has come to dominate chemical production, the market for triglycerides and their derivatives has remained strong; this is partly to the combination of availability and functionality/structure that is not easily replicated by fossil-derived equivalents. Triglycerides have three distinct routes for application as building blocks for other chemicals and materials (Figure 4.17). The first option is to process triglycerides in their intact form, targeting the functionality on the fatty acid chains for chemical transformations to give modified triglycerides. The second and third options are linked as the ester bonds yields fatty acids and glycerol that can be utilised separately as platform molecules. Intermediate examples also exist where the target is mono- or diglycerides, that is, the cleavage of only one or two of the ester bonds.

Modification of native triglycerides, while maintaining the triester moiety, is typically used for the synthesis of monomers for free radical, cationic, olefin metathesis and condensation polymerisations and also to produce additives for polymeric materials such as cross-linkers and plasticisers [105, 106, 206–209]. Additionally, hydrogenation, hydration and oxidation of the chains are used for the synthesis of lubricants base oils or for components of personal care and cosmetic products [210]. All of the modifications highlighted in Table 4.6 are applicable to free fatty acids or the alkyl esters.

**Table 4.6** Triglyceride platform products from the chemical modification of triglycerides.

| Starting oil | Modification | Modified triglyceride product |
|---|---|---|
| Linseed oil (same reaction applicable to all unsaturated vegetable oils) [211] | Epoxidation | |
| Soybean oil (same reaction applicable to all unsaturated vegetable oils) [212] | Epoxidation Methanolysis | |
| Soybean oil (same reaction applicable to all unsaturated vegetable oils) [213, 214] | Hydroformylation Hydrogenation | |
| High oleic sunflower oil (same reaction applicable to other high mono-unsaturated oils) [215] | Metathesis (acrylate) | |
| Soybean oil (same reaction applicable to all unsaturated vegetable oils) [216, 217] | Pericyclic reaction (ene) | |
| Castor oil (same reaction possible on hydroxylated vegetable oils) [207] | Esterification | $R = CO_2H$ (maleate) or H (acrylate) |

G: glycerol moiety and two more fatty acid chains, these fatty acids either being the same or different from the fatty acid shown in detail

Historically, surfactants are the chemicals most widely produced from triglycerides, though the increasing demand for biodiesel over the last few decades has seen this become the most predominant product from vegetable oils [218]. Both soaps and biodiesel are examples of the route to chemicals from triglycerides via the cleavage of the fatty acid chain from the glycerol. Several free fatty acids and their esters are of value as oleochemicals, including ricinoleic acid, oleic acid, lauric acid and palmitic acid, and the chemicals formed from these acids are referred to as oleochemicals. The longer-chained saturated acids (lauric and palmitic) are mainly used for surfactant production or additives to personal care formulations. The carbon–carbon double bonds in oleic and linoleic acids are used for further functionalisation of the chain, such as epoxidation, hydration and alkylation, but can also be used for chain scission, typically via ozonolysis (forming azelaic acid, used for formation of nylon-6,9) or ethenolysis (Figure 4.18) [219]. Unsaturation can be used to connect two fatty acid chains together, the most common being a $C_{36}$ dimer acid used in adhesives, polyesters and polyurethanes [220]. These are often further modified by reduction to the dimer diol, which is also useful for polymer applications [116]. Ricinoleic acid is another fatty acid of high value for chemical production, both as a result of its β-hydroxy alkene moiety and its high abundance in castor oil triglycerides, typically between 85 and 90% [221]. The most important products of ricinoleic acid are the result of chain scission (alkali fusion or pyrolysis), giving sebacic acid, 10-hydroxydecanic acid or 10-undecenoic acid (Figure 4.18) depending upon the conditions used, while also producing one equivalent of 2-octanol, 2-octanone or heptanal, respectively [222, 223]. The larger scission products are typically used for the production of various nylons (nylon-4,10, nylon-6,10, nylon-10,10, nylon-11) [116]. Additionally, ricinoleic acid or its esters can be hydrogenated or dehydrated to various useful derivatives (Figure 4.18) [224]. The carboxylic acid function can

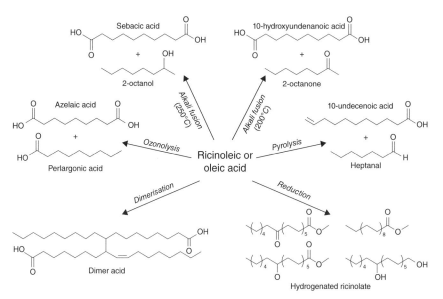

**Figure 4.18** *Triglyceride platform: example oleochemicals derivable from oleic and ricinoleic acid.*

also be used as the target for chemical modification. Examples include reduction to fatty alcohols or esterification with other alcohols including vinyl alcohols for monomer synthesis [104, 225, 226].

In other instances triglycerides are used as a feedstock for fuel production [227], either via MeOH transesterification to give FAMEs, or decarboxylation to give long chain hydrocarbons (Figure 4.19). It is from the production of biodiesel where the glycerol by-product is typically envisaged to be sourced as a platform molecule and the commodity chemicals that can be derived from glycerol are very diverse, some examples of which are shown in Figure 4.19 [227–235].

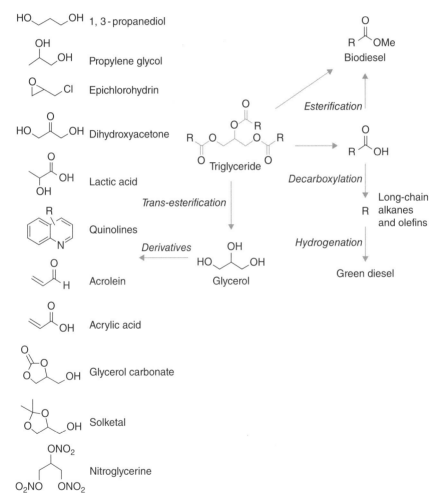

**Figure 4.19** *Triglyceride platform: fuels from fatty acids and commodity chemicals from glycerol.*

Hydrogenations are therefore used to form 1,3-propandiol and propylene glycol while additional oxidation and dehydration steps yield acrolein, lactic acid and acrylic acid [236]. Oxidation of glycerol is also possible, forming a range of hydroxy acids, aldehydes and ketones (such as dihydroxyacetone). Deoxygenation and dehydroxylation of glycerol can also be used to remove functionality, eventually yielding alkenes and alkanes which are useful as drop-in replacements for fossil equivalents [235, 237, 238]. Reaction with dimethyl carbonate or urea is used to produce glycerol carbonate, useful as a solvent or for a variety of further transformations [239], while reaction of glycerol with HCl is used to produce epichlorohydrin. The ketal, solketal, is formed from the reaction of acetone with glycerol and this protected glycerol is useful in forming mono-, di- and triglycerides of controlled composition [240]. The Skraup reaction between glycerol and anilines can be used in the production of substituted quinolones, finding use in pharmaceuticals, dyes and chelating agents, for example [241]. Glycerol has also been used for many years in the production of the commercial explosive nitroglycerine (glycerol trinitrate).

## 4.9    Conclusion

As the age of cheap oil nears its end and environmental concerns relating to the extraction of non-sustainable fossil resources apply pressure to the petrochemical industry, new sustainable resources for chemicals and materials will be required. The traditional chemical industry relies heavily on a small set of base chemical building blocks that are produced globally on an enormous scale. Within this chapter it has been demonstrated how a new set of bio-derived building blocks, so call platform molecules, can be used to supplant these base chemicals and form the cornerstone of a more sustainable chemical industry that is less reliant on fossil resources. First-hand knowledge from the authors and a thorough review of recent scientific literature has been used to compile a list of the most promising platform molecules, giving the reader an appreciation of the diversity of chemicals readily obtainable from biomass. The processing technologies involved in the conversion of biomass to chemicals has been reviewed, as has the different constituent parts of biomass and how these can be used to access different platform molecules. The higher heteroatom content of platform molecules versus base chemicals has also been discussed, and the effects of downstream chemistry considered. Finally, four example platforms or platform molecules, one from each of the core processing technologies, were reviewed in detail, highlighting how each can be used as a versatile building block that can lead to many other valuable chemicals. Although drop-in replacements for petrochemicals are in many cases derivable from biomass, there is also great potential for new chemistry, new materials and new products to be developed within the context of the biorefinery, and platform molecules form an essential part of these facilities. Evident from the discussions within this chapter is the great promise, growing interest and certain economic and environmental benefits of a move towards a chemical industry more reliant on renewable, bio-derived feedstocks for sustainable chemical production.

# References

1. Levy, P.F., Sanderson, J.E., Kispert, R.G. and Wise, D.L. (1981) Biorefining of biomass to liquid fuels and organic-chemicals. *Enzyme and Microbial Technology*, **3** (3), 207–215.
2. Lipinsky, E.S. (1978) Fuels from biomass–integration with food and materials systems. *Science*, **199** (4329), 644–651.
3. Werpy, T. and Pedersen, G. (2005) *Top Value Added Chemicals From Biomass*, vol. **1**. US Department of Energy, Oak Ridge, USA.
4. Werpy, T., Bozell, J.J., White, J.F. and Johnson, D. (2007) *Top Value Added Chemicals From Biomass. Volume 2: Results of Screening for Potential Candidates From Biorefinery Lignin*. US Department of Energy, Oak Ridge, USA.
5. Lane, J. (2013) Top Molecules: The DOE's 12 Top Biobased List–What's Worked Out?. Available at http://www.biofuelsdigest.com/biobased/2013/01/08/top-molecules-the-does-12-top-biobased-list-whats-worked-out/ (accessed 28 August 2014).
6. OPX Biotechnologies Inc. (2014) Available at http://www.opxbio.com/ (accessed 28 August 2014).
7. Metabolix (2014) Sustainable Biobased Chemicals. Available at http://www.metabolix.com/Products/Biobased-Chemicals (accessed 28 August 2014).
8. BASF (2014) Cargill and Novozymes Achieve Milestone in Bio-Based Acrylic Acid Process. Available at http://www.basf.com/group/pressrelease/P-13-356 (accessed 28 August 2014).
9. Biofine Inc. (2014) Biofine Technology. Available at http://biofinetechnology.com/ (accessed 28 August 2014).
10. Itaconix (2014) Itaconix Corporation. Available at http://www.itaconix.com/ (accessed 28 August 2014).
11. Davis, S. (2011) *Chemical Economics Handbook Product Review: Petrochemical Industry Overview*. SRI Consulting, Englewood, CO, USA.
12. Massey, R., Jacobs, M., Gallager, L.A. *et al.* (2013) *Global Chemicals Outlook–Towards Sound Management of Chemicals*. UNEP, Nairobi, Kenya.
13. Braskem (2014) Green Polyethylene. Available at http://www.braskem.com.br/site.aspx/green-products-USA (accessed 28 August 2014).
14. Chemicals-technology.com (2014) Braskem Ethanol-to-Ethylene Plant, Brazil. Available at http://www.chemicals-technology.com/projects/braskem-ethanol/ (accessed 28 August 2014).
15. Gotro, J. (2013) Bio-Based Polypropylene; Multiple Synthetic Routes under Investigation. Available at http://polymerinnovationblog.com/bio-based-polypropylene-multiple-synthetic-routes-under-investigation/ (accessed 28 August 2014).
16. Genomatica (2014) Extensive Butadiene Development Program. Available at http://www.genomatica.com/products/butadiene/ (accessed 28 August 2014).
17. Anellotech Inc. (2013) Anellotech Announcs Ability to Produce Large Volume Product Development Samples of Biomass-Derived Benzene and Toluene. Available at http://www.anellotech.com/downloads/Anellotech-release03-06-2013.pdf (accessed 28 August 2014).

18. Broeren, M. (2013) *Production of Bio-Methanol*. International Renewable Energy Agency, Abu Dhabi, UAE.

19. Clark, J.H., Farmer, T.J., Macquarrie, D.J. *et al* (2012) Microwave heating for the rapid conversion of sugars and polysaccharides to 5-chloromethyl furfural. *Green Chemistry*, **15** (1), 72–75.

20. Alonso, D.M., Wettstein, S.G. and Dumesic, J.A. (2013) Gamma-valerolactone, a sustainable platform molecule derived from lignocellulosic biomass. *Green Chemistry*, **15** (3), 584–595.

21. Skibar, W., Grogan, G., McDonald, J. and Pitts, M. (2009) *UK Expertise for Exploitation of Biomass-Based Platform Chemicals*, The FROPTOP Group, Runcorn, UK.

22. Serrano-Ruiz, J.C., Luque, R. and Sepulveda-Escribano, A. (2011) Transformations of biomass-derived platform molecules: from high added-value chemicals to fuels via aqueous-phase processing. *Chemical Society Reviews*, **40** (11), 5266–5281.

23. Sanders, J.P.M., Clark, J.H., Harmsen, G.J. *et al.* (2012) Process intensification in the future production of base chemicals from biomass. *Chemical Engineering and Processing*, **51**, 117–136.

24. European Renewable Resources and Materials Association (2007) Accelerating the development of the market for bio-based products in Europe. Report of the Taskforce of Bio-Based Products. Available at http://www.errma.com/wp-content/uploads/2014/02/Annex-5-prep_bio.pdf (accessed 15 September 2014).

25. Bozell, J.J., Moens, L., Elliott, D.C. *et al.* (2000) Production of levulinic acid and use as a platform chemical for derived products. *Resources Conservation and Recycling*, **28** (3–4), 227–239.

26. Rackemann, D.W. and Doherty, W.O.S. (2011) The conversion of lignocellulosics to levulinic acid. *Biofuels Bioproducts and Biorefining-Biofpr*, **5** (2), 198–214.

27. Yuan, J., Li, S.S., Yu, L. *et al.* (2013) Copper-based catalysts for the efficient conversion of carbohydrate biomass into gamma-valerolactone in the absence of externally added hydrogen. *Energy and Environmental Science*, **6** (11), 3308–3313.

28. Weusthuis, R.A., Aarts, J.M.M.J. and Sanders, J.P.M. (2011) From biofuel to bioproduct: is bioethanol a suitable fermentation feedstock for synthesis of bulk chemicals? *Biofuels Bioproducts and Biorefining-Biofpr*, **5** (5), 486–494.

29. Rabemanolontsoa, H. and Saka, S. (2013) Comparative study on chemical composition of various biomass species. *RSC Advances*, **3** (12), 3946–3956.

30. Melero, J.A., Iglesias, J. and Garcia, A. (2012) Biomass as renewable feedstock in standard refinery units. feasibility, opportunities and challenges. *Energy and Environmental Science*, **5** (6), 7393–7420.

31. Assary, R.S., Kim, T., Low, J.J. *et al.* (2012) Glucose and fructose to platform chemicals: understanding the thermodynamic landscapes of acid-catalysed reactions using high-level Ab initio methods. *Physical Chemistry Chemical Physics*, **14** (48), 16603–16611.

32. Tuck, C.O., Perez, E., Horvath, I.T. *et al.* (2012) Valorization of biomass: deriving more value from waste. *Science*, **337** (6095), 695–699.

33. Dodson, J.R., Cooper, E.C., Hunt, A.J. *et al.* (2013) Alkali silicates and structured mesoporous silicas from biomass power station wastes: the emergence of bio-MCMs. *Green Chemistry*, **15** (5), 1203–1210.

34. Kobayashi, H. and Fukuoka, A. (2013) Synthesis and utilisation of sugar compounds derived from lignocellulosic biomass. *Green Chemistry*, **15** (7), 1740–1763.

35. Patel, M., Crank, M., Dornsburg, V. *et al.* (2006) *Medium and Long-Term Opportunities and Risks of the Biotechnological Production of Bulk Chemicals From Renewable Resources*, The BREW Project, Utrecht, Netherlands.
36. Balat, M. (2011) Production of bioethanol from lignocellulosic materials via the biochemical pathway: a review. *Energy Conversion and Management*, **52** (2), 858–875.
37. Balat, M., Balat, H. and Oz, C. (2008) Progress in bioethanol processing. *Progress in Energy and Combustion Science*, **34** (5), 551–573.
38. Limayem, A. and Ricke, S.C. (2012) Lignocellulosic biomass for bioethanol production: current perspectives, potential issues and future prospects. *Progress in Energy and Combustion Science*, **38** (4), 449–467.
39. Viikari, L., Vehmaanpera, J. and Koivula, A. (2012) Lignocellulosic ethanol: from science to industry, *Biomass and Bioenergy*, **46** 13–24.
40. Maki-Arvela, P., Salmi, T., Holmbom, B. *et al.* (2011) Synthesis of sugars by hydrolysis of hemicelluloses-a review. *Chemical Reviews*, **111** (9), 5638–5666.
41. Zaldivar, J., Nielsen, J. and Olsson, L. (2001) Fuel ethanol production from lignocellulose: a challenge for metabolic engineering and process integration. *Applied Microbiology and Biotechnology*, **56** (1–2), 17–34.
42. Xu, Y.X. and Hanna, M.A. (2010) Optimum conditions for dilute acid hydrolysis of hemicellulose in dried distillers grains with solubles. *Industrial Crops and Products*, **32** (3), 511–517.
43. Cara, P.D., Pagliaro, M., Elmekawy, A. *et al.* (2013) Hemicellulose hydrolysis catalysed by solid acids. *Catalysis Science and Technology*, **3** (8), 2057–2061.
44. Ormsby, R., Kastner, J.R. and Miller, J. (2012) Hemicellulose hydrolysis using solid acid catalysts generated from biochar. *Catalysis Today*, **190** (1), 89–97.
45. Zhuang, J.P., Liu, Y., Wu, Z. *et al.* (2009) Hydrolysis of wheat straw hemicellulose and detoxification of the hydrolysate for xylitol production. *Bioresources*, **4** (2), 674–686.
46. Zhou, L.P., Shi, M.T., Cai, Q.Y. *et al.* (2013) Hydrolysis of hemicellulose catalyzed by hierarchical H-USY zeolites – the role of acidity and pore structure. *Microporous and Mesoporous Materials*, **169**, 54–59.
47. Dutta, S., De, S., Saha, B. and Alam, M.I. (2012) Advances in conversion of hemicellulosic biomass to furfural and upgrading to biofuels. *Catalysis Science and Technology*, **2** (10), 2025–2036.
48. Moller, M. and Schroder, U. (2013) Hydrothermal production of furfural from xylose and xylan as model compounds for hemicelluloses. *RSC Advances*, **3** (44), 22253–22260.
49. W. De Jong and G. Marcotullio, Overview of biorefineries based on co-production of furfural, existing concepts and novel developments, *International Journal of Chemical Reactor Engineering*, **8**, 1–24 (2010).
50. Wang, L., Fan, X.G., Tang, P.W. and Yuan, Q.P. (2013) Xylitol fermentation using hemicellulose hydrolysate prepared by acid pre-impregnated steam explosion of corncob. *Journal of Chemical Technology and Biotechnology*, **88** (11), 2067–2074.
51. Kautola, H. (1990) Itaconic acid production from xylose in repeated-batch and continuous bioreactors. *Applied Microbiology and Biotechnology*, **33** (1), 7–11.
52. Kobayashi, H., Komanoya, T., Guha, S.K. *et al.* (2011) Conversion of cellulose into renewable chemicals by supported metal catalysis. *Applied Catalysis A-General*, **409**, 13–20.
53. Wilson, D.B. (2011) Microbial diversity of cellulose hydrolysis. *Current Opinion in Microbiology*, **14** (3), 259–263.

54. Huang, Y.B. and Fu, Y. (2013) Hydrolysis of cellulose to glucose by solid acid catalysts. *Green Chemistry*, **15** (5), 1095–1111.

55. Geboers, J.A., Van de Vyver, S., Ooms, R. *et al.* (2011) Chemocatalytic conversion of cellulose: opportunities, advances and pitfalls. *Catalysis Science and Technology*, **1** (5), 714–726.

56. Mohanram, S., Amat, D., Choudhary, J. *et al.* (2013) Novel perspectives for evolving enzyme cocktails for lignocellulose hydrolysis in biorefineries. *Sustainable Chemical Processes*, **1**, 15.

57. Dadi, A.P., Varanasi, S. and Schall, C.A. (2006) Enhancement of cellulose saccharification kinetics using an ionic liquid pretreatment step. *Biotechnology and Bioengineering*, **95** (5), 904–910.

58. Moller, M., Harnisch, F. and Schroder, U. (2013) Hydrothermal liquefaction of cellulose in subcritical water-the role of crystallinity on the cellulose reactivity. *RSC Advances*, **3** (27), 11035–11044.

59. Benoit, M., Rodrigues, A., Zhang, Q.H. *et al.* (2011) Depolymerization of cellulose assisted by a nonthermal atmospheric plasma. *Angewandte Chemie-International Edition*, **50** (38), 8964–8967.

60. Orozco, A., Ahmad, M., Rooney, D. and Walker, G. (2007) Dilute acid hydrolysis of cellulose and cellulosic bio-waste using a microwave reactor system. *Process Safety and Environmental Protection*, **85** (B5), 446–449.

61. Fan, J.J., De Bruyn, M., Budarin, V.L. *et al.* (2013) Direct microwave-assisted hydrothermal depolymerization of cellulose. *Journal of the American Chemical Society*, **135** (32), 11728–11731.

62. Dutta, S. and Pal, S. (2014) Promises in direct conversion of cellulose and lignocellulosic biomass to chemicals and fuels: combined solvent-nanocatalysis approach for biorefinery. *Biomass and Bioenergy*, **62**, 182–197.

63. Song, J.L., Fan, H.L., Ma, J. and Han, B.X. (2013) Conversion of glucose and cellulose into value-added products in water and ionic liquids. *Green Chemistry*, **15** (10), 2619–2635.

64. Liu, B., Zhang, Z.H. and Zhao, Z.K. (2013) Microwave-assisted catalytic conversion of cellulose into 5-hydroxymethylfurfural in ionic liquids. *Chemical Engineering Journal*, **215**, 517–521.

65. Kaldstrom, M., Kumar, N. and Murzin, D.Y. (2011) Valorization of cellulose over metal supported mesoporous materials. *Catalysis Today*, **167** (1), 91–95.

66. Kusema, B.T., Faba, L., Kumar, N. *et al.* (2012) Hydrolytic hydrogenation of hemicellulose over metal modified mesoporous catalyst. *Catalysis Today*, **196** (1), 26–33.

67. Oltmanns, J.U., Palkovits, S. and Palkovits, R. (2013) Kinetic investigation of sorbitol and xylitol dehydration catalyzed by silicotungstic acid in water. *Applied Catalysis A-General*, **456**, 168–173.

68. Li, N. and Huber, G.W. (2010) Aqueous-phase hydrodeoxygenation of sorbitol with pt/SiO2-Al2O3: identification of reaction intermediates. *Journal of Catalysis*, **270** (1), 48–59.

69. Menon, V. and Rao, M. (2012) Trends in bioconversion of lignocellulose: biofuels, platform chemicals and biorefinery concept. *Progress in Energy and Combustion Science*, **38** (4), 522–550.

70. Varanasi, P., Singh, P., Auer, M. *et al.* (2013) Survey of renewable chemicals produced from lignocellulosic biomass during ionic liquid pretreatment. *Biotechnology for Biofuels*, **6**, 14.

71. Krivorotova, T. and Sereikaite, J. (2014) Seasonal changes of carbohydrates composition in the tubers of jerusalem artichoke. *Acta Physiologiae Plantarum*, **36** (1), 79–83.

72. Ricca, E., Calabro, V., Curcio, S. and Iorio, G. (2009) Fructose production by chicory inulin enzymatic hydrolysis: a kinetic study and reaction mechanism. *Process Biochemistry*, **44** (4), 466–470.

73. Shen, X., Wang, Y.X., Hu, C.W. *et al.* (2012) One-pot conversion of inulin to furan derivatives catalyzed by sulfated TiO2/mordenite solid acid. *Chemcatchem*, **4** (12), 2013–2019.

74. Ajavakom, A., Supsvetson, S., Somboot, A. and Sukwattanasinitt, M. (2012) Products from microwave and ultrasonic wave assisted acid hydrolysis of chitin. *Carbohydrate Polymers*, **90** (1), 73–77.

75. Chen, X., Chew, S.L., Kerton, F.M. and Yan, N. (2014) Direct conversion of chitin into a N-containing furan derivative. *Green Chemistry*, **16** (4), 2204–2212.

76. Anastasakis, K., Ross, A.B. and Jones, J.M. (2011) Pyrolysis behaviour of the main carbohydrates of brown macro-algae. *Fuel*, **90** (2), 598–607.

77. Ruiz, H.A., Rodriguez-Jasso, R.M., Fernandes, B.D. *et al.* (2013) Hydrothermal processing, as an alternative for upgrading agriculture residues and marine biomass according to the biorefinery concept: a review. *Renewable and Sustainable Energy Reviews*, **21**, 35–51.

78. Pan, X.J., Arato, C., Gilkes, N. *et al.* (2005) Biorefining of softwoods using ethanol organosolv pulping: preliminary evaluation of process streams for manufacture of fuel-grade ethanol and co-products. *Biotechnology and Bioengineering*, **90** (4), 473–481.

79. Higson, A. and Smith, C. (2011) *Lignin. NNFCC Renewable Chemicals Factsheet* The National Non-Food Crops Centre, York, UK.

80. Baker, D.A. and Rials, T.G. (2013) Recent advances in low-cost carbon fiber manufacture from lignin. *Journal of Applied Polymer Science*, **130** (2), 713–728.

81. Ruiz-Duenas, F.J. and Martinez, A.T. (2009) Microbial degradation of lignin: how a bulky recalcitrant polymer is efficiently recycled in nature and how we can take advantage of this. *Microbial Biotechnology*, **2** (2), 164–177.

82. Azadi, P., Inderwildi, O.R., Farnood, R. and King, D.A. (2013) Liquid fuels, hydrogen and chemicals from lignin: a critical review. *Renewable and Sustainable Energy Reviews*, **21**, 506–523.

83. Wang, H., Tucker, M. and Ji, Y. (2013) Recent development in chemical depolymerization of lignin: a review, *Journal of Applied Chemistry*, **2013**, 1–9.

84. Kang, S.M., Li, X.L., Fan, J. and Chang, J. (2013) Hydrothermal conversion of lignin: a review. *Renewable and Sustainable Energy Reviews*, **27**, 546–558.

85. Zhang, X., Tu, M.B. and Paice, M.G. (2011) Routes to potential bioproducts from lignocellulosic biomass lignin and hemicelluloses. *Bioenergy Research*, **4** (4), 246–257.

86. Ye, Y.Y., Zhang, Y., Fan, J. and Chang, J. (2012) Novel method for production of phenolics by combining lignin extraction with lignin depolymerization in aqueous ethanol. *Industrial and Engineering Chemistry Research*, **51** (1), 103–110.

87. Iyayi, C.B. and Dart, R.K. (1982) The degradation of p-coumaryl alcohol by Aspergillus Flavus. *Microbiology*, **128**, 1473–1482.

88. Dale, B.E., Allen, M.S., Laser, M. and Lynd, L.R. (2009) Protein feeds coproduction in biomass conversion to fuels and chemicals. *Biofuels Bioproducts and Biorefining-Biofpr*, **3** (2), 219–230.

89. Hayashi, Y., Yamaguchi, J., Sumiya, T. and Shoji, M. (2004) Direct proline-catalyzed asymmetric alpha-aminoxylation of ketones. *Angewandte Chemie-International Edition*, **43** (9), 1112–1115.

90. Hou, C., Zhu, H., Li, Y.J. and Li, Y.F. (2012) Immobilized proline and its derivatives employed in the catalysis of asymmetric organic synthesis. *Progress in Chemistry*, **24** (9), 1729–1741.

91. Opalka, S.M., Longstreet, A.R. and McQuade, D.T. (2011) Continuous proline catalysis via leaching of solid proline. *Beilstein Journal of Organic Chemistry*, **7**, 1671–1679.

92. Konst, P.M., Turras, P.M.C.C., Franssen, M.C.R. *et al.* (2010) Stabilized and immobilized bacillus subtilis arginase for the biobased production of nitrogen-containing chemicals. *Advanced Synthesis and Catalysis*, **352** (9), 1493–1502.

93. Konst, P.M., Scott, E.L., Franssen, M.C.R. and Sanders, J.P.M. (2011) Acid and base catalyzed hydrolysis of cyanophycin for the biobased production of nitrogen containing chemicals. *Journal of Biobased Materials and Bioenergy*, **5** (1), 102–108.

94. Sanders, J., Scott, E., Weusthuis, R. and Mooibroek, H. (2007) Bio-refinery as the bio-inspired process to bulk chemicals. *Macromolecular Bioscience*, **7** (2), 105–117.

95. Zhang, Y.X., Kumar, A., Vadlani, P.V. and Narayanan, S. (2013) Production of nitrogen-based platform chemical: cyanophycin biosynthesis using recombinant Escherichia coli and renewable media substitutes. *Journal of Chemical Technology and Biotechnology*, **88** (7), 1321–1327.

96. Spekreijse, J., Le Notre, J., van Haveren, J. *et al.* (2012) Simultaneous production of biobased styrene and acrylates using ethenolysis. *Green Chemistry*, **14** (10), 2747–2751.

97. van Haveren, J., Scott, E.L. and Sanders, J. (2008) Bulk chemicals from biomass. *Biofuels Bioproducts and Biorefining-Biofpr*, **2** (1), 41–57.

98. Lammens, T.M., Franssen, M.C.R., Scott, E.L. and Sanders, J.P.M. (2012) Availability of protein-derived amino acids as feedstock for the production of bio-based chemicals. *Biomass and Bioenergy*, **44**, 168–181.

99. Becker, E.W. (2007) Micro-algae as a source of protein. *Biotechnology Advances*, **25** (2), 207–210.

100. Casado, F.J., Montano, A., Spitzner, D. and Carle, R. (2013) Investigations into acrylamide precursors in sterilized table olives: evidence of a peptic fraction being responsible for acrylamide formation. *Food Chemistry*, **141** (2), 1158–1165.

101. Buhlert, J., Carle, R., Majer, Z. and Spitzner, D. (2006) Thermal degradation of peptides and formation of acrylamide. *Letters in Organic Chemistry*, **3** (5), 356–357.

102. Buhlert, J., Carle, R., Majer, Z. and Spitzner, D. (2007) Thermal degradation of peptides and formation of acrylamide, part 2. *Letters in Organic Chemistry*, **4** (5), 329–331.

103. Arribas-Lorenzo, G. and Morales Navas, F.J. (2012) Recent insights in acrylamide as carcinogen in foodstuffs. *Advances in Molecular Toxicology*, **6**, 163–193.

104. C. Scrimgeour, *Chemistry of fatty acids*. In: Bailey's Industrial Oil and Fat Products (ed. F. Shahidi), John Wiley & Sons, Chichester (2005), pp. 1–43.

105. Khot, S.N., Lascala, J.J., Can, E. *et al.* (2001) Development and application of triglyceride-based polymers and composites. *Journal of Applied Polymer Science*, **82** (3), 703–723.

106. Guner, F.S., Yagci, Y. and Erciyes, A.T. (2006) Polymers from triglyceride oils. *Progress in Polymer Science*, **31** (7), 633–670.

107. Tan, H.W., Aziz, A.R.A. and Aroua, M.K. (2013) Glycerol production and its applications as a raw material: a review. *Renewable and Sustainable Energy Reviews*, **27**, 118–127.

108. Farris, R.D. (1979) Methyl-esters in the fatty-acid industry. *Journal of the American Oil Chemists Society*, **56**, A770–A773.

109. Schwab, W., Fuchs, C. and Huang, F.C. (2013) Transformation of terpenes into fine chemicals. *European Journal of Lipid Science and Technology*, **115** (1), 3–8.

110. Harvey, B.G., Wright, M.E. and Quintana, R.L. (2010) High-density renewable fuels based on the selective dimerization of pinenes. *Energy and Fuels*, **24**, 267–273.

111. Meylemans, H.A., Quintana, R.L. and Harvey, B.G. (2012) Efficient conversion of pure and mixed terpene feedstocks to high density fuels. *Fuel*, **97**, 560–568.

112. Swift, K.A.D. (2004) Catalytic transformations of the major terpene feedstocks. *Topics in Catalysis*, **27** (1–4), 143–155.

113. Stanciulescu, M. and Ikura, M. (2007) Limonene ethers from tire pyrolysis oil – part 2: continuous flow experiments. *Journal of Analytical and Applied Pyrolysis*, **78** (1), 76–84.

114. Stanciulescu, M. and Ikura, M. (2006) Limonene ethers from tire pyrolysis oil – part 1: batch experiments. *Journal of Analytical and Applied Pyrolysis*, **75** (2), 217–225.

115. Sannita, E., Aliakbarian, B., Casazza, A.A. *et al.* (2012) Medium-temperature conversion of biomass and wastes into liquid products, a review. *Renewable and Sustainable Energy Reviews*, **16** (8), 6455–6475.

116. de Jong, E., Higson, A., Walsh, P. and Wellisch, M. (2011) *Bio-Based Chemicals: Value Added Products From Biorefineries*, IEA Bioenergy. Task 42 Biorefinery, Wageningen, Netherlands.

117. Sheldrake, G.N. and Schleck, D. (2007) Dicationic molten salts (ionic liquids) as re-usable media for the controlled pyrolysis of cellulose to anhydrosugars. *Green Chemistry*, **9** (10), 1044–1046.

118. Long, J.X., Guo, B., Li, X.H. *et al.* (2011) One step catalytic conversion of cellulose to sustainable chemicals utilizing cooperative ionic liquid pairs. *Green Chemistry*, **13** (9), 2334–2338.

119. Jang, Y.S., Kim, B., Shin, J.H. *et al.* (2012) Bio-based production of C2–C6 platform chemicals. *Biotechnology and Bioengineering*, **109** (10), 2437–2459.

120. Fenouillot, F., Rousseau, A., Colomines, G. *et al.* (2010) Polymers from renewable 1,4:3,6-dianhydrohexitols (isosorbide, isomannide and isoidide): a review. *Progress in Polymer Science*, **35** (5), 578–622.

121. Mascal, M. and Nikitin, E.B. (2010) High-yield conversion of plant biomass into the key value-added feedstocks 5-(hydroxymethyl)furfural, levulinic acid, and levulinic esters via 5-(chloromethyl)furfural. *Green Chemistry*, **12** (3), 370–373.

122. Xiao, Z.J., Wang, X.M., Huang, Y.L. *et al.* (2012) Thermophilic fermentation of acetoin and 2,3-butanediol by a novel geobacillus strain. *Biotechnology for Biofuels*, **5**, 88.

123. Gao, C., Zhang, L.J., Xie, Y.J. *et al.* (2013) Production of (3S)-acetoin from diacetyl by using stereoselective NADPH-dependent carbonyl reductase and glucose dehydrogenase. *Bioresource Technology*, **137**, 111–115.

124. Curran, K.A., Leavitt, J., Karim, A. and Alper, H.S. (2013) Metabolic engineering of muconic acid production in Saccharomyces cerevisiae. *Metabolic Engineering*, **15**, 55–66.

125. Frost, J.W., Miermont, A., Draths Corporation, Schweitzer, D. and Bui, V. (2010) Preparation of trans,trans muconic acid and trans,trans muconates, Patent No. WO 2010/148049 A2.

126. Wright, M.M. and Brown, R.C. (2007) Comparative economics of biorefineries based on the biochemical and thermochemical platforms. *Biofuels, Bioproducts and Biorefining-Biofpr*, **1** (1), 49–56.

127. Cheiky, M., Sills, R.A. and Jarand, M.L. (2012) Method for enhancing soil growth using bio-char, Cool Planet Biofuels Inc., US Patent 8,236,085.

128. Novak, J.M., Cantrell, K.B., Watts, D.W. *et al.* (2014) Designing relevant biochars as soil amendments using lignocellulosic-based and manure-based feedstocks. *Journal of Soils and Sediments*, **14** (2), 330–343.

129. Mullen, C.A., Boateng, A.A., Goldberg, N.M. *et al.* (2010) Bio-oil and bio-char production from corn cobs and stover by fast pyrolysis. *Biomass and Bioenergy*, **34** (1), 67–74.

130. McHenry, M.P. (2009) Agricultural bio-char production, renewable energy generation and farm carbon sequestration in Western Australia: certainty, uncertainty and risk. *Agriculture Ecosystems and Environment*, **129** (1–3), 1–7.

131. Abnisa, F., Daud, W.M.A.W., Husin, W.N.W. and Sahu, J.N. (2011) Utilization possibilities of palm shell as a source of biomass energy in Malaysia by producing bio-oil in pyrolysis process. *Biomass and Bioenergy*, **35** (5), 1863–1872.

132. Wang, Z.X., Dong, T., Yuan, L.X. *et al.* (2007) Characteristics of bio-oil-syngas and its utilization in fischer-tropsch synthesis. *Energy and Fuels*, **21** (4), 2421–2432.

133. Sarkar, S. and Kumar, A. (2010) Large-scale biohydrogen production from bio-oil. *Bioresource Technology*, **101** (19), 7350–7361.

134. Xiu, S.N. and Shahbazi, A. (2012) Bio-oil production and upgrading research: a review. *Renewable and Sustainable Energy Reviews*, **16** (7), 4406–4414.

135. Mortensen, P.M., Grunwaldt, J.D., Jensen, P.A. *et al.* (2011) A review of catalytic upgrading of bio-oil to engine fuels. *Applied Catalysis A-General*, **407** (1–2), 1–19.

136. Bridgwater, A.V. (2012) Review of fast pyrolysis of biomass and product upgrading. *Biomass and Bioenergy*, **38**, 68–94.

137. C. A. Mullen, A. A. Boateng and K. B. Hicks, *et al.* Analysis and comparison of bio-oil produced by fast pyrolysis from three barley biomass/byproduct streams, *Energy and Fuels*, **24**, 699–706 (2010).

138. Knoef, H. (2012) *Handbook of Biomass Gasification*, Biomass Technology Group (BTG), Enschede.

139. Goransson, K., Soderlind, U., He, J. and Zhang, W.N. (2011) Review of syngas production via biomass DFBGs. *Renewable and Sustainable Energy Reviews*, **15** (1), 482–492.

140. Mondal, P., Dang, G.S. and Garg, M.O. (2011) Syngas production through gasification and cleanup for downstream applications – recent developments. *Fuel Processing Technology*, **92** (8), 1395–1410.

141. Puig-Arnavat, M., Bruno, J.C. and Coronas, A. (2010) Review and analysis of biomass gasification models. *Renewable and Sustainable Energy Reviews*, **14** (9), 2841–2851.

142. Woolcock, P.J. and Brown, R.C. (2013) A review of cleaning technologies for biomass-derived syngas. *Biomass and Bioenergy*, **52**, 54–84.

143. Borg, O., Hammer, N., Enger, B.C. *et al.* (2011) Effect of biomass-derived synthesis gas impurity elements on cobalt fischer-tropsch catalyst performance including in situ sulphur and nitrogen addition. *Journal of Catalysis*, **279** (1), 163–173.

144. Spath, P.L. and Dayton, D.C. (2003) *Preliminary Screening - Technical and Economic Assessment of Synthesis Gas to Fuels and Chemicals With Emphasis on the Potential for Biomass-Derived Syngas*, National Renewable Energy Laboratory NREL/TP-510-34929, Golden, CO, USA.

145. Yin, X.L., Leung, D.Y.C., Chang, J. *et al.* (2005) Characteristics of the synthesis of methanol using biomass-derived syngas. *Energy and Fuels*, **19** (1), 305–310.

146. Olsbye, U., Svelle, S., Bjorgen, M. *et al.* (2012) Conversion of methanol to hydrocarbons: how zeolite cavity and pore size controls product selectivity. *Angewandte Chemie-International Edition*, **51** (24), 5810–5831.

147. Li, Y.W., He, D.H., Ge, S.H. *et al.* (2008) Effects of CO2 on synthesis of isobutene and isobutane from CO2/CO/H-2 reactant mixtures over zirconia-based catalysts. *Applied Catalysis B-Environmental*, **80** (1–2), 72–80.

148. Semelsberger, T.A., Borup, R.L. and Greene, H.L. (2006) Dimethyl ether (DME) as an alternative fuel. *Journal of Power Sources*, **156** (2), 497–511.

149. Arcoumanis, C., Bae, C., Crookes, R. and Kinoshita, E. (2008) The potential of di-methyl ether (DME) as an alternative fuel for compression-ignition engines: a review. *Fuel*, **87** (7), 1014–1030.

150. Kanda, H., Li, P. and Makino, H. (2013) Production of decaffeinated green tea leaves using liquefied dimethyl ether. *Food and Bioproducts Processing*, **91** (C4), 376–380.

151. Khandan, N., Kazemeini, M. and Aghaziarati, M. (2008) Determining an optimum catalyst for liquid-phase dehydration of methanol to dimethyl ether. *Applied Catalysis A-General*, **349** (1–2), 6–12.

152. Bae, J.W., Potdar, H.S., Kang, S.H. and Jun, K.W. (2008) Coproduction of methanol and dimethyl ether from biomass-derived syngas on a Cu-ZnO-Al2O3/gamma-Al2O3 hybrid catalyst. *Energy and Fuels*, **22** (1), 223–230.

153. Haynes, A. (2010) Catalytic methanol carbonylation. *Advances in Catalysis*, **53**, 1–45.

154. Haynes, A. (2006) Acetic acid synthesis by catalytic carbonylation of methanol. *Catalytic Carbonylation Reactions*, **18**, 179–205.

155. Liu, Y.Y., Murata, K., Inaba, M. and Takahara, I. (2013) Synthesis of ethanol from methanol and syngas through an indirect route containing methanol dehydrogenation, DME carbonylation, and methyl acetate hydrogenolysis. *Fuel Processing Technology*, **110**, 206–213.

156. Gupta, M., Smith, M.L. and Spivey, J.J. (2011) Heterogeneous catalytic conversion of dry syngas to ethanol and higher alcohols on cu-based catalysts. *ACS Catalysis*, **1** (6), 641–656.

157. Spivey, J.J. and Egbebi, A. (2007) Heterogeneous catalytic synthesis of ethanol from biomass-derived syngas. *Chemical Society Reviews*, **36** (9), 1514–1528.

158. Herman, R.G. (2000) Advances in catalytic synthesis and utilization of higher alcohols. *Catalysis Today*, **55** (3), 233–245.

159. Verkerk, K.A.N., Jaeger, B., Finkeldei, C.H. and Keim, W. (1999) Recent developments in isobutanol synthesis from synthesis gas. *Applied Catalysis A-General*, **186** (1–2), 407–431.

160. Zaman, S. and Smith, K.J. (2012) A review of molybdenum catalysts for synthesis gas conversion to alcohols: catalysts, mechanisms and kinetics. *Catalysis Reviews-Science and Engineering*, **54** (1), 41–132.

161. Surisetty, V.R., Dalai, A.K. and Kozinski, J. (2011) Alcohols as alternative fuels: an overview. *Applied Catalysis A-General*, **404** (1–2), 1–11.

162. Daniell, J., Kopke, M. and Simpson, S.D. (2012) Commercial biomass syngas fermentation. *Energies*, **5** (12), 5372–5417.

163. Okabe, K., Murata, K., Nakanishi, M. *et al.* (2009) Fischer-tropsch synthesis over ru catalysts by using syngas derived from woody biomass. *Catalysis Letters*, **128** (1–2), 171–176.

164. Fenton, H.J.H. and Gostling, M. (1901) Derivatives of methylfurfural. *Journal of the Chemical Society*, **79**, 807–816.

165. Fenton, H.J.H. and Gostling, M. (1899) Bromomethylfurfuraldehyde. *Journal of the Chemical Society, Transactions*, **75**, 423–433.

166. Hill, H.B. and Jennings, W.L. (1893) On certain products of the dry distillation of wood: methylfurfurol and methylpyromucic acid. *American Chemical Journal*, **15**, 159–185.

167. Hamada, K., Suzukamo, G. and Nagase, T. (1978) Furaldehydes, Sumitomo Chemical Co. Ltd., DE 2745743.

168. Szmant, H.H. and Chundury, D.D. (1981) The preparation of 5-chloromethylfurfuraldehyde from high fructose corn syrup and other carbohydrates. *Journal of Chemical Technology and Biotechnology*, **31** (4), 205–212.

169. Fischer, E. and von Neyman, H. (1914) Notiz Über Ω-Chlormethyl- Und Äthoxymethyl-Furfurol. *Berichte Der Deutschen Chemischen Gesellschaft*, **47**, 973–977.

170. Haworth, W.N. and Jones, W.G.M. (1944) The conversion of sucrose into furan compounds. part I. 5-hydroxymethylfurfuraldehyde and some derivatives. *Journal of the Chemical Society*, 667–670.

171. Timko, J.M. and Cram, D.J. (1974) Furanyl unit in host compounds. *Journal of the American Chemical Society*, **96** (22), 7159–7160.

172. Mascal, M. and Nikitin, E.B. (2008) Direct, High-yield conversion of cellulose into bio-fuel, *Angewandte Chemie-International Edition*, **47**(41), 7924–7926.

173. Mascal, M. and Nikitin, E.B. (2009) Towards the efficient, total glycan utilization of biomass. *Chemsuschem*, **2** (5), 423–426.

174. Mascal, M. and Nikitin, E.B. (2009) Dramatic advancements in the saccharide to 5-(chloromethyl)furfural conversion reaction. *Chemsuschem*, **2** (9), 859–861.

175. Brasholz, M., von Kanel, K., Hornung, C.H. *et al.* (2011) Highly efficient dehydration of carbohydrates to 5-(chloromethyl)furfural (CMF), 5-(hydroxymethyl)furfural (HMF) and levulinic acid by biphasic continuous flow processing. *Green Chemistry*, **13** (5), 1114–1117.

176. van Putten, R.J., van der Waal, J.C., de Jong, E. *et al.* (2013) Hydroxymethylfurfural, a versatile platform chemical made from renewable resources. *Chemical Reviews*, **113** (3), 1499–1597.

177. Yang, W.R. and Sen, A. (2010) One-step catalytic transformation of carbohydrates and cellulosic biomass to 2,5-dimethyltetrahydrofuran for liquid fuels. *Chemsuschem*, **3** (5), 597–603.

178. Thananatthanachon, T. and Rauchfuss, T.B. (2010) Efficient production of the liquid fuel 2,5-dimethylfuran from fructose using formic acid as a reagent. *Angewandte Chemie-International Edition*, **49** (37), 6616–6618.

179. Simmie, J.M. (2012) Kinetics and thermochemistry of 2,5-dimethyltetrahydrofuran and related oxolanes: next next-generation biofuels. *Journal of Physical Chemistry A*, **116** (18), 4528–4538.

180. Gruter, G.J.M. and Dautzenberg, F. (2007) Preparation of 5-?alkoxymethylfurfural ethers for use as fuel or monomer in polymerization. EU Patent 1834950.

181. Gruter, G.J.M. and Manzer, L.E. (2009) Hydroxymethylfurfural ethers with mixed alcohols as gasoline and diesel fuel additives, World Patent WO 2009030508.

182. Gruter, G.J.M. (2009) (Alkoxymethyl)furfural ethers with branched alcohols as gasoline and diesel fuel additives. World Patent WO 2009030511.

183. Windom, B.C., Lovestead, T.M., Mascal, M. *et al.* (2011) Advanced distillation curve analysis on ethyl levulinate as a diesel fuel oxygenate and a hybrid biodiesel fuel. *Energy and Fuels*, **25** (4), 1878–1890.

184. Leibig, C.M., Mullen, B. and Rieth, L., *et al.* (2010) Cellulosic-derived levulinic ketal esters: a new building block, Abstracts of Papers of the American Chemical Society, **239**, 111–116.

185. Lange, J.P., Price, R., Ayoub, P.M. *et al.* (2010) Valeric biofuels: a platform of cellulosic transportation fuels. *Angewandte Chemie-International Edition*, **49** (26), 4479–4483.

186. Fenton, H.J.H. and Robinson, F. (1909) Homologues of furfuraldehyde. *Journal of the Chemical Society*, **95**, 1334–1340.

187. Zhou, X.Y. and Rauchfuss, T.B. (2013) Production of hybrid diesel fuel precursors from carbohydrates and petrochemicals using formic acid as a reactive solvent. *Chemsuschem*, **6** (2), 383–388.

188. Zhong, S.H., Daniel, R., Xu, H. *et al.* (2010) Combustion and emissions of 2,5-dimethyl-furan in a direct-injection spark-ignition engine. *Energy and Fuels*, **24**, 2891–2899.

189. Williams, C.L., Chang, C.C., Do, P. *et al.* (2012) Cycloaddition of biomass-derived furans for catalytic production of renewable p-xylene. *ACS Catalysis*, **2** (6), 935–939.

190. Brandvold, T.A. (2010) Carbohydrate route to para-xylene and terephthalic acid. World Patent WO 2010151346.

191. Shiramizu, M. and Toste, F.D. (2011) On the diels-alder approach to solely biomass-derived polyethylene terephthalate (PET): conversion of 2,5-dimethylfuran and acrolein into p-xylene. *Chemistry-A European Journal*, **17** (44), 12452–12457.

192. Masuno, M.N., Cannon, D. and Bissel, J., *et al.* (2013) Method of producing p-xylene by reacting 2,5-dimethylfuran with ethylene in the presence of lewis acid catalyst and oxidation of p-xylene to terephthalic acid. World Patent WO 2013040514.

193. Mikochik, P. and Cahana, A. (2012) Conversion of 5-(chloromethyl)-2-furaldehyde into 5-methyl-2-furoic acid and derivatives thereof. EU Patent 2606039 A1.

194. Silks, L.A., Gordon, J.C., Wu, R. and Hanson, S.K. (2011) Process for preparation of furan derivatives by carbon chain extension through aldol reaction. US Patent 20110040109.

195. Jira, R. and Braunling, H. (1987) Synthesis of polyarenemethines, a new class of conducting polymers. *Synthetic Metals*, **17** (1–3), 691–696.

196. Cooper, W.F. and Nuttall, W.H. (1912) Furan-2:5-dialdehyde. *Journal of the Chemical Society, Transactions*, **101**, 1074–1081.

197. Durre, P. (2008) Fermentative butanol production – bulk chemical and biofuel. *Incredible Anaerobes: From Physiology to Genomics to Fuels*, **1125**, 353–362.

198. Xue, C., Zhao, X.Q., Liu, C.G. *et al.* (2013) Prospective and development of butanol as an advanced biofuel. *Biotechnology Advances*, **31** (8), 1575–1584.

199. Ramey, D. E. (1998) Continuous two stage, dual path anaerobic fermentation of butanol and other organic solvents using two different strains of bacteria, Environmental Energy Inc., US Patent 5,753,474.

200. Ramey, D.E. and Yang, S.T. (2004) *Production of Butyric Acid and Butanol From Biomass*, US Department of Energy, Morgantown, West Virginia, USA.

201. Xue, C., Zhao, J.B., Liu, F.F. *et al.* (2013) Two-stage in situ gas stripping for enhanced butanol fermentation and energy-saving product recovery. *Bioresource Technology*, **135**, 396–402.

202. Mascal, M. (2012) Chemicals from biobutanol: technologies and markets. *Biofuels Bioproducts and Biorefining-Biofpr*, **6** (4), 483–493.

203. Harvey, B.G. and Meylemans, H.A. (2011) The role of butanol in the development of sustainable fuel technologies. *Journal of Chemical Technology and Biotechnology*, **86** (1), 2–9.

204. ICIS Chemical Business (2010) *ICIS, Soaps and Detergents Oleochemicals*, ICIS Chemical Business, London, UK.

205. D. C. Taylor, M. A. Smith and P. Fobert, *et al.* Plant systems – metabolic engineering of higher plants to produce bio-industrial oils, in *Comprehensive Biotechnology* 2nd edn (ed. M. Moo-Young), Pergamon Press, Oxford, pp. 67–85 (2011).

206. Xia, Y. and Larock, R.C. (2010) Vegetable oil-based polymeric materials: synthesis, properties, and applications. *Green Chemistry*, **12** (11), 1893–1909.

207. Sharma, V. and Kundu, P.P. (2006) Addition polymers from natural oils – a review. *Progress in Polymer Science*, **31** (11), 983–1008.

208. Bailosky, L.C., Bender, L.M., Bode, D. *et al.* (2013) Synthesis of polyether polyols with epoxidized soy bean oil. *Progress in Organic Coatings*, **76** (12), 1712–1719.

209. Raquez, J.M., Deleglise, M., Lacrampe, M.F. and Krawczak, P. (2010) Thermosetting (bio) materials derived from renewable resources: a critical review. *Progress in Polymer Science*, **35** (4), 487–509.

210. Salimon, J., Salih, N. and Yousif, E. (2012) Industrial development and applications of plant oils and their biobased oleochemicals. *Arabian Journal of Chemistry*, **5** (2), 135–145.

211. de Espinosa, L.M. and Meier, M.A.R. (2011) Plant oils: the perfect renewable resource for polymer science?!. *European Polymer Journal*, **47** (5), 837–852.

212. Petrovic, Z.S., Javni, I., and Guo, A. (2002) Method of making natural oil-based polyols and polyurethanes there from. US Patent 6,433,121.

213. Guo, A., Zhang, W. and Petrovic, Z.S. (2006) Structure-property relationships in polyurethanes derived from soybean oil. *Journal of Materials Science*, **41** (15), 4914–4920.

214. Petrovic, Z.S. and Javni, I. (2002) Process for the synthesis of epoxidised natural oil-based isocyanate prepolymers for applications in polyurethanes. US Patent 6,399,698.

215. Biermann, U., Metzger, J.O. and Meier, M.A.R. (2010) Acyclic triene metathesis oligo- and polymerization of high oleic sun flower oil. *Macromolecular Chemistry and Physics*, **211** (8), 854–862.

216. Thames, S.F., Smith, O.W. and Evans, J.M., *et al.* (2014) Functionalised vegetable oil derivatives, latex compositions and coatings, L. Southern diversified products. US Patent 7,361,710.

217. Eren, T., Kusefoglu, S.H. and Wool, R. (2003) Polymerization of maleic anhydride-modified plant oils with polyols. *Journal of Applied Polymer Science*, **90** (1), 197–202.

218. Gunstone, F.D. (2004) *The Chemistry of Oils and Fats: Sources, Composition, Properties and Uses*, Blackwell Publishing, Oxford.

219. Meier, M.A.R. (2009) Metathesis with oleochemicals: new approaches for the utilization of plant oils as renewable resources in polymer science. *Macromolecular Chemistry and Physics*, **210** (13–14), 1073–1079.

220. Di Franco, E. (2004) Method of isolating oleic acid and producing lineloic dimer/trimer acids via selective reactivity. Arizona Chemical Company, US Patent 6,835,324.

221. Binder, R.G., Kohler, G.O., Goldblatt, L.A. and Applewhite, T.H. (1962) Chromatographic analysis of seed oils – fatty acid composition of castor oil, *Journal of the American Oil Chemists Society*, **39** (12), 513.

222. Azcan, N. and Demirel, E. (2008) Obtaining 2-octanol, 2-octanone, and sebacic acid from castor oil by microwave-induced alkali fusion. *Industrial and Engineering Chemistry Research*, **47** (6), 1774–1778.

223. Vasishtha, A.K., Trivedi, R.K. and Das, G. (1990) Sebacic acid and 2-octanol from castor-oil. *Journal of the American Oil Chemists Society*, **67** (5), 333–337.

224. Mutlu, H. and Meier, M.A.R. (2010) Castor oil as a renewable resource for the chemical industry. *European Journal of Lipid Science and Technology*, **112** (1), 10–30.

225. Alam, S., Kalita, H., Jayasooriya, A. *et al.* (2014) 2-(vinyloxy)ethyl soyate as a versatile platform chemical for coatings: an overview. *European Journal of Lipid Science and Technology*, **116** (1), 2–15.

226. Chernykh, A., Alam, S., Jayasooriya, A. *et al.* (2013) Living carbocationic polymerization of a vinyl ether monomer derived from soybean oil, 2-(vinyloxy)ethyl soyate. *Green Chemistry*, **15** (7), 1834–1838.

227. Smith, B., Greenwell, H.C. and Whiting, A. (2009) Catalytic upgrading of tri-glycerides and fatty acids to transport biofuels. *Energy and Environmental Science*, **2** (3), 262–271.

228. Yang, F.X., Hanna, M.A. and Sun, R.C. (2012) Value-added uses for crude glycerol-a byproduct of biodiesel production. *Biotechnology for Biofuels*, **5** (13), 1–10.

229. Lin, Y.C. (2013) Catalytic valorization of glycerol to hydrogen and syngas. *International Journal of Hydrogen Energy*, **38** (6), 2678–2700.

230. Mattam, A.J., Clomburg, J.M., Gonzalez, R. and Yazdani, S.S. (2013) Fermentation of glycerol and production of valuable chemical and biofuel molecules. *Biotechnology Letters*, **35** (6), 831–842.

231. Corma, A., Huber, G.W., Sauvanauda, L. and O'Connor, P. (2008) Biomass to chemicals: catalytic conversion of glycerol/water mixtures into acrolein, reaction network. *Journal of Catalysis*, **257** (1), 163–171.

232. Zhou, C.H.C., Beltramini, J.N., Fan, Y.X. and Lu, G.Q.M. (2008) Chemoselective catalytic conversion of glycerol as a biorenewable source to valuable commodity chemicals. *Chemical Society Reviews*, **37** (3), 527–549.

233. Behr, A., Eilting, J., Irawadi, K. *et al.* (2008) Improved utilisation of renewable resources: new important derivatives of glycerol. *Green Chemistry*, **10** (1), 13–30.

234. Zhou, C.H., Zhao, H., Tong, D.S. *et al.* (2013) Recent advances in catalytic conversion of glycerol. *Catalysis Reviews-Science and Engineering*, **55** (4), 369–453.

235. Di Mondo, D., Ashok, D., Waldie, F. *et al.* (2011) Stainless steel as a catalyst for the total deoxygenation of glycerol and levulinic acid in aqueous acidic medium. *ACS Catalysis*, **1** (4), 355–364.

236. Almeida, J.R.M., Favaro, L.C.L. and Quirino, B.F. (2012) Biodiesel biorefinery: opportunities and challenges for microbial production of fuels and chemicals from glycerol waste. *Biotechnology for Biofuels*, **5**, 48.

237. ten Dam, J. and Hanefeld, U. (2011) Renewable chemicals: dehydroxylation of glycerol and polyols. *Chemsuschem*, **4** (8), 1017–1034.

238. Busch, R. and Crawford, A.R. (2014) Biofuel composition and manufacturing process. World Patent WO 2009/035689.

239. Sonnati, M.O., Amigoni, S., Taffin de Givenchy, E.P. *et al.* (2013) Glycerol carbonate as a versatile building block for tomorrow: synthesis, reactivity, properties and applications. *Green Chemistry*, **15** (2), 283–306.

240. De Torres, M., Jimenez-Oses, G., Mayoral, J.A. *et al.* (2012) Glycerol ketals: synthesis and profits in biodiesel blends. *Fuel*, **94** (1), 614–616.

241. Manske, R.H.F., Ledingham, A.E. and Ashford, W.R. (1949) The preparation of quinolines by a modified skraup reaction. *Canadian Journal of Research*, **27** (9), 359–367.

# 5

# Monomers and Resulting Polymers from Biomass

**James A. Bergman[1] and Michael R. Kessler[2]**

[1] *Materials Science and Engineering Department, Iowa State University, USA*

[2] *School of Mechanical and Materials Engineering, Washington State University, USA*

## 5.1 Introduction

This chapter discusses biobased monomers, some of which are promising as comonomers to increase bio-content even though they do not provide synthetic pathways to 100% biorenewable polymers, while other systems already illustrate the possibility of providing 100% biorenewable materials. Optimization of these systems so that they match the properties and costs of fossil-fuel-based polymers is an ongoing emphasis of the research discussed here. Some biobased polymers already display better properties than their fossil fuel counterparts, leaving only production hurdles to be overcome in order to enter high-volume markets.

Each section will introduce a new class of 'monomer', discussing relevant material sources and feedstocks, necessary chemical modifications, and production methods. This includes monomers from vegetable oils, furans, terpenes, rosin, tannins, and α-hydroxy acids (AHAs). The first two material sources are of growing interest as polymer synthesis techniques have matured over the past

*Introduction to Chemicals from Biomass*, Second Edition. Edited by James Clark and Fabien Deswarte.
© 2015 John Wiley & Sons, Ltd. Published 2015 by John Wiley & Sons, Ltd.

**Figure 5.1** *Thermoset plastic resin products from biobased feedstocks developed by the Biopolymers & Biocomposites Research Team at Iowa State University. Reproduced with permission from [1] http://www.biocom.iastate.edu/newsroom/gallery.html.*

decade, offering vegetable oils (or the fatty acids they contain) and furans the potential to become platform monomers that can be tailored to any number of material types and applications. The chemistries of terpenes, rosin, and tannins have been established for some time, but with the renewed interest in alternative feedstocks and advances in chemistry, these monomers have seen a renaissance recently. Lastly, AHAs, in particular lactic acid, have been used in biomedical applications for decades and have recently shown to be competitive in high-volume applications such as packaging. Figure 5.1 shows a variety of products made from thermosetting, moldable plastic materials derived from biofeedstocks, demonstrating the broad field of applications for these materials.

This chapter focuses on currently available materials, most of which are available on the commodities scale rather than on the scale provided by genetic engineering with regard to tailor-made monomers from bacteria, plants, and other sources. It should however be recognized that while some researchers continuously push to optimize the utilization of the available natural materials, others focus on how to manipulate nature to provide the materials needed.

## 5.2    Polymers from Vegetable Oils

This section briefly highlights some important contributions to the research of triglyceride (vegetable oil) -based thermoplastics and thermosets. The authors recognize the brevity of this review and therefore want to draw attention to several in-depth discussions on various topics discussed here [2–8].

During the first decade of the twenty-first century the use of vegetable oils as feedstock for renewable materials was increasingly pursued. Vegetable oils have a number of inherent properties providing them with the potential to replace petroleum feedstocks for the production of polymeric materials, including ready availability, potential biodegradability, low toxicity, and a variety of functional groups [7]. The industrial consumption of soybean oil for non-food products was approximately 15% between 2001 and 2005, leaving much room for growth in this field [5].

In general, vegetable oils consist of triglycerides of glycerol and various fatty acids [4]. Table 5.1 provides an overview of the chemical structures of common fatty acids, which are generally long, straight-chain molecules with an even number of carbon atoms. A variety of fatty acids are present in a single vegetable oil, which is typically named for the biomass of origin (i.e. soy bean oil derived from soy beans). Table 5.2 provides an overview of the fatty acid contents in several commonly used vegetable oils. While the use of plant-based oils in commodity goods is not novel, the recent interest has focused on vegetable oils or modified vegetable oils that can be used to produce materials with properties comparable to those synthesized from fossil-derived feedstocks [9–11]. Research has shown that vegetable oils and modified vegetable oils can be utilized with traditional polymer chemistry routes including, but not limited to, free-radical, polycondensation, cationic, anionic, and metal-catalyzed polymerizations. In general, researchers have taken advantage of the unsaturation sites found in many fatty acids to achieve monomer-like reactivity. The double bonds can be conjugated, epoxidized, or hydrolyzed in order to afford them the reactivity necessary for traditional polymer chemistry techniques. While several naturally occurring fatty acids contain some of these functional groups, as seen in Table 5.1, the majority of research and the focus of this discussion is on commodity oils, produced in the millions of tons per year, that contain simple fatty acids such as palmitic, stearic, oleic, linoleic, and linoleic acid [2, 4, 12]. Commodity oils such as soybean, canola, and corn typically contain between three and five double bonds per triglyceride [5].

Vegetable oils are categorized as drying, semi-drying, and non-drying depending on their degree of unsaturation: drying oils have an iodine value higher than 130; semi-drying oils have an iodine value between 90 and 130; and non-drying oils have an iodine value below 90 [5]. The iodine value is defined as the amount of iodine (in milligrams) that is required to consume all the double bonds in 100 g of the specific oil [5]. Most commodity oils are semi-drying and after modification contain multiple reactive sites per triglyceride, which makes them good candidates for cross-linked materials and thermoset plastics.

**Table 5.1** Structures of common fatty acids [5, 7].

| Fatty acid | Structure |
| --- | --- |
| Caprylic | |
| Capric | |
| Lauric | |
| Myristic | |
| Palmitic | |

Palmitoleic

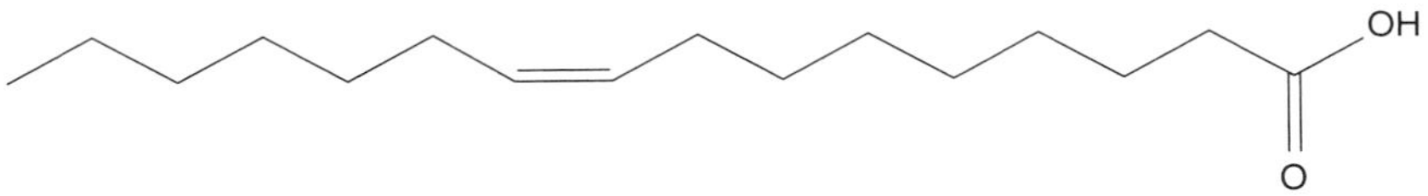

Stearic

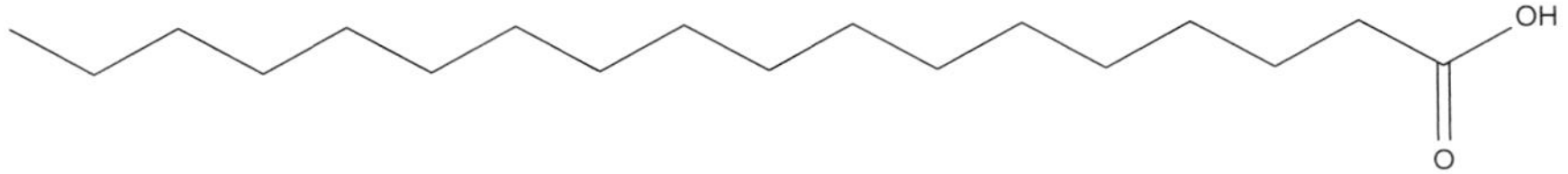

Oleic

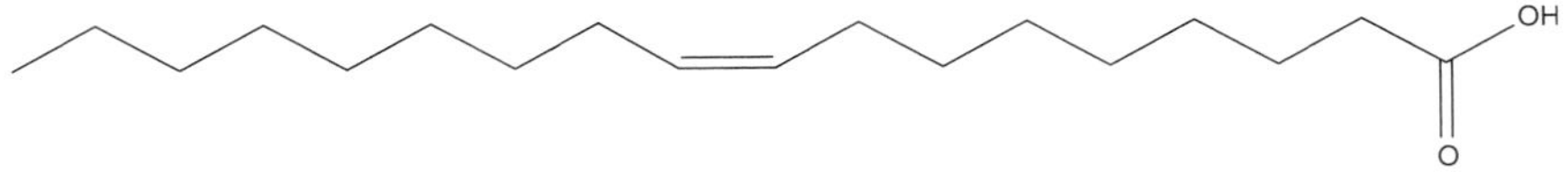

Linoleic

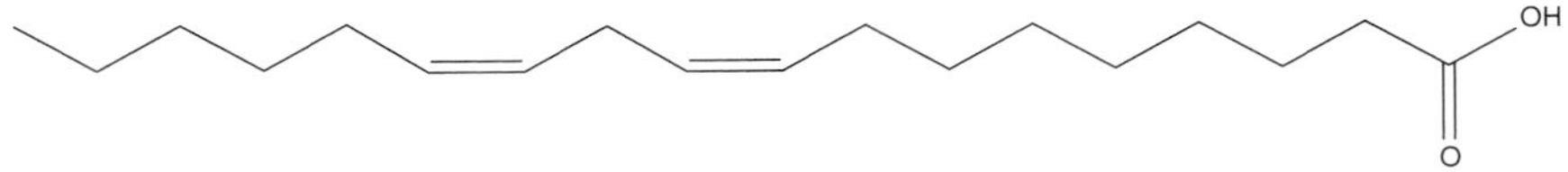

Linolenic

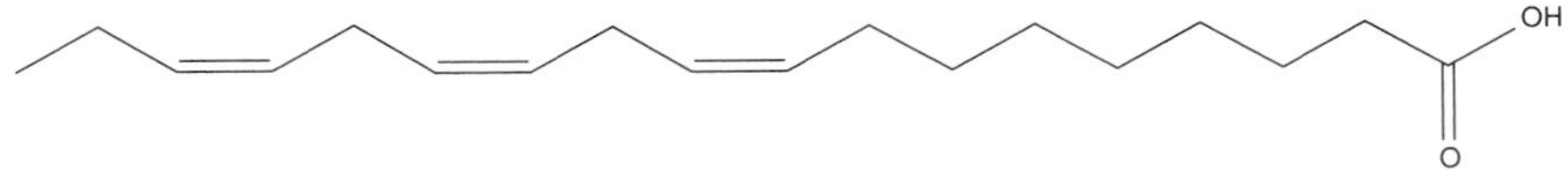

α-Eleostearic

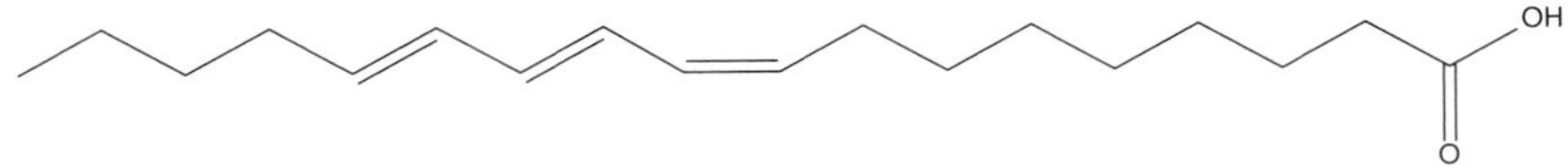

(Continued)

**Table 5.1** (Continued)

| Fatty acid | Structure |
| --- | --- |
| Ricinoleic | |
| Vernolic | |
| Licanic | |

**Table 5.2**  *Fatty acid compositions of common oils [2, 5].*

| Oil | Palmitic acid | Stearic acid | Oleic acid | Linoleic acid | Linolenic acid |
|---|---|---|---|---|---|
| Canola | 4.1 | 1.8 | 60.9 | 21.0 | 8.8 |
| Corn | 10.9 | 2.0 | 25.4 | 59.6 | 1.2 |
| Cottonseed | 21.6 | 2.6 | 18.6 | 54.4 | 0.7 |
| Linseed | 5.5 | 3.5 | 19.1 | 15.3 | 56.6 |
| Olive | 13.7 | 2.5 | 71.1 | 10.0 | 0.6 |
| Soybean | 11.0 | 4.0 | 23.4 | 53.3 | 7.8 |
| Tung | — | 4.0 | 8.0 | 4.0 | — |
| Fish | — | — | 18.2 | 1.1 | 0.99 |
| Castor | 1.5 | 0.5 | 5.0 | 4.0 | 0.5 |
| Palm | 39 | 5 | 45 | 9 | — |
| Oiticica | 6 | 4 | 8 | 8 | — |
| Rapeseed | 4 | 2 | 56 | 26 | 10 |
| Sunflower | 6 | 4 | 42 | 47 | 1 |
| High oleic[a] | 6.4 | 3.1 | 82.6 | 2.3 | 3.7 |

[a] Genetically engineered soybean oil from DuPont [2].

### 5.2.1   Isolation of Vegetable Oil

The two commonly used methods for extracting vegetable oil from biomass are mechanical separation and solvent extraction [13]. Mechanical extraction utilizes mechanical pressure such as shearing to break the cells of the biomass, liberating the oil. The main benefit to this method is low capital costs. Unfortunately, the forces used to break the cells also generate a large amount of heat, which can degrade the oil. Also, the mechanical processes reduce the oil content in the biomass to only 5–10% by mass [13]. The more efficient solvent extraction has therefore become the method of choice for vegetable oil recovery. Solvent extraction requires higher capital investment but provides significantly higher oil yields than mechanical separation [13]. The process involves soaking the biomass in an organic solvent, typically hexane, which is able to permeate the cell walls and oil bodies. The oil solubilizes in the hexane and is free to diffuse into the surrounding environment. By circulating fresh hexane into the environment and removing the oil-laden solvent, the oil is carried away from the biomass [13]. By washing the biomass several times using this process, solvent extraction reduces the oil content of the biomass to below 1% by mass [13]. The oil is then recovered through distillation.

### 5.2.2   Thermosets of Vegetable Oils and Comonomers

Vegetable oils rich in double bonds, specifically those with conjugated double bonds, lend themselves to cross-linking via polymerization with monomers such as styrene, α-methyl styrene, or divinyl benzene [5]. Richard Larock's group extensively studied the copolymerization of soybean, linseed, corn, tung, and fish oils with vinyl monomers through cationic and free-radical polymerization to produce a broad range of thermosetting materials [14–18]. Under appropriate conditions, virgin oils are cross-linked with styrene and divinyl benzene to produce thermosets comparable to thermosets made from petroleum-based monomers with glass transition temperatures ranging from 0 to 105°C, and moduli from 6 to 2000 MPa. These materials display characteristic behaviors, such as dynamic mechanical properties, tensile stress-strain behavior, damping behavior, and shape memory, comparable to materials from fossil-based monomers [19–22]. These results emphasize that vegetable-oil-based materials are able to compete directly with and replace materials derived from fossil-based monomers.

### 5.2.3    Epoxidized and Acrylated Epoxidized Vegetable Oil

Modifications to the double bonds in the vegetable oil triglycerides provide sites with higher reactivity, offering more pathways to vegetable oil polymerization. Epoxidized soybean oil and acrylated epoxidized soybean oil (AESO) are two commercially available modified oils. Scheme 5.1 shows their modification

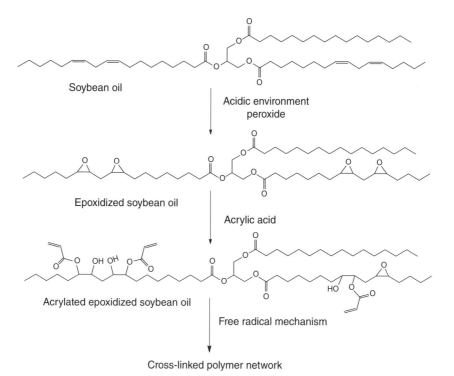

**Scheme 5.1**    *Modification pathway for AESO, which can be used to produce thermosetting polymers [3, 7].*

pathway. Epoxidized vegetable oils are obtained through a standard epoxidation reaction, where the double bonds react with hydrogen peroxide in an acidic environment [23]. Epoxidized vegetable oils can be used as plasticizers in polyvinyl chloride or as toughening agents, and as a platform for further modifications [3].

Reacting epoxidized vegetable oil with acrylic acid produces acrylated epoxidized vegetable oil. Acrylated epoxidized vegetable oils are used for surface coating and ink applications [24–29]. The properties of cross-linked AESO and styrene were extensively studied by R. Wool's group [2, 30]; they observed a plasticizing effect in thermosets of 100% AESO which they contributed to the inherent structure of the vegetable oil. The storage modulus for the 100% AESO material was similar to that of a 50/50 mix of AESO and styrene at low temperatures, while at high temperatures the storage modulus was inversely proportional to the styrene content [2]. The behavior at elevated temperatures was explained by the increased cross-linking density attainable in 100% AESO. However, the tensile properties significantly increased after incorporation of styrene; the tensile modulus of 100% AESO was approximately 440 MPa, while for a mix of 60% AESO and 40% styrene the modulus was 1.6 GPa. Ultimate tensile strengths also increased from approximately 6 MPa for 100% AESO to approximately 21 MPa for the 60/40 AESO/styrene mix [2]. Replacing the styrene with another biorenewable monomer (acrylated epoxidized fatty methyl ester) provided a route to increase the renewability of materials produced from AESO [31]. The fact that the properties of the material can be tuned by changing the ratio of AESO and monomer (styrene or acrylated epoxidized fatty methyl ester) illustrates the possibility that these materials may replace fossil-based polymers for structural applications [2, 32].

Recent work with AESO utilized facile-controlled free-radical polymerization techniques, such as atom transfer radical polymerization (ATRP) and reversible addition fragmentation chain transfer (RAFT), to synthesize block copolymers with triglyceride-based monomers [33]. This work reveals a possible pathway to replacing the fossil-based thermoplastics by renewable materials.

### 5.2.4    Polyurethanes from Vegetable Oil

Epoxidized vegetable oils can be reacted to form polyols, commonly by acid-catalyzed epoxide ring opening by water or by acid-catalyzed alcoholysis with a mono-alcohol, typically methanol [8]. These vegetable-oil-based polyols can then be reacted with diisocyanates to form polyurethanes, as shown in Scheme 5.2 [6, 34].

Generally, polyurethanes are used in applications such as coatings, adhesives, sealants, and elastomers, but the major application is as flexible or rigid foams [6]. When used in flexible foams, polyols typically have a molecular weight ranging from 3,000 to 6,000 Da with a hydroxyl functionality near 3 [6]. In rigid foams, the polyols have molecular weights below 1,000 Da and a higher hydroxyl functionality (between 3 and 6) [6]. Because polyols from vegetable oils have functionalities between 3 and 5, with molecular weights below 1,000 Da, they are suitable for rigid foam applications [35].

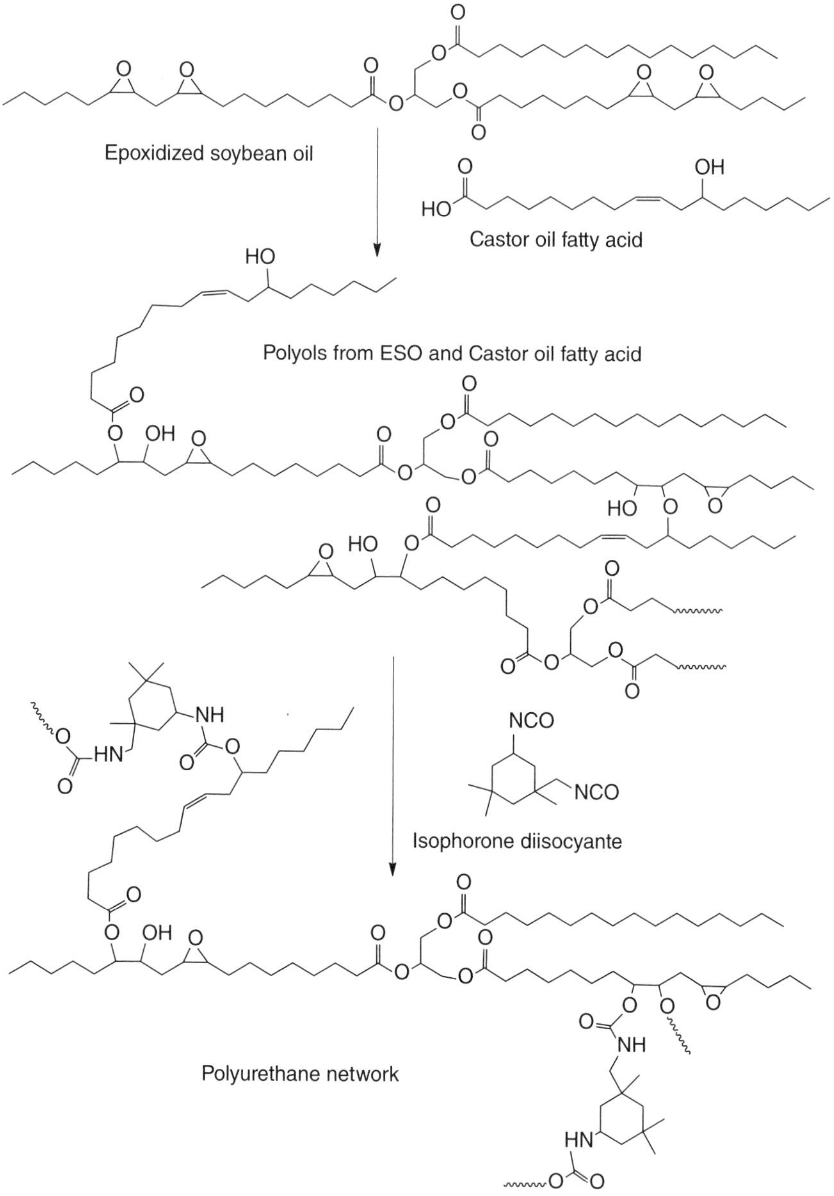

**Scheme 5.2** *Example of a polyurethane produced from vegetable oil, with a fatty acid polyol intermediate and cross-linking with isophorone diisocyanate [34].*

To increase the renewable content of the polyurethanes produced with vegetable-oil-based polyols, diisocyanates have been synthesized from renewable sources. Here, the fatty-acid-derived 1,7-heptamethylene diisocyanate and the isosorbide-based diisocyanates are of interest [36, 37]. Both diisocyanates were used to react with vegetable-based polyols to produce polyurethanes with 100% renewable carbon content. The 1,7-heptamethylene diisocyanate was compared to the fossil-based 1,6-hexamethlylene diisocyanate and displayed similar properties [36]. In addition, it was shown that 1,7-heptamethylene could be used to synthesize thermoplastic polyurethanes from completely renewable sources [38].

In an effort to remove volatile organic compounds from the polyurethane systems, water-borne polyurethane dispersions have grown in popularity. These systems are used in coatings and adhesives [39–42]. Combining emulsion polymerization of acrylic monomers with the polyurethane dispersions to form urethane-acrylic latexes significantly increased the thermal and mechanical properties of the resulting polyurethane films [43]. Other work showed methoxylated soybean oil polyols with hydroxyl numbers ranging from 2.4 to 4.0 could be used to vary the cross-linking density of the polyurethanes, producing materials ranging from elastomeric polymers, ductile plastics, to rigid plastics, thus illustrating the versatility of the material for a wide range of possible applications [44].

### 5.2.5   Polyesters

Alkyd resins are polyesters formed by esterification of polyhydroxy alcohols, such as glycerol, with polybasic acids and fatty acids. Alkyd resins are one of the oldest materials produced by vegetable oils, with reports of various systems for well over 100 years [4]. The resins can be designed to exhibit good film properties with high viscosity and good drying and hardness behavior [4]. The nature of the components (fatty acids and glycerol) and the ester linkages provide a higher level of biodegradability to these resins, making them suitable for possible medical uses [4, 45].

There are three main reaction schemes to produce polyesters: polycondensation of hydroxy acids; polycondensation of a diacid and a diol; or ring-opening polymerization of lactones [4]. Monoglycerides, obtained by the alcoholysis of vegetable oils, can be directly reacted with anhydrides, such as glutaric, phthalic, maleic, and succinic anhydride, to produce polyester resins [46, 47]. Polyester resins from vegetable oil enjoy growing popularity because they are economical and easy to implement [4]. The resins are classified based on their 'oil length', referring to the percentage of oil contained in the resin. Short-oil resins contain less than 40% oil, medium-oil lengths range from 40 to 60% oil, and long-oil alkyds contain more than 60% oil [4]. The oil content alters the final properties of, for example, coatings; short-oil polyester resins are used for baked finishes on automobiles and appliances, while long-oil resins are used in brushing enamels [4].

It is also possible to produce biobased polyester thermoplastics [48, 49]; Petrovic *et al.* produced a high–molecular-weight linear polyester utilizing a self-transesterification reaction of 9-hydroxynonanoic acid, a hydroxyl acid produced from castor oil [49]. The polymer produced from 9-hydroxynonanoic acid (PHNME) is analogous to polycaprolactone (PCL), but has a longer hydrocarbon chain between the ester groups [49]. PCL is a conventional polyester with wide use in the biomedical industry. The longer carbon chain in PHNME results in a higher melting point and better thermal stability than PCL [49]. Polyesters formed with ricinoleic acid (from castor oil) and lactic acid also show promise for biomedical applications [45]. It was shown that by manipulating the composition of ricinoleic acid, the polymers produced could be tuned from solid to liquid at room temperature [50, 51].

As with polyurethanes, polyesters from vegetable oils are well suited for ecologically friendly, water-based coatings. These systems typically use alkyd resins with high acid numbers, neutralized with amines [52]. The alkyd resins can be used in water-based coatings and paints [53, 54].

### 5.2.6  Polyamides

Polyamides based on vegetable oils are used in the ink and paint industry [55, 56]. Polyamides are thixotropic, which improves film appearance and allows for easy application of coatings [4]. Vegetable-oil-based polyamides can be produced by condensation reaction of fatty-acid-based dimer acids and diamines [56]. An important polyamide from a vegetable oil is Polyamide 11 (PA 11), derived from castor oil [57, 58]. PA 11 exhibits good dimensional stability, electrical properties, a wide range of flexibility, low cold brittleness temperature, and good chemical resistance [7].

### 5.2.7  Vegetable Oil Conclusion

Although this was by no means a comprehensive review of the topic of polymers based on vegetable oils, the reader may have gained an appreciation of their potential; vegetable oils have the ability to replace fossil-based feedstocks in a large capacity using familiar chemistries and technologies. Virgin vegetable oils can be cross-linked with vinyl monomers to produce thermosets comparable to fossil-based materials in a number of ways [19–22]. Modified vegetable oils obtain reactivities similar to traditional monomers and are capable of producing materials suitable for structural applications [2, 30]. Polyurethanes produced from vegetable oils have shown promise for foam and coating applications as well as a route to renewable thermoplastics [35, 38–42]. Polyesters from vegetable oils have applications in coatings and biomedical products [4, 45]. Polyamides from vegetable oils can be used for the synthesis of PA 11, and have been shown to have properties suitable for paints [55, 56, 58].

Moving forward, the versatility of vegetable oil feedstocks combined with relatively new polymerizations methods, such as acyclic metathesis polymerizations (ADMET), ring-opening metathesis polymerization (ROMP), ATRP, and RAFT, provides opportunities to meet the need for specialized and novel thermosets and thermoplastics [33, 59–62]. Lastly, as research in genetically modified organisms continues to develop, there is potential for an upstream increase in purity of fatty acid contents, for an increase in functionalities of fatty acids, and for an increase in production of vegetable oils [3, 63].

## 5.3  Furan Chemistry

This section introduces the established but still growing area of polymers developed from furans. Two furan derivatives are of particular interest: furfural and 5-hydroxymethylfurfural (HMF). Furans and furan derivatives are a promising field in renewable chemistry as they are platform chemicals that can be used to produce fine chemicals, pharmaceuticals, and agrochemicals, as well as polymers [64]. Several excellent in-depth reviews on furan chemistry for biorenewable polymers are available [64–68].

Furan is a heterocyclic diene. The first-generation derivatives – furfural and 5-HMF, shown in Figure 5.2 – can be produced from mono, oligo, and polysaccharide sources using several methods [67]. Straightforward modifications of furfural and HMF lead to the production of several monomers suitable for polymerization [68]. Figure 5.2 depicts some of these, including 2,5-furancarboxydialdehyde (FCDA), 2,5-furandicarboxylic acid (FDCA), 2,5-furandicarboxylic acid dichloride (FDCC), and isopropylidene *bis*-(2,5-furandiylmethylene)diisocyanate. Furan derivatives show promise for the production of polymers, generally following condensation routes to polyesters, polyamides, polyurethanes, and others [67]. Limited success has been reported with regard to free radical, cationic, and anionic polymerizations because of difficulties mostly contributed to the central furan molecule, suggesting limited room for growth [67].

### 5.3.1  Production of Furfural and HMF

Furfural and 5-HMF are commonly produced by acid-catalyzed dehydration of pentose and hexose sugars, respectively [67]. Carbohydrate feedstocks for the production of these chemicals include many by-products, such as corn cobs, rice hulls, and sugarcane bagasse. Furfural has been a commodity product since the early twentieth century; production typically consists of a pretreatment, hydrolysis, and refining [69]. During the pretreatment, the feedstock (corn cobs, oat hulls, etc.) are crushed and mixed with sulfuric acid [69]. The hydrolysis is carried out at high temperatures, and the escaping steam from the reaction mixture carries out water and furfural; subsequently the steam is filtered and condensed [69]. The

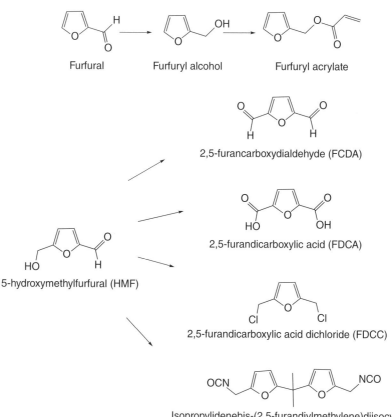

**Figure 5.2**    *Furanic monomers; a sample of the variety afforded by furan based chemistry [67, 68].*

condensate is then separated into various fractions, including furfural, by distillation; typical yields reach only about 15% of dry matter [67, 69]. Purified, furfural is a clear liquid.

Unlike furfural, HMF is solid at room temperature. The production of HMF has not yet moved beyond the pilot scale. A recent review details the chemistry of the production of HMF from mechanism to process technology [64]. The three general methods for pilot-scale production of HMF differ in reaction medium: an aqueous process; a process with non-aqueous solvents; and a two-phase process with both aqueous and non-aqueous solutions [64, 70–72]. The starting reagent for these processes is typically fructose, and all three processes have drawbacks such as the use of homogeneous acid catalysts and low yields; these challenges will need to be met before HMF production can meet industrial-scale demand [64].

### 5.3.2    Second-Generation Derivatives

Furfural and 5-HMF benefit from the ease of adding functional groups to their structures. As illustrated in Figure 5.2, 2,5-FCDA, 2,5-FDCA, 2,5-FDCC, and isopropylidene *bis*-(2,5-furandiylmethylene)diisocyanate can be produced from the first-generation derivative HMF. These monomers are bi-functional and well suited for the condensation polymerizations to produce polyesters, polyamides, and polyurethanes. Modifications to furfural produce mono-functional monomers such as furfuryl acrylates, which are precursors for chain reaction polymerizations such as free radical and cationic polymerization [67]. The major exception is furfuryl alcohol (FA), which is the hydrogenation product of furfural; FA is a commonly used monomer in polycondensation reactions [67]. There are numerous modifications to furfural and HMF cataloged, many of which follow oxidation or reduction chemistry route. In-depth information on these modifications can be found elsewhere [65, 68, 73].

### 5.3.3    Addition Polymerizations

The resonance effect of the furan heterocycle prevents furan derivatives that are not sufficiently substituted from participation in free radical polymerizations [65]. In fact, a whole class of free radical inhibitors has been developed on the basis of the stabilizing effect of some conjugated furan derivatives [65, 67]. Furyl acrylates and furfuryl methacrylates do display appropriate reactivity to undergo free radical polymerization, though it was reported that at high conversion rates, cross-linking and chain branching occurred in modest amounts, suggesting some furyl radical participation [67, 74].

Furan derivatives bearing 2-alkenyl structures, such as 2-vinylfuran and 2-isoprophenyl-5-methylfuran, can be activated for polymerization under mild cationic conditions. In the case of 2-vinylfuran, two types of additions have been reported – one through normal vinyl growth and the other through alkylation at C5 – which gives rise to an irregular pattern on the polymer backbone [67]. Side reactions can be eliminated by using the more substituted 2-isoprophenyl-5-methylfuran which only participates through vinyl additions, producing a traditional thermoplastic homopolymer [67].

The efficiency of anionic polymerization is also limited in furan derivatives. The epoxy substituted, 2-furyloxirane is susceptible to anionic polymerization, though care has to be taken in selecting an initiator. Conventional initiators can force both the $\alpha$ and $\beta$ opening of the epoxide ring, producing an irregular polymer structure [67]. Regiospecific $Al(iPrO)_3$ acts on the $\alpha$ position only, producing a regular polymer [65].

### 5.3.4    Furfuryl Alcohol

In light of the fact that most furfural derivatives used for the described polymer chain reactions have low efficiency, it may seem odd that furfural is a commodity product [69]. However, furfural is almost exclusively produced for FA. The condensation of FA results in a very useful cross-linked material. What should be a straightforward acid-catalyzed condensation reaction is complicated by two unavoidable side reactions, the scope of which is discussed elsewhere [65, 67, 75]. As a result of the side reactions, the product is a dark thermoset instead of a clear thermoplastic [67]. The resins produced by condensation of FA have found a number of applications, including metal casting cores, corrosion-resistant coatings, polymer concretes, wood adhesives, low flammability materials, and porous carbon [67, 76–80]. They are also used in hybrid materials such as carbon-silica, nanocomposite carbons, PFA-silica, Nafion-PFA, and sol-gel-based nanocomposites [67, 80–87].

### 5.3.5    Polyesters

Polyesters are one of the most promising markets for furan-based polymers. Despite the fact that a patent for polyesters from furan-based monomers was filed as early as in the middle of the twentieth century [88], the potential for these materials has only recently been realized [64, 89–92]. A specific target is poly(ethylene terephthalate) (PET), which is a fossil-based polyester with a market volume of 50 million tons per year [93]. Several groups have investigated furan analogs to PET using the 2,5-FDCA furan derivative [89, 91, 92, 94–97]. There are several synthetic routes to polyesters based on FDCA, including solution polycondensation, polytransesterification, and solid-state postcondensation [91, 92]. Each route requires the participation of a comonomer with FDCA. Here, ethylene glycol is of particular interest as it can also be derived from renewable resources, offering an alternative to PET with a completely renewable carbon content [89].

Gomes *et al.* investigated a variety of FDCA-based polyesters using different diols, including biobased ethylene glycol, propylene glycol, isosorbide, isoidide, and bis(2,5-hydroxymethyl)furan [91]. Scheme 5.3 provides an overview of the reaction steps. The work investigated both solution polycondensation and polytransesterification synthesis of polyesters; the molecular weights of the produced polymers were studied by size exclusion chromatography and perfluorinated end-groups [91]. The molecular weights of the polymers produced ranged from low (2 kDa) to moderately high (45 kDa). Transesterification was the preferred technique because it allowed for the production of the higher-molecular-weight polymers [91]. The thermal properties of the polyesters were studied using thermal gravimetric analysis (TGA), differential scanning calorimetry (DSC), and X-ray diffraction and it was shown that poly(ethylene 2,5-furandicarboxylate) (PEF) exhibited a glass transition temperature $T_g = 80°C$, a melting

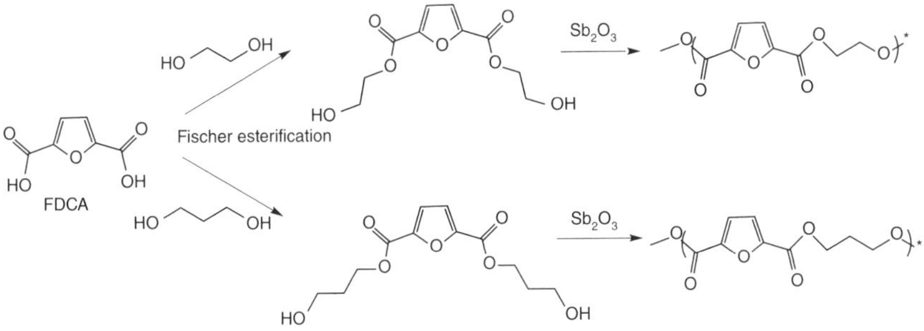

**Scheme 5.3** *Pathway to polyester from FDCA, using ethylene glycol and propylene glycol [91].*

temperature $T_m = 215°C$, and an onset thermal decomposition $T_{di} = 300°C$, suggesting that PEF could successfully replace PET [91].

Knoop *et al.* followed up the work by Gomes *et al.* investigating PEF synthesized using a solid-state postcondensation technique [92]. The work examined PEF, poly(1,3-propylenefuranoate) (PPF), poly(1,4-butylenefuranoate) (PBF), and their terephthalic counterparts. It was shown that the furan-based polymers had similar $T_g$ but lower $T_m$, suggesting similar material behavior in specific applications but lower processing temperatures [92]. The mechanical properties of both amorphous and annealed PEF were studied using tensile testing and dynamic mechanical thermal analysis (DMTA). PEF had a Young's modulus of 2450 MPa, comparable to that of PET (2000 MPa) [92, 98], which is another indication that PEF may be a viable alternative to PET.

Polyesters from furans can also be used to produce cross-linked thermosets. The first-generation furan derivative, 5-HMF, can be substituted to give a conjugated double bond next to the furan [99]. This double bond is photoreactive at 308 nm and can be used to cross-link with similar moieties [99]. This can be exploited using low $T_g$ aliphatic hydroxyesters as comonomers to produce cross-linked elastomers, with $T_g$ below room temperature [67, 99].

### 5.3.6   Polyamides

Polycondensation reactions can be used to produce polyamides based on furan moieties, an approach primarily explored by Gandini's research group [100–104]. Aliphatic and aromatic diamines can be used to react with 2,5-FDCA and 3,4-FDCA to produce polymers ranging from amorphous to semi-crystalline materials [67]. Gharbi *et al.* studied a series of polyamides based on a diacid chloride (2,2′-*bis*[2-(5-chloroformylfuryl)]propane), which can be obtained by coupling of FDCA as seen in Scheme 5.4 [103–105]. Diacid chloride was reacted with various aliphatic and aromatic diamines, and the thermal properties of the resulting polymers were examined using DSC and TGA. The polyamides with aliphatic amide

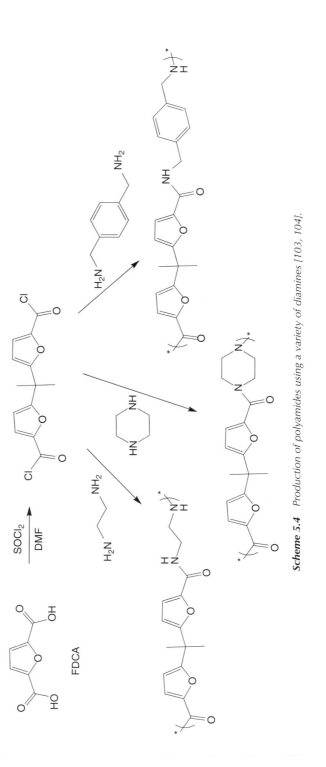

**Scheme 5.4** Production of polyamides using a variety of diamines [103, 104].

linkages were mostly amorphous, with glass transition temperatures significantly higher than their polyester analogs [103]. The furan-aromatic polyamides displayed semi-crystalline behavior, with melting temperatures above 200°C [104]. In all cases the decomposition temperature was above 300°C with some aromatic polyamides exhibiting decomposition temperatures above 400°C, indicating good thermal stability of these materials [67, 103, 104].

Polyester amides are an emerging class of materials with many attractive features; they combine the excellent mechanical properties of polyamides and the biodegradability of polyesters [106]. Recent investigations by Triki *et al.* used co-polycondensation of 5,5′-(propane-2,2-diyl)-*bis*(furan-2-carboxylic acid) (BFA), hexamethylene diamine, and ethane-1,2-diol to produce a series of furan-based polyester amides with varying amounts of amines [107], and studied them using DSC and TGA. The furan-based polymers were all amorphous and an increase in glass transition temperature was observed as the amine content was increased [107]. This underscores the versatility of furan-based monomers used in copolymerization reactions, and identifies another method of tuning the physical properties of furan-based polymers.

Despite relatively sparse interest in academic literature, furan-based polyamides have seen growing industrial interest. Recent patents by Benecke *et al.* utilize FDCA and aromatic diamines to produce curatives for polyureas, hybrid urethanes, chain extenders for polyurethanes, and for reaction injection molding [108, 109]. In addition, companies have announced plans to bring FDCA-based polyamides to markets such as carpets, textiles, and engineering plastics for automotive and electronic applications [110].

### 5.3.7   Other Polymers

Polyurethanes can also be produced using furan-based diols, furan-based diisocyanates, or both for a completely furan-based polyurethane. Again, the majority of this work has been advanced in Gandini's lab, and a series of papers outlines the synthesis, kinetics, and property characterization of polyurethanes from furans [111–115]. Their work explored the stability of the furan-based monomers, with the notable finding that 2,5-furan diisocyanate is not stable [113]. In general, furan diisocyanates are more reactive than their aromatic counterparts; however, furan alcohols are less reactive than their aliphatic complements [73]. Thermal analysis by DSC and TGA suggested that polyurethanes tend to crystallize and exhibit moderate thermal stability, with onset of degradation temperatures at around 200°C [114]. Their work also showed that it is possible to produce thermoplastic elastomers by incorporation of poly(tetramethylene oxide) glycol, PCL glycol, or poly(butadiene) glycol as diol chain extenders [115]. Furan-based thermoplastic elastomers showed lower moduli and thermal transitions compared to their aromatic/aliphatic commercial counterparts, leaving only niche applications for their use [115].

As detailed elsewhere, furan-based monomers can be used to produce even more types of polymers [68]. Two types of particular interest in electronic applications are poly-Schiff bases and polyhydrazides [73]. The poly-Schiff bases produced by furanic chemistry are typically oligomers, with degrees of polymerization reaching only approximately 10 [116]. Despite the low degree of polymerization, poly-Schiff bases between furanic dialdehyde and aromatic diamine have sufficient conjugations to achieve a conductivity of $10^{-8}$ S cm$^{-1}$ for undoped materials and $10^{-4}$ S cm$^{-1}$ for iodine-doped materials [116]. In addition, polyhydrazides with their highly polar macromolecules are potentially useful for electronic and optical applications [73].

### 5.3.8    Furan Conclusion

This introduction to furan-based chemistry illustrates the vast opportunities afforded by this special class of chemicals. It is possible to consider furan chemistry as an entirely new realm of polymer chemistry, capable of producing macromolecular materials comparable to or surpassing those derived from fossil fuel sources [67]. Both FA from furfural and 2,5-FDCA from 5-HMF have already secured a spot on the commodities market. As other furan chemistries are optimized, the market applications for furan-based polymers will grow to include the applications currently serviced by fossil-based polymers.

## 5.4    Terpenes

Terpenes are volatile organics, produced mainly by plants as secondary metabolites. The basic structure of terpenes is based on 5C isoprene with a hydrocarbon chain containing one or more carbon–carbon double bonds. Typically, monoterpenes have a general formula of $C_{10}H_{16}$ with a large range of available structures as illustrated in Figure 5.3. Two important features of monoterpenes are the vast number of stereoisomers available to a given structure accessible via facile isomerization and the wide variety of oxygenated derivatives that can be produced from the basic structures [117].

This section will briefly touch on the industrial production of turpentine, a mixture of monoterpenes, cationic polymerization of α-pinene and β-pinene, both homopolymerization and copolymerization, the polymerization of non-pinene terpenes, and conclude with a mention of terpenoids. Outside the scope of this discussion, terpenes find wide use in the production of fragrances and flavors [120]. In addition, terpenes hold promise as environmentally friendly solvents; camphene, which has been shown to be harmless, has been tested as a solvent for polypropylene [121]. Also not covered is isoprene, a hemiterpene. One of the most-studied monomers, isoprene has traditionally been produced as a by-product of the ethylene industry but recent advances in fermentation techniques have allowed for a renewable pathway to the production of isoprene [122].

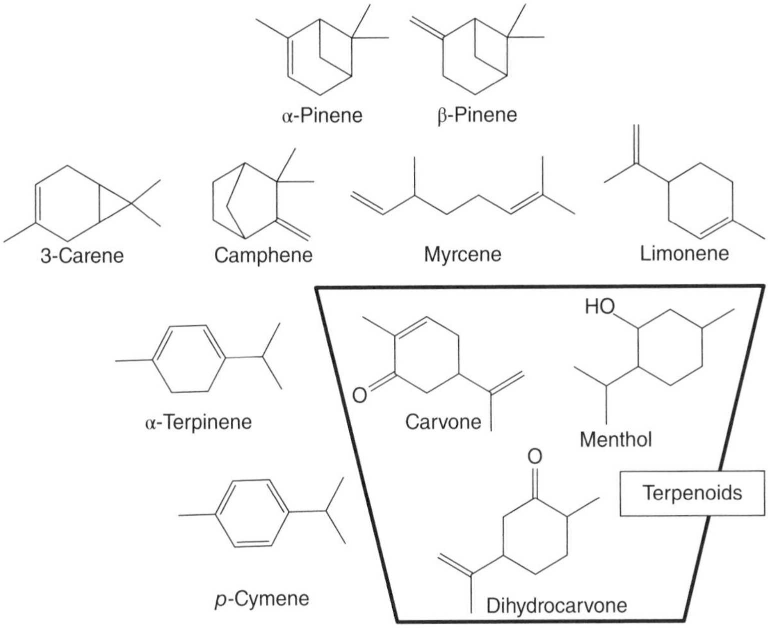

**Figure 5.3** *Structures of common monoterpenes and terpenoids, all produced by manipulation of pinenes [117–119].*

### 5.4.1 Production of Turpentine

Turpentine is a mixture of mainly α-pinene and β-pinene and other monoterpenes that constitute only a small fraction of the mixture [117]. Turpentine is the volatile fraction of pine resin and is obtained by steam-distillation of the resin; the non-volatile residue left after this process is rosin (Section 5.5). Turpentine is classified by the method used to isolate the pine resin. The two main classes are gum pine resin, acquired by tapping living trees, and sulphate pine resin, obtained during the Kraft pulping of pine wood [117]. The majority of the pine resin is isolated using the latter technique [117]. The composition of turpentine is dependent on a number of factors, including tree species, geographic location, and isolation method. Typical compositions range from α-pinene between 45 and 95% and β-pinene between 0.5 and 28% [117]. Approximately 300,000 tons of turpentine are produced annually, though currently a large portion of the turpentine produce is not used as a monomeric material [117]. A number of interesting terpenes can be isolated from the volatile fractions of plant oils, such as: caryophyllene and humulene from clove or hop oils; carvone, a terpenoid, from spearmint and caraway oils; and limonene from citrus peel oil [123–125]. These oils are produced on an industrial scale of $10^4$ tons per year [123, 124].

The majority of terpenes produced annually are α-pinene and β-pinene. Through thermal or acidic isomerization, any number of monoterpenes can be produced from these feedstocks. A review by Corma *et al.* describes the vast array of manipulations possible and the final products obtained from these pinene starting materials. Isomerization, hydrogenation, epoxidation, oxidation, and other techniques are utilized to produce a large variety of compounds, including terpenoids and terpenols [119]. More recently, thiol addition to limonene and β-pinene has been explored, which allows the modified terpene to participate in the facile thiol-ene click reaction [125].

### 5.4.2    Cationic Polymerization of Pinenes

It is generally accepted that monoterpenes do not participate in radical homopolymerization [117]. Monoterpenes can undergo cationic polymerization as long as an electron-donating group is on the double bond so that the cation is obtainable. Despite the prevalence of α-pinene, the majority of the research over the past century has centered on β-pinene because of its reactive, exocyclic double bond [117]. Only oligomers of α-pinene are obtained using cationic polymerization because of the steric hindrance of the necessary endocyclic addition [126].

Early work with β-pinene showed that Lewis acid metal halides such as $AlCl_3$ and $ZrCl_4$ are able to initiate homopolymerization reactions, though low temperatures (–40°C) are needed to achieve polymers with molecular weights of a few thousand [127]. Of the initiators explored, the most efficient is $EtAlCl_2$ which has been used to produce poly(β-pinene)s with molecular weights up to 40,000 at low temperatures and of a few thousands at room temperature [128, 129]. In addition, living cationic polymerization systems were developed for β-pinene, allowing for the synthesis of block copolymers [130]. Lu *et al.* reported a system of 2-chloroethylvinyl ether, isopropoxytitanium chloride and tetra-*n*-butylammonium chloride in dichloromethane at low temperatures that enabled a living cationic homopolymerization of β-pinene [131].

### 5.4.3    Copolymerization of Pinenes

The copolymerization of α-pinene and β-pinene with several synthetic monomers has been investigated, including styrene, α-methylstyrene, and isobutene [117, 132–135]. Increased polymerization efficiency was observed in the copolymerized systems to the point copolymers between α-pinene and isobutene, with molecular weights ranging from a few thousands up to 29,000 reported [133]. While this system eludes the goal of a completely renewable backbone, it is evidence that these monomer systems can be incorporated into diverse polymer backbones in order to increase the renewable carbon content.

Copolymerization was reported with the living cationic systems. It was reported that the systems containing β-pinene/styrene, β-pinene/*p*-methylstyrene and β-pinene/isobutylene all proceeded with high efficiency of β-pinene consumption [130, 136]. These systems were capable of producing polymers with molecular weights of several thousands. Extending this technique to incorporate chemical functionalities capable of participating in free radical polymerizations allowed for higher-order molecular architecture designs. Lu *et al.* reported an example where homo- and copolymers of β-pinene were end-capped by a methacrylic group, from which free radical polymerization of methyl methacrylate could be carried out to give a diblock polymer [137].

### 5.4.4   Polymerization of Non-Pinene Terpenes

Myrcene and limonene are two terpenes that have recently gained growing attention because of possible manipulations that afford facile polymerization. For myrcene, an isomer of pinene obtainable by pyrolysis, it was shown that ring closing metathesis can produce monomeric 3-methylenecyclopentene, which can undergo radical, anionic, and cationic polymerization [138]. The study showed that although feasible, free radical polymerization was not optimal. Both anionic and cationic polymerization were efficient, and a cationic system of *i*BuOCH(Cl) Me/ZnCl$_2$/Et$_2$O in toluene was capable of producing polymers with molecular weights higher than 20,000 [138]. DSC characterization showed both $T_g$ (11°C) and $T_m$ (65 and 105°C), suggesting the polymer was semi-crystalline [138].

Limonene, another promising terpene, was the focus of several recent studies by Firdaus *et al.* While also obtainable by isomerization of pinene, the most abundant source of limonene is citrus peel oil which is produced on an industrial scale [125]. Firdaus utilized thio-alcohol (2-mercaptoethanol) and thio-ester (methyl thioglycolate) to react via a free radical mechanism with the double bonds in the terpene [125]. The reactivity differences between the terminal double bond and the endocyclic double bond allowed manipulation of where the thiols reacted [125]. Both double bonds could be substituted to allow difunctional monomers to participate in facile condensation reactions to produce a linear polyester with molecular weights around 10,000 Da [125]. Higher-molecular-weight polymers were achievable via copolymerization between the difunctional limonene derivative, long-chain fatty-acid-based diesters, and diols [125]. Most of the polyesters showed semi-crystalline behavior, displaying both $T_g$ (−45°C) and $T_m$ (−15 to 50°C) [125].

In a follow-up study, Firdaus and Meier showed how this technique could be used to produce both polyamides and polyurethanes [139]. The polyamides produced had molecular weights up to 12,000 Da and thermal properties ranging from amorphous to semi-crystalline [139]. The polyurethanes were obtained via an isocyanate-free method and also had molecular weights up to 12,000 Da with similar thermal properties ranging from amorphous to semi-crystalline [139].

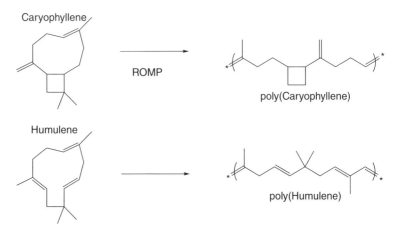

**Scheme 5.5**  *Examples of linear thermoplastics achievable by ROMP of the sesquiterpenes caryophyllene and humulene [123].*

More complex terpenes are also available, and caryophyllene and humulene are two examples of sesquiterpenes that contain three isoprene units and are found in both clove oil and hop oil [123]. A recent study showed the ability to use ring-opening metathesis on these sesquiterpenes to produce thermoplastic polymers with molecular weights of the order $10^4$ Da. Scheme 5.5 illustrates the complex structure of these sesquiterpenes and the resulting thermoplastic backbones [123]. The resulting polymers had unsaturation sites that can be targeted for cross-linking or can be hydrogenated to ensure thermoplastic behavior [123]. The polymers form soft materials with low $T_g$ (–15 to –50°C), which makes them attractive for film and coating applications [123].

### 5.4.5    Terpenoids

Terpenoids are similar to terpenes in that their structure is based on the C5 iso-prene building block. However, while terpenes are hydrocarbons, terpenoids also contain functional groups. Terpenoids occur naturally in plant oils or can be produced via chemical modification of terpenes. Two common terpenoids are carvone and menthol. Recent work by Lowe *et al.* focused on carvone, a terpenoid found in spearmint and caraway oils [124]. The work demonstrated a pathway to poly-esters through ring-opening polymerization of lactone [124]. The process utilized a Baeyer–Villiger oxidation to modify carvone (a cyclic ketone) into a lactone, which was then polymerized through a facile ring-opening mechanism [124]. The resulting polymers had molecular weights in the tens of thousands and had available double bonds for post-polymerization functionalization [124]. This work illustrates a pathway to thermoset materials from the terpenoid/terpene platform.

### 5.4.6  Terpene Conclusion

Currently, the main industrial use of polyterpene resins is to impart tack in adhesive components, provide good gloss, moisture resistance, and flexibility in wax coatings, and to provide viscosity and density controls in casting wax [117]. With a supply of more than $10^5$ ton per year and a growing number of synthetic routes to higher molecular polymers, growing applications for terpene-based polymers can be expected. With the ability to incorporate biodegradable ester linkages through polycondensation reactions or ring-opening polymerizations, these polymers become good candidates for biomedical applications [140].

## 5.5  Rosin

Rosin is the non-volatile residue left after steam distillation of pine resin, where turpentine is the volatile fraction produced during this process (Section 5.4). Rosin is a mixture of resin acids, which are typically tricyclic carboxylic acids with the generic formula $C_{19}H_{29}COOH$ [141]. The use of rosin to take advantage of its hydrophobic and binding properties is at least as old as the production of wooden ships, and there are several excellent reviews on the topic [141–144]. This section introduces the common production methods and basic chemical structure of rosin, discuss the role of rosin in epoxy resins and in thermosetting polymers, and concludes with recent reports of incorporating rosin in thermoplastics.

### 5.5.1  Production and Chemistry of Rosin

Rosin is the residue left after the steam distillation of pine resin. Rosin is classified by the technique used to isolate the pine resin; the two main classes are gum pine rosin, where pine resin is acquired by tapping living trees, and tall oil rosin, also known as sulfate rosin, where pine resin is obtained during the Kraft pulping of pine wood [141]. Rosin is composed of a mixture of resin acids (90–95%) and neutral compounds [142]. The resin acid composition of rosin is dependent on the pine resin production, the tree species, and geographic location, along with other factors [141, 144]. Several common resin acids are shown in Figure 5.4; it should be noted that for the purpose of our discussion, no distinction will be made between stereoisomers [118, 144]. As illustrated in Figure 5.4 many acids are isomers of each other, where the location of the double bonds distinguishes each type. The acid species can be categorized as abietic-type (including abietic, neoabietic, palustric, and levopimaric) and pimaric-type (including pimaric, isopimaric, and sandaracopimaric) [141, 144]. There are two lesser isomeric groups that are outside the scope of this treatment, isoprimarane-type and labdane-type rosin acids [141]. Abietic-type acids possess chemically useful conjugated double bonds, while pimaric-type acids do not. As a monomeric mixture, rosin is a semi-transparent solid with melting temperatures dependent on the acid

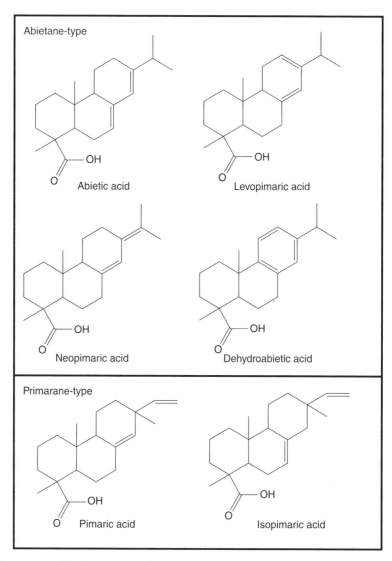

***Figure 5.4*** *Molecular structures of common abietic-type and pimaric-type rosins [118, 141].*

compositions; acid melt temperatures range from 150°C for levopimaric acid to 217°C for pimaric acid [145].

Rosin is produced on an industrial scale at over 1 million tons a year, the majority of which is gum rosin followed by tall oil rosin [146]. The quality of rosin and its derivatives is judged based on four basic parameters [141, 142, 146]: acid number, saponification number, color, and softening point. The acid number is a measure of the free carboxylic groups; as the carboxylic acids are functionalized, the acid number decreases. The saponification number is a measure of the total carboxylic groups. The color of the rosin is often a detrimental factor in many applications; a color intensity increase correlates to a decrease in quality [141]. The softening point is associated with the glass transition temperature of the rosin, which directly influences potential applications.

### 5.5.2   Epoxy Resins from Rosin

The chemical functionality, in particular of the abietic-type acids, lends rosin to a number of functionalizations that afford appropriate reactivity for use in epoxy resins. Gum rosins contain approximately 90% abietic-type acids, and many chemical modifications can be efficiently designed around this chemistry [145]. Accessing isomers can be facilitated by simple acid-catalyzed isomerization [141]. Outside the scope of this discussion, dehydrogenation of abietic-type acids produces dehydroabietic acid with an aromatic ring capable of typical aromatic substitutions, such as acylation, sulphonation, and nitration [141]. The Diels–Alder product of abietic-type acids, in particular levopimaric acid, with maleic anhydride has received much attention as a curing agent for epoxy resins. This Diels–Alder reaction is shown in Scheme 5.6 [118, 147–151]. The cure kinetics and thermal mechanical properties of these agents are comparable to their fossil-based counterparts, and they exhibit slightly better $T_g$ [147, 150]. However, the degradation temperatures for these systems are slightly below those of their fossil-based counterparts [147, 150]. A two-step process to produce epoxy curing agents from

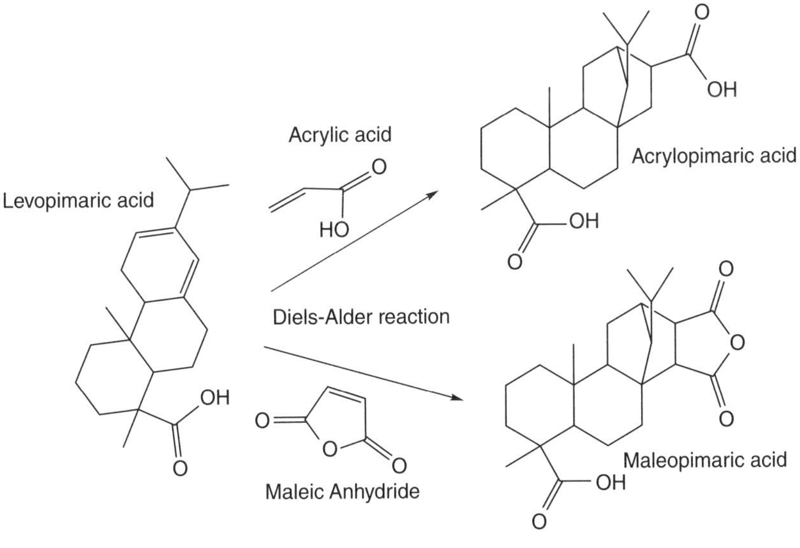

**Scheme 5.6**   *Diels–Alder products of reaction between levopimaric acid with acrylic acid and maleic anhydride [118].*

rosin utilizes diols to couple two abietic-type acids which are then functionalized with a Diels–Alder addition of maleic anhydride [118, 147, 148, 151]. This is a promising route, because changing the length of the coupling diol chain provides the opportunity to regulate the flexibility of the resulting thermoset [148].

The production of epoxy resin binders from rosin typically follows a Diels–Alder reaction with acrylic acid or maleic anhydride, again targeting the levopimaric acid isomer, or a reaction with formaldehyde [141, 145, 152, 153]. The modified rosin is then further functionalized with a base-catalyzed esterification with epichlorohydrin to produce a functional analog to fossil-based epoxies [145, 153]. A system studied by Atta *et al.* exhibited good mechanical properties and high chemical and solvent resistance, suggesting the potential for rosin-based epoxies for a wide range of protective coating applications [145, 153].

### 5.5.3   Polyesters and Polyurethanes from Rosin

The ability to produce polyesters and polyurethanes from rosin has been demonstrated [154–156]. Jin *et al.* used maleic-modified rosin to produce two kinds of rosin-based polyester polyols by esterification of the anhydride species with either diethylene glycol or ethylene glycol [154]. The resulting polyols were reacted with a commercial isocyanate and compared with an industrially available polyester polyol. It was shown that the foaming properties of the rosin-based polyols were comparable to its fossil-based counterparts, and the thermal decomposition temperature of the rosin-based polyols was slightly higher than that of the fossil-based polyol [154]. Unfortunately, little work has followed up on the early study, leaving questions about the mechanical properties of the foam.

Unsaturated polyesters were produced by Atta *et al.* using acrylic-acid-substituted rosin, maleic anhydride, and propylene or ethylene glycol [156]. The resulting polymers had only moderate molecular weight (5 kDa), but could be cross-linked [156]. Combining the unsaturated polyesters with styrene and curing the system led to materials with good mechanical properties and good resistance to both acid and alkali environments [156]. This work illustrates the viability of using modified rosin for a number of membrane applications.

The above examples illustrate that rosin can be functionalized and reacted using polycondensation methods. Although more exhaustive reviews of the versatility of rosin are available, this industrially available platform molecule has received relatively little attention, indicating a promising area for future research [118, 141, 157].

### 5.5.4   Thermoplastic Polymers from Rosin: Controlled Radical Techniques

Polymerization of thermoplastics from rosin has typically followed condensation routes, achieving only low to moderate molecular weights. Recent work by the Tang group at the University of South Carolina, Columbia, SC, US explored the polymerization of modified rosin using relatively new controlled free-radical polymerization techniques of ATRP and RAFT [157–160]. The work utilized an acrylated dehydroabietic acid, where the acylation occurred on the carboxylic acid and the dehydroabietic acid form was used for the stabilized aromatic ring. The acrylic functionality can be further removed from the rosin by using an aliphatic spacer between the acrylate and the carboxylic acid [158]. Polymers with molecular weights ranging from 10 to 100 kDa were produced and a relationship between the length of the aliphatic spacer and the steric effect observed was established [118, 158]. The Tang group has reported the production of interesting antimicrobial polymers from similar thermoplastics [161].

The Tang group explored the combination of poly(dehydroabietic acid) (PDA) and PCL. PCL is a well-known biodegradable polyester, synthesized by ring-opening polymerization (ROP) of ε-caprolactone. By combining ROP and ATRP, Wilbon *et al.* reported the first rosin-containing block copolymer of PDA and PCL [160]. As expected, the composition of the blocks dictated the thermal properties of the final polymer and a decrease in $T_m$ was associated with a higher content of PDA, because the PDA interfered with the crystallization of PCL [160].

An interesting follow up to the PDA-PCL block copolymer study investigated the effects of rosin acids as side chain pendants to PCL. Yao *et al.* synthesized a PCL polymer backbone and, using azide-alkyne click chemistry, coupled alkyne-modified dehydroabietic acid to the PCL backbone [162]. The $T_g$ of PCL prior to coupling ranged between –40 and –60°C, and after coupling the $T_g$ increased to 55–85°C [162]. The resulting polymer had improved thermal properties while maintaining the excellent ability of PCL to biodegrade, suggesting promise for a number of applications including food packaging, drug delivery, and other biomedical applications [118, 162].

### 5.5.5   Rosin Conclusion

The use of rosin acids as a renewable monomer has, like many biorenewable chemistries, received renewed interest over the past decade. The tricyclic, multifunctional acids, illustrated in Figure 5.4, are capable of competing with their fossil-based counterparts in a number of areas, including as epoxy resins and as thermosetting polyesters [147, 148, 153, 156]. More importantly, new polymerization techniques have shown that rosin acids are useful in the design of novel biomedical materials and biodegradable thermoplastics [118, 162]. These biorenewable monomers have the potential to be used in a number of film and coating applications. With continued research into the behavior of their block copolymers, these stiff monomers may be the renewable answer to styrene in the search for an all-natural substitute for the ubiquitous styrene–butadiene–styrene triblock polymer used in so many applications.

## 5.6    The Potential of Tannins

We provide a brief introduction to the potential of tannins. Despite a promising aromatic structure with viable reactive sites, the use of tannins in polymeric materials has been limited. On a broader scale, tannins have some interesting properties that fit well with the goal of creating chemicals from biomass. Outside the scope of this chapter, tannins are used in the manufacture of leather, as wine, beer, and juice additives, as ore flotation agents, in cement as a superplasticizer, and more recently in pharmaceutical applications [163]. As a polymer tannins mostly find use in adhesives, in particular as a binder for wood where the condensation reaction between formaldehyde and tannin is exploited to form a thermosetting resin. The use of tannins in wood adhesives is well established and is not discussed here [163–167]. Recent work by Tondi *et al.* has produced interesting results with regards to the condensation reaction between tannin and FA to produce a near completely biorenewable thermoset foam for applications in insulation, which is the focus of this discussion [168]. This section will introduce the oligomeric structure of condensed tannins and use the tannin-based foams as an example for the potential of this widely available chemical.

A generalized structure of a condensed tannin molecule is shown in Figure 5.5. The two types of tannins are condensed and hydrolyzable; the latter constitutes less than 10% of the tannins produced and is therefore not discussed here [164]. Condensed tannins are natural products of the oligomerization of aromatic flavonoids, where the degree of oligomerization depends on the source of the flavonoids [163]. In mimosa and quebracho tannins, the number of flavonoid repeat units ranges between 2 and 11 with an average degree of polymerization between 4 and 5; for pine tannins the average degree of polymerization is higher (between 6 and 7) [163]. Tannins are produced by a large number of woody plants and can be obtained through simple water extraction; common sources are the wood of quebracho, chestnut, and mangrove and the bark of oak, black wattle, black mimosa, and several species of pines and firs. Tannin production reached a historic high after World War II and, after several decades of decline, it is currently approximately 200 000 tons per year [163]. Today, countries with high tannin productions include Brazil, South Africa, India, Zimbabwe, Tanzania (mimosa tannin), Argentina (quebracho tannin), and Indonesia (mangrove tannin) [163].

As illustrated in Figure 5.5, flavonoids are composed of a tricyclic monomer unit with two phenolic rings, A and B. Using the reactivity of these species with formaldehyde and other aldehydes to form cross-linked polymers is well established [163–167]. However, it should be pointed out that crude tannins are approximately 70–80% of these active polyflavonoid species, with the remaining ingredients acting as inert impurities [163]. Purification of tannins has proven to be difficult; tannins are therefore used with these impurities, which act as plasticizers. Resin additives are often needed to increase the cross-linking density of the tannin matrix [163]. In addition, the bulkiness and steric hindrance of the tannin oligomers can cause incomplete consumption of reactive sites when reacted with formaldehyde,

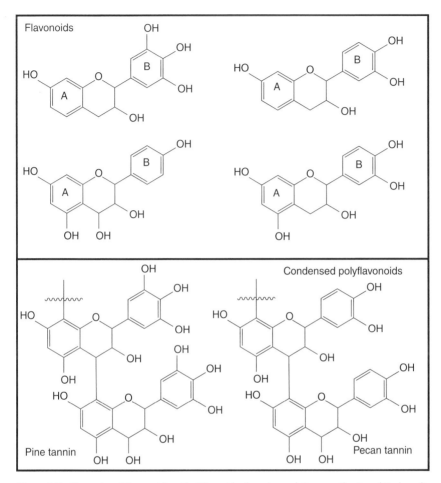

**Figure 5.5**  *Examples of flavonoids with different hydroxyl populations on the A and B phenolic rings, and representative condensed polyflavonoids from pine and pecan sources [163].*

suggesting the use of an additional linker to help close the gaps the methylene bridge cannot close [163]. Despite its industrial availability, the reactivity of its phenolic ring A, and the success of tannins in wood adhesives, the chemical has received little attention in current research which is instead focused on the production of biobased polymers.

### 5.6.1   Recent Work with Tannin Polycondensation

The production of cross-linked foams using tannin-based systems was first reported in the early 1990s, but it has only recently gained attention in a series of publications by the Pizzi group at the University of Lorraine, Epinal, France

[168–177]. The foams are produced using a mixture of tannin extract, formaldehyde, FA (Section 5.3), and a blowing agent, utilizing a three-step process [170]. In the first step, the mixture is stirred vigorously [170]. A strong acid, acting as the catalyst, is then added, initiating the second step [170]. During the second step, the exothermic condensation reaction between the FA, tannin, and formaldehyde causes the blowing agent to boil; the evaporation of the blowing agent causes the reaction mixture to foam and cure instantaneously [170]. During the last step, the cross-linked foam is stabilized and hardened [170]. This approach successfully produced foams from mimosa, quebracho, and pine tannins, with slightly different cure kinetics for each tannin [170]. The investigation showed that an important balance between the evaporation rate of the blowing agent and the cure kinetics had to be met and that a mixture of tannins, including pine tannin, contributed to this balance [170].

Mechanical compression tests of the tannin-based foams showed compressive strengths ranging from 0.03 to 3.97 MPa, depending on relative density and direction of failure. These values compare favorably to the compressive strength of fossil-based phenol-based foams, which range from 0.06 to 5.4 MPa for the same relative densities [171]. In addition, it was determined that the foams had very low thermal conductivities ($0.024–0.030 \, \mathrm{W \, m^{-1} \, K^{-1}}$) and were fire resistant and self-extinguishing [171]. The combination of these properties, together with the low material costs, suggests possible applications as insulation materials [171].

Another study examined the pyrolysis of tannin-based foams at 900°C to produce carbon foams [172]. These carbon foams possessed the good mechanical strength and low thermal conductivity observed in the organic foams [172]. The study showed electrical conductivities for the carbon foams of 1.47 and 1.13 S cm$^{-1}$ along the $z$- and $xy$-directions, respectively, which represent good conductivities for the low bulk density [172]. Studies by the Pizzi group suggested applications for tannin-based foams other than insulation, including catalyst supports and wastewater filters [178, 179].

Most recently, the Pizzi group showed that tannin-FA chemistries can be implemented for wood adhesives avoiding the use of formaldehyde, for foams without using blowing agents or formaldehyde, and for thermosetting plastics [176, 180, 181]. The first two methods are important because they reduce fossil-based formaldehyde and represent a greener approach to the production of solvent-free foams [176]. The thermosetting plastics produced by condensation reactions between tannin and FA could be molded and exhibited good post-curing mechanical and thermal properties [181]. Figure 5.6 shows the cross-linked networked formed by these materials; the incorporated aromatic and cyclic species indicate good material strength. The break strength of tannin-furan plastics was above 190 MPa and the Young's modulus was 2.16 GPa, rendering these materials competitive with a number of fossil-based materials [181]. Although the synthesis of tannin-furfuryl thermosets is tedious, once the procedure is optimized the properties of the resulting moldable plastics make them a competitive option for many applications.

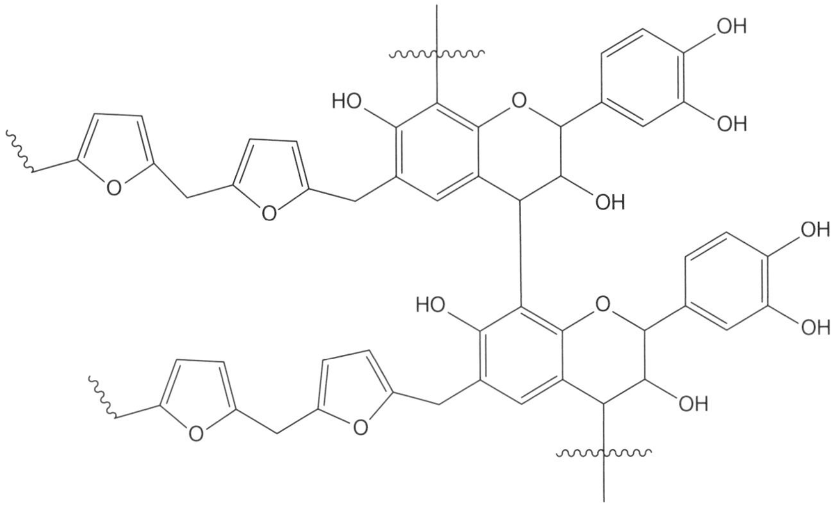

**Figure 5.6** *Representative cross-linked network of the moldable plastics produced by condensation reaction between tannin and FA [181].*

### 5.6.2 Tannins Conclusion

The use of tannins in wood adhesives is well established, but surprisingly little attention has been given to the application of tannins in other areas. Recent efforts by the Pizzi group have shown a number of possible applications for tannins. These include uses in insulation, as catalyst supports, and for active filtration [168, 178, 179]. In addition, it was shown that it is possible to produce moldable thermosetting plastics from tannins, with material strengths competitive with traditional fossil-based thermoset plastics [181].

## 5.7 Alpha-Hydroxy Acids

Certain AHAs have gained attention over the past decade as potential competitors for conventional polyesters, such as PET for high-volume uses in packaging and textiles. This discussion focuses on two biorenewable AHAs: polylactic acid (PLA) and polyglycolic acid (PGA). PLA was discovered in the early part of the twentieth century and has gained industrial relevance over the past decade. Currently, several companies worldwide are producing PLA [182–186]. PLA is a polyester with stereoisomers, which can be used to tune the properties of the bulk polymer from amorphous to semi-crystalline. This section describes the production of PLA, its properties and current applications in packing and biomedical applications. Detailed reviews of PLA have been prolific over the past decade [187–195].

### 5.7.1  Production of PLA

The structures of glycolic acid and lactic acid with their respective homopolymers are shown in Figure 5.7. Both glycolic acid and lactic acid are produced naturally. Glycolic acid is commonly produced from fossil feedstocks; however, it can be recovered directly from biowaste with high sugar content. Currently, research into an economically feasible, fermentation-style production of glycolic acid is ongoing [196]. Lactic acid is produced by fermentation. This section describes the production of lactic acid, the process to produce lactide, and the polymerization of lactide to PLA, which is analogous to the production of PGA.

Lactic acid is typically produced via bacterial fermentation using carbohydrate sources such as corn cobs and stalks, lignocellulose/hemicellulose hydrolyzates, cassava bagasse, and other industrial bio-by-product streams [189]. The most widely used bacteria are *Lactobacilli* with several common strains: *L. amylophilus*, *L. bavaricus*, *L. casei*, *L. delbrueckii*, *L. jensenii*, and *L. acidophilus* [195]. Lactic acid comes in two optically active configurations – L and D isomers – and specific bacteria will predominately produce one of the two stereoisomers. L-lactic acid is produced by *L. amylophilus*, *L. bavaricus*, or *L. casei*; D-lactic acid or a mixture of L- and D-lactic acid is produced by *L. delbrueckii*, *L. jensenii*, or *L. acidophilus* [190]. After fermentation, lactic acid is separated by a neutralization of the fermentation bath, typically with a basic salt. The soluble lactic acid solution

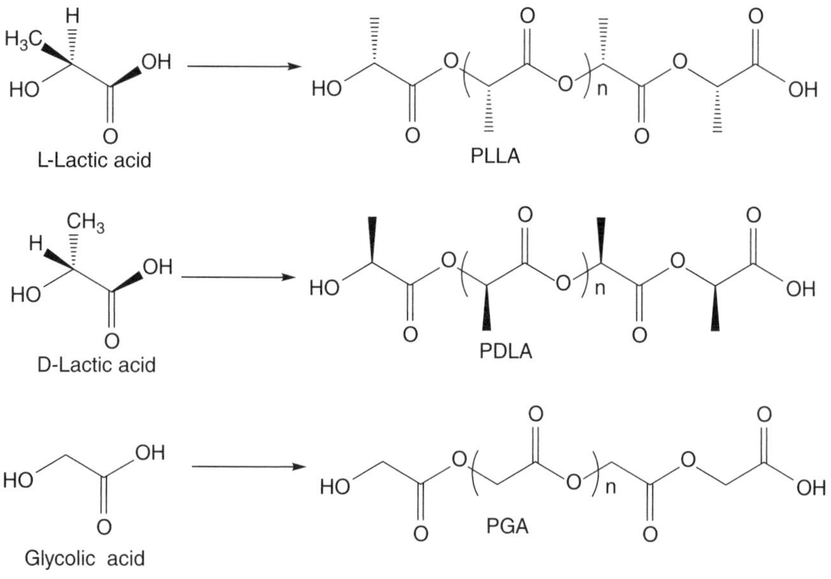

**Figure 5.7**  Molecular structures for L-lactic acid, D-lactic acid, and glycolic acid and the representative polymer backbone for poly(L-lactic acid) (PLLA), poly(D-lactic acid) (PDLA), and PGA.

is then filtered, and the acid evaporated, crystallized, and acidified [190]. Further purification steps are taken for pharmaceutical or food applications.

Direct polycondensation polymerization of lactic acid is possible, though this typically results in only low-molecular-weight polymers [194]. To produce polymers with higher molecular weight, an intermediate (lactide) is typically formed. This cyclic dimer of lactic acid can then be polymerized by ring-opening polymerization to form high-molecular-weight polymers, as shown in Scheme 5.7 [194]. In order to produce lactide, lactic acid is oligomerized by polycondensation,

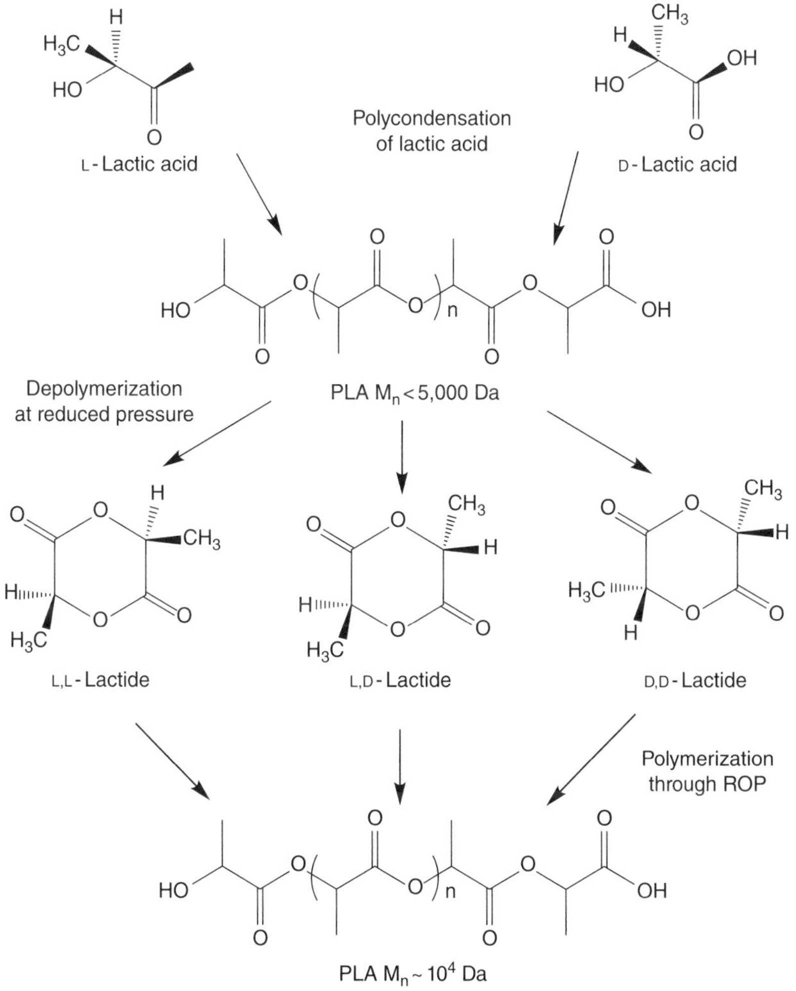

**Scheme 5.7**  *Production of high-molecular-weight PLA polymers by ring-opening polymerization of lactide. Lactide, in various stereoisomers, is produced by depolymerization of low-molecular-weight PLA at reduced pressure [194].*

the reaction temperature is raised and the pressure reduced. This initiates a depolymerization process that results in lactide, which in turn is evaporated *in situ* [194]. Lactide, like lactic acid, comes in different stereoisomeric forms: L- and D-lactides derived from homo lactic acids and rac-lactide (meso-lactide) derived from a combination of L- and D-lactic acids. Ring-opening polymerization of lactide is possible via cationic, anionic, or coordination-insertion mechanisms; the most common approach uses stannous octoate as catalyst with a coordination-insertion mechanism [190].

The production of PGA is analogous to that of PLA; here, glycolic acid is used first to produce glycolide, which then undergoes ROP to produce PGA. In fact, the systems are similar enough to allow the copolymerization of glycolide and lactide to produce copolymers with properties between those of the homopolymers. In addition, the monomers are often copolymerized in ROP systems with ε-caprolactone, a fossil-based biodegradable monomer [190].

### 5.7.2  Properties of PLA

The properties of PLA match those of other thermoplastic polyesters. Homopolymers consisting of 93% or more L-lactide are semi-crystalline, potentially reaching 40% crystallinity [190]. When the polymer composition contains between 50 and 93% L-lactide, the system is amorphous [190]. The melt transition of the homopolymer is between 130 and 180°C and glass transition is typically observed between 50 and 80°C [190]. A racemic mixture of poly(L-lactide) and poly(D-lactide) produces stereo-complexation that can increase the mechanical and thermal properties of the material; in this case the $T_m$ can be as high as 230°C [190].

Because one of the most important applications of PLA is in flexible packaging, the physical and barrier properties of the material have been extensively studied [192]. PLA films exhibit carbon dioxide permeability of $1.76 \times 10^{-17}$ kg m m$^{-2}$ s$^{-1}$ Pa$^{-1}$, which ranges between that of polystyrene (PS) and PET [197]. Both PET and PLA are hydrophobic, that is, they absorb very little water and have similar low water vapor permeability coefficients [192].

The mechanical properties of PLA depend on a number of parameters, including crystallinity, polymer structure, molecular weight, and processing, which can be tuned to meet the requirements of a given application. PLA is a brittle polymer, with a typical elastic modulus of 2.1 GPa and elongation at break of 9% [192, 194, 198]. Plasticization, mixing, or copolymerization of lactide can be used to increase elasticity of the resulting polymer. The importance of the production method was demonstrated by a study where fiber tensile modulus was increased from between 6.5 and 9.3 GPa to between 9.6 and 16 GPa by changing from a melt spinning system to a solution spinning system for poly(L-lactide) [194].

Degradation properties are important defining qualities of both PLA and PGA. Both polymers undergo thermal and hydrolytic degradation. PLA begins to decompose between 230 and 260°C [190]. The hydrolytic degradation of PLA begins with the slow uptake of water, which leads to the hydrolysis of ester linkages, breaking the polymer backbone into fragments. Additional water uptake allows for further degradation of the ester bonds [190]. This degradation process proceeds faster in amorphous polymers, which allows for a faster water uptake. The hydrolytic decomposition renders these materials compostable. Their biological compatibility and biodegradability make PLA and PGA attractive for biomedical applications.

### 5.7.3   Applications of PLA

Both PLA and PGA and their copolymers offer strength and biodegradability, required for biomedical applications. Sutures made from PGA/PLA copolymer fibers were one of the first bioresorbable medical products and, for decades, these copolymers have been used in various medical devices [199–202]. The mechanical properties of PLA allow for fracture fixation devices, such as screws and plates, that are absorbed by the body within a few weeks [203]. This approach has the benefit that the affected area in the body is slowly reintroduced to stress as the mechanical properties of the PLA degrade gradually over time [203]. Drug delivery platforms were also developed, with drug release profiles tuned by copolymerization of PLA with PGA or meso-lactide [204]. Recent work showed that it is possible to build porous PLA scaffolds in order to culture different cell types for cell-based gene therapy [187].

PLA offers better mechanical properties than PS for commercial packaging applications [192]. A number of companies are producing PLA commercially with PLA in most of the world's major consumer markets (Europe, Japan, and the US) in medical devices and food packing [183–186, 190]. Unfortunately, for the use of PLA to expand into new applications, several hurdles still need to be overcome including brittleness, limited barrier properties, and low heat resistance.

## 5.8   Conclusion

This chapter provided an overview of the relevant research and industrial applications of monomers from biomass and their resulting polymers, highlighting their versatility and the fact that monomers from renewable sources can compete with fossil-based monomers. As scaling-up hurdles are overcome and chemistries are optimized, the future offers many opportunities for these materials.

## References

1. Iowa State University (2013) Photo Gallery: Biopolymers & Biocomposites Research Team. Available at http://www.biocom.iastate.edu/newsroom/gallery.html (accessed 28 August 2014).
2. Khot, S.N., Lascala, J.J., Can, E. *et al.* (2001) Development and application of triglyceride-based polymers and composites. *Journal of Applied Polymer Science*, **82** (3), 703–723.

3. Wool, R. and Sun, XS. (2005) *Polymers and Composite Resins from Plant Oils. Bio-Based Polymers and Composites.* Academic Press, Burlington, MA, USA, pp. 56–113.

4. Guner, F.S., Yagci, Y. and Erciyes, A.T. (2006) Polymers from triglyceride oils. *Progress in Polymer Science*, **31** (7), 633–670.

5. Belgacem MN and Gandini A (2008) Materials from vegetable oils: major sources, properties and applications. In: *Monomers, Polymers and Composites from Renewable Resources*, first edition (eds Belgacem MN and Gandini A). Elsevier Ltd., Great Britain, pp. 39–66.

6. Petrovic, Z.S. (2008) Polyurethanes from vegetable oils. *Polymer Reviews*, **48** (1), 109–155.

7. Xia, Y. and Larock, R.C. (2010) Vegetable oil-based polymeric materials: synthesis, properties, and applications. *Green Chemistry*, **12** (11), 1893–1909.

8. Desroches, M., Escouvois, M., Auvergne, R. *et al.* (2012) From vegetable oils to polyurethanes: synthetic routes to polyols and main industrial products. *Polymer Reviews*, **52** (1), 38–79.

9. Cunningham, A.L. and Yapp, W.J. (1974) Liquid polyol compositions. The Sherwin-Williams Company. US Patent 3,827,993.

10. Bussell, G.W., inventor; Inmont Corporation, assignee (1974) Maleinized fatty acid esters of 9-oxatetracyclo-4.4.1 O O undecan-4-ol. US Patent 3,855,163 A.

11. Harris, E.B., Hodakowski, L.E., Osborn, C.L., inventors; Union Carbide Corporation, assignee (1978) Polymerizable epoxide-modified compositions. US Patent 4,119,640.

12. Gunstone, F. (1996) *Fatty Acid and Lipid Chemistry*, 1st edn, Blackie Academic & Professional, Great Britain, p. 252.

13. Kemper, T. (2005) Oil extraction, in *Bailey's Industrial Oil and Fat Products* (eds A.E. Bailey and F. Shahidi), John Wiley & Sons, Hoboken New Jersey, pp. 57–98.

14. Li, F., Marks, D.W., Larock, R.C. and Otaigbe, J.U. (2000) Fish oil thermosetting polymers: synthesis, structure, properties and their relationships. *Polymer*, **41** (22), 7925–7939.

15. Li, F.K. and Larock, R.C. (2000) Thermosetting polymers from cationic copolymerization of tung oil: Synthesis and characterization. *Journal of Applied Polymer Science*, **78** (5), 1044–1056.

16. Li, F.K. and Larock, R.C. (2001) New soybean oil-styrene-divinylbenzene thermosetting copolymers. I. Synthesis and characterization. *Journal of Applied Polymer Science*, **80** (4), 658–670.

17. Li, F.K., Perrenoud, A. and Larock, R.C. (2001) Thermophysical and mechanical properties of novel polymers prepared by the cationic copolymerization of fish oils, styrene and divinylbenzene. *Polymer*, **42** (26), 10133–10145.

18. Valverde, M., Andjelkovic, D., Kundu, P.P. and Larock, R.C. (2008) Conjugated low-saturation soybean oil thermosets: Free-radical copolymerization with dicyclopentadiene and divinylbenzene. *Journal of Applied Polymer Science*, **107** (1), 423–430.

19. Li, F.K. and Larock, R.C. (2000) New soybean oil-styrene-divinylbenzene thermosetting copolymers. II. Dynamic mechanical properties. *Journal of Polymer Science Part B— Polymer Physics*, **38** (21), 2721–2738.

20. Li, F.K. and Larock, R.C. (2001) New soybean oil-styrene-divinylbenzene thermosetting copolymers. III. Tensile stress–strain behavior. *Journal of Polymer Science Part B—Polymer Physics*, **39** (1), 60–77.

21. Li, F.K. and Larock, R.C. (2002) New soybean oil-styrene-divinylbenzene thermosetting copolymers. Good damping properties. IV. *Polymers for Advanced Technologies*, **13** (6), 436–449.

22. Li, F.K. and Larock, R.C. (2002) New soybean oil-styrene-divinylbenzene thermosetting copolymers. V. Shape memory effect. *Journal of Applied Polymer Science*, **84** (8), 1533–1543.

23. Eckwert, K., Gutsche, B., Jeromin, L., *et al.*, inventors; Henkel Kommanditgesellschaft auf Aktien, assignee (1987) Process for the epoxidation of olefinically unsaturated hydrocarbon compounds with peracetic acid. US Patent 4,647,678 A.

24. Harris, E.B., Hodakowski, L.E., Osborn, C.L., inventors; Union Carbide Corporation, assignee (1978) Polymerizable epoxide-modified compositions. US Patent 4,119,640.

25. Trecker, D.J., Borden, G.W., Smith, O.W., inventors; Union Carbide Corporation, assignee (1976) Method for curing acrylated epoxidized soybean oil amine compositions. US Patent 3,979,270.

26. Trecker, D.J., Borden, G.W., Smith, O.W., inventors; Union Carbide Corporation, assignee (1976) Acrylated epoxidized soybean oil amine compositions and method. US Patent 3,931,075.

27. Borden, G.W., Smith, O.W., Trecker, D.J., inventors; Union Carbide Corporation, assignee (1977). Acrylated epoxidized soybean oil urethane derivatives. US Patent 4,025,477.

28. Pashley, R.M., Senden, T.J., Morris R.A., *et al.*, inventors; The Australian National University. The Acton and Memtec Ltd., assignee (1994) Polymerizable porphyrins. US Patent 5,360,880.

29. Likavec, W.R., Bradley, C.R., inventors; Day-Glo Color Corporation, assignee (1999) Ultraviolet and electron beam radiation curable fluorescent printing ink concentrates and printing inks. US Patent 5,886,628.

30. Khot, S., Kusefoglu, S., Giuseppe, P., *et al.*, inventors; University of Delaware, assignee (2000) High modulus polymers and composites from plant oils. US Patent 6,121,398.

31. Campanella, A., La Scala, J.J. and Wool, R.P. (2009) The use of acrylated fatty acid methyl esters as styrene replacements in triglyceride-based thermosetting polymers. *Polymer Engineering and Science*, **49** (12), 2384–2392.

32. La Scala, J. and Wool, R.P. (2005) Property analysis of triglyceride-based thermosets. *Polymer*, **46** (1), 61–69.

33. Cochran, E., Williams, R., Hernandez, N., Cascione, A., inventors. (2013) Thermoplastic elastomers via atom transfer radical polymerization of plant oil. US application 13/744,733.

34. Zhang, C., Xia, Y., Chen, R. *et al.* (2013) Soy-castor oil based polyols prepared using a solvent-free and catalyst-free method and polyurethanes therefrom. *Green Chemistry*, **15** (6), 1477–1484.

35. Zlatanic, A., Lava, C., Zhang, W. and Petrovic, Z.S. (2004) Effect of structure on properties of polyols and polyurethanes based on different vegetable oils. *Journal of Polymer Science Part B—Polymer Physics*, **42** (5), 809–819.

36. Hojabri, L., Kong, X. and Narine, S.S. (2009) Fatty acid-derived diisocyanate and biobased polyurethane produced from vegetable oil: synthesis, polymerization, and characterization. *Biomacromolecules*, **10** (4), 884–891.

37. Zenner, M.D., Xia, Y., Chen, J.S. and Kessler, M.R. (2013) Polyurethanes from isosorbide-based diisocyanates. *Chemsuschem*, **6** (7), 1182–1185.

38. Hojabri, L., Kong, X. and Narine, S.S. (2010) Functional thermoplastics from linear diols and diisocyanates produced entirely from renewable lipid sources. *Biomacromolecules*, **11** (4), 911–918.

39. Lee, S.Y., Lee, J.S. and Kim, B.K. (1997) Preparation and properties of water-borne polyurethanes. *Polymer International*, **42** (1), 67–76.

40. Rahman, M.M. and Kim, H.-D. (2006) Synthesis and characterization of waterborne polyurethane adhesives containing different amounts of ionic groups (I). *Journal of Applied Polymer Science*, **102** (6), 5684–5691.

41. Chattopadhyay, D.K. and Raju, K.V.S.N. (2007) Structural engineering of polyurethane coatings for high performance applications. *Progress in Polymer Science*, **32** (3), 352–418.

42. Park, S.H., Chung, I.D., Hartwig, A. and Kim, B.K. (2007) Hydrolytic stability and physical properties of waterborne polyurethane based on hydrolytically stable polyol. *Colloids and Surfaces A—Physicochemical and Engineering Aspects*, **305** (1–3), 126–131.

43. Lu, Y. and Larock, R.C. (2007) New hybrid latexes from a soybean oil-based waterborne polyurethane and acrylics via emulsion polymerization. *Biomacromolecules*, **8** (10), 3108–3114.

44. Lu, Y. and Larock, R.C. (2008) Soybean-oil-based waterborne polyurethane dispersions: effects of polyol functionality and hard segment content on properties. *Biomacromolecules*, **9** (11), 3332–3340.

45. Slivniak, R., Ezra, A. and Domb, A.J. (2006) Hydrolytic degradation and drug release of ricinoleic acid–lactic acid copolyesters. *Pharmaceutical Research*, **23** (6), 1306–1312.

46. Aydin, S., Akcay, H., Ozkan, E. *et al.* (2004) The effects of anhydride type and amount on viscosity and film properties of alkyd resin. *Progress in Organic Coatings*, **51** (4), 273–279.

47. Dutta, N., Karak, N. and Dolui, S.K. (2004) Synthesis and characterization of polyester resins based on Nahar seed oil. *Progress in Organic Coatings*, **49** (2), 146–152.

48. Kozlowska, A. and Ukielski, R. (2004) New type of thermoplastic multiblock elastomers-poly(ester-block-amide)s-based on oligoamide 12 and oligoester prepared from dimerized fatty acid. *European Polymer Journal*, **40** (12), 2767–2772.

49. Petrovic, Z.S., Milic, J., Xu, Y.J. and Cvetkovic, I. (2010) A chemical route to high molecular weight vegetable oil-based polyhydroxyalkanoate. *Macromolecules*, **43** (9), 4120–4125.

50. Slivniak, R. and Domb, A.J. (2005) Lactic acid and ricinoleic acid based copolyesters. *Macromolecules*, **38** (13), 5545–5553.

51. Slivniak, R., Langer, R. and Domb, A.J. (2005) Lactic and ricinoleic acid based copolyesters stereocomplexation. *Macromolecules*, **38** (13), 5634–5639.

52. Ostberg, G., Hulden, M., Bergenstahl, B. and Holmberg, K. (1994) Alkyd emulsions. *Progress in Organic Coatings.*, **24**, 281–297.

53. Nakayama, Y. (1998) Polymer blend systems for water-borne paints. *Progress in Organic Coatings*, **33**, 108–116.

54. Aigbodion, A.I., Okieimen, F.E., Obazee, E.O. and Bakare, I.O. (2003) Utilisation of maleinized rubber seed oil and its alkyd resin as binders in water-borne coatings. *Progress in Organic Coatings.*, **46**, 28–31.

55. Fan, X.D., Deng, Y.L., Waterhouse, J. *et al.* (1999) Development of an easily deinkable copy toner using soy-based copolyamides. *Journal of Applied Polymer Science*, **74** (6), 1563–1570.

56. Deng, Y.L., Fan, X.D. and Waterhouse, J. (1999) Synthesis and characterization of soy-based copolyamides with different alpha-amino acids. *Journal of Applied Polymer Science*, **73** (6), 1081–1088.

57. Koch, R. (1977) Nylon 11. *Polymer News*, **3** (6), 302–307.

58. Nayak, P.L. (2000) Natural oil-based polymers: opportunities and challenges. *Journal of Macromolecular Science—Reviews in Macromolecular Chemistry and Physics*, **C40** (1), 1–21.

59. Rybak, A., Fokou, P.A. and Meier, M.A.R. (2008) Metathesis as a versatile tool in oleochemistry. *European Journal of Lipid Science and Technology*, **110** (9), 797–804.

60. Henna, P. and Larock, R.C. (2009) Novel thermosets obtained by the ring-opening metathesis polymerization of a functionalized vegetable oil and dicyclopentadiene. *Journal of Applied Polymer Science*, **112** (3), 1788–1797.

61. Xia, Y. and Larock, R.C. (2010) Castor oil-based thermosets with varied crosslink densities prepared by ring-opening metathesis polymerization (ROMP). *Polymer*, **51** (12), 2508–2514.

62. Alemdar, N., Erciyes, A.T. and Bicak, N. (2012) Styrenated sunflower oil polymers from raft process for coating application. *Journal of Applied Polymer Science*, **125** (1), 10–18.

63. Zhang, B.X., Chen, H.Q., Li, M. *et al.* (2013) Genetic engineering of Yarrowia lipolytica for enhanced production of trans-10, cis-12 conjugated linoleic acid. *Microbial Cell Factories*, **12**.

64. van Putten, R.J., van der Waal, J.C., de Jong, E. *et al.* (2013) Hydroxymethylfurfural, a versatile platform chemical made from renewable resources. *Chemical Reviews*, **113** (3), 1499–1597.

65. Gandini, A. and Belgacem, M.N. (1997) Furans in polymer chemistry. *Progress in Polymer Science*, **22** (6), 1203–1379.

66. Gandini, A. and Belgacem, M.N. (2002) Furfural and furanic polymers. *Actualite Chimique.*, **11–12**, 56–61.

67. Gandini, A. and Belgacem, M. (2008) Furan derivatives and furan chemistry at the service of macromolecular materials, in *Monomers, Polymers and Composites from Renewable Resources*, First edn (eds M.N. Belgacem and A. Gandini), Elsevier Ltd, London, pp. 115–152.

68. Amarasekara, A.S. (2012) 5-Hydroxymethylfurfural based polymers, in *Renewable Polymers: Synthesis, Processing, and Technology* (ed V. Mittal), Scrivener Publishing LLC, Somerset, NJ, pp. 381–428.

69. Win, D.T. (2005) Furfural—gold from garbage. *AU Journal of Technology*, **8** (4), 185–190.

70. Fleche, G., Gaset, A., Gorrichon, J-P., *et al.*, inventors; Roquette Freres, assignee (1982) Process for manufacturing 5-hydroxymethylfurfural. US Patent 4,339,387.

71. Rapp, K.M., inventor; Süddeutsche Zucker-Aktiengesellschaft, assignee (1988) Process for preparing pure 5-hydroxymethylfurfuraldehyde. US Patent 4,740,605.

72. Bazoa, C., Raymond, F., Rigal, L., Gaset, L., inventors; Furchim, assignee (1992) Process for the manufacture of high purity hydroxymethylfurfural. France 2,669,635.

73. Moreau, C., Belgacem, M.N. and Gandini, A. (2004) Recent catalytic advances in the chemistry of substituted furans from carbohydrates and in the ensuing polymers. *Topics in Catalysis*, **27** (1–4), 11–30.

74. Lange, J., Davidenko, N., Rieumont, J. and Sastre, R. (2002) Study of network formation in furfuryl methacrylate photopolymerisation at different temperatures. *The Tobita method applied to the polymerisation at low conversions. Polymer.*, **43** (3), 1003–1011.

75. McKillip, W.J. (1989) Chemistry of furan polymers. *ACS Symposium Series*, **385**, 408–423.

76. Muthukumar, M. and Mohan, D. (2005) Studies on furan polymer concrete. *Journal of Polymer Research*, **12** (3), 231–241.

77. Lande, S., Westin, M. and Schneider, M. (2004) Properties of furfurylated wood. *Scandinavian Journal of Forest Research.*, **19**, 22–30.

78. Lande, S., Westin, M., Schneider, M.H. (2004) Environmental friendly wood protection: furfurylated wood as alternative to traditional wood preservation. Proceedings of ICECFOP1: 1st International Conference on Environmentally-Compatible Forest Products, Oporto, Portugal, 22–24 September 2004, pp. 163–175.

79. Kawashima, D., Aihara, T., Kobayashi, Y. *et al.* (2000) Preparation of mesoporous carbon from organic polymer/silica nanocomposite. *Chemistry of Materials*, **12** (11), 3397–3401.

80. Zarbin, A.J.G., Bertholdo, R. and Oliveira, M. (2002) Preparation, characterization and pyrolysis of poly(furfuryl alcohol)/porous silica glass nanocomposites: novel route to carbon template. *Carbon*, **40** (13), 2413–2422.

81. Müller, H., Rehak, P., Jager, C. *et al.* (2000) A concept for the fabrication of penetrating carbon/silica hybrid materials. *Advanced Materials*, **12** (22), 1671–1675.

82. Yao, J.F., Wang, H.T., Chan, K.Y. *et al.* (2005) Incorporating organic polymer into silica walls: A novel strategy for synthesis of templated mesoporous silica with tunable pore structure. *Microporous and Mesoporous Materials*, **82** (1–2), 183–189.

83. Yao, J.F., Wang, H.T., Liu, J. *et al.* (2005) Preparation of colloidal microporous carbon spheres from furfuryl alcohol. *Carbon*, **43** (8), 1709–1715.

84. Principe, M., Suarez, H., Jimenez, G.H. *et al.* (2007) Composites prepared from silica gel and furfuryl alcohol with p-toluenesulfphonic acid as the catalyst. *Polymer Bulletin*, **58** (4), 619–626.

85. Yi, B., Rajagopalan, R., Foley, H.C. *et al.* (2006) Catalytic polymerization and facile grafting of poly(furfuryl alcohol) to single-wall carbon nanotube: Preparation of nanocomposite carbon. *Journal of the American Chemical Society*, **128** (34), 11307–11313.

86. Liu, J., Wang, H.T., Cheng, S.A. and Chan, K.Y. (2004) Naflon-polyfurfuryl alcohol nanocomposite membranes with low methanol permeation. *Chemical Communications*, **6**, 728–729.

87. Grund, S., Kempe, P., Baumann, G. *et al.* (2007) Nanocomposites prepared by twin polymerization of a single-source monomer. *Angewandte Chemie-International Edition*, **46** (4), 628–632.

88. Lincoln, J., Drewit, J.G.N., inventors; Celanese Corporation, assignee (1951) Polyesters from heterocyclic components. US Patent 2,551,731.

89. Gandini, A., Silvestre, A.J.D., Neto, C.P. *et al.* (2009) The furan counterpart of poly(ethylene terephthalate): An alternative material based on renewable resources. *Journal of Polymer Science Part A—Polymer Chemistry*, **47** (1), 295–298.

90. Gandini, A. (2011) The irruption of polymers from renewable resources on the scene of macromolecular science and technology. *Green Chemistry*, **13** (5), 1061–1083.

91. Gomes, M., Gandini, A., Silvestre, A.J.D. and Reis, B. (2011) Synthesis and characterization of poly(2,5-furan dicarboxylate)s based on a variety of diols. *Journal of Polymer Science Part A—Polymer Chemistry*, **49** (17), 3759–3768.

92. Knoop, R.J.I., Vogelzang, W., van Haveren, J. and van Es, D.S. (2013) High molecular weight poly(ethylene-2,5-furanoate); critical aspects in synthesis and mechanical property determination. *Journal of Polymer Science Part A—Polymer Chemistry*, **51** (19), 4191–4199.

93. Jong, E.D., Dam, M.A., Sipos, L., Gruter, G.J.M. (2012) Furandicarboxylic acid (FDCA), a versatile building block for a very interesting class of polyesters. In *Biobased Monomers, Polymers, and Materials* (eds Smith PB and Gross RA). American Chemical Society, ACS Symposium Series no. 1105, pp. 1–13.

94. Gruter, G.-J.M., Sipos, L. and Dam, M.A. (2012) Accelerating research into bio-based FDCA-polyesters by using small scale parallel film reactors. *Combinatorial Chemistry & High Throughput Screening*, **15** (2), 180–188.

95. Ma, J., Yu, X., Xu, J. and Pang, Y. (2012) Synthesis and crystallinity of poly(butylene 2,5-furandicarboxylate). *Polymer*, **53** (19), 4145–4151.

96. Ma, J., Pang, Y., Wang, M. *et al.* (2012) The copolymerization reactivity of diols with 2,5-furandicarboxylic acid for furan-based copolyester materials. *Journal of Materials Chemistry*, **22** (8), 3457–3461.

97. Zhu, J., Cai, J., Xie, W. *et al.* (2013) Poly(butylene 2,5-furan dicarboxylate), a biobased alternative to PBT: synthesis, physical properties, and crystal structure. *Macromolecules*, **46** (3), 796–804.

98. Lepoittevin, B. and Roger, P. (2011) *Handbook of Engineering and Specialty Thermoplastics*, vol. **3**, Scrivener Publishing LLC, Salem, MA/Wiley, Hoboken, NJ, pp. 97–126.

99. Lasseuguette, E., Gandini, A., Belgacem, M.N. and Timpe, H.J. (2005) Synthesis, characterization and photocross-linking of copolymers of furan and aliphatic hydroxyethylesters prepared by transesterification. *Polymer*, **46** (15), 5476–5483.

100. Mitiakoudis, A., Gandini, A. and Cheradame, H. (1985) Polyamides containing furanic moieties. *Polymer Communications*, **26** (8), 246–249.

101. Mitiakoudis, A. and Gandini, A. (1991) Synthesis and characterization of furanic polyamides. *Macromolecules*, **24** (4), 830–835.

102. Gharbi, S. and Gandini, A. (1999) Polyamides incorporating furan moieties. I. Interfacial polycondensation of 2,2′-bis(5-chloroformyl-2-furyl)propane with 1,6-diaminohexane. *Acta Polymerica*, **50** (8), 293–297.

103. Gharbi, S., Afli, A., El Gharbi, R. and Gandini, A. (2001) Polyamides incorporating furan moieties. 4. Synthesis, characterisation and properties of a homologous series. *Polymer International*, **50** (5), 509–514.

104. Abid, S., El Gharbi, R. and Gandini, A. (2004) Polyamides incorporating furan moieties. 5. Synthesis and characterisation of furan-aromatic homologues. *Polymer*, **45** (17), 5793–5801.

105. Gharbi, S., Andreolety, J.P. and Gandini, A. (2000) Polyesters bearing furan moieties IV. Solution and interfacial polycondensation of 2,2′-bis(5-chloroformyl-2-furyl)propane with various diols and bisphenols. *European Polymer Journal*, **36** (3), 463–472.

106. Rodriguez-Galan, A., Franco, L. and Puiggali, J. (2011) Degradable Poly(ester amide)s for Biomedical Applications. *Polymers*, **3** (1), 65–99.

107. Triki, R., Abid, M., Tessier, M. *et al.* (2013) Furan-based poly(esteramide)s by bulk copolycondensation. *European Polymer Journal*, **49** (7), 1852–1860.

108. Benecke, H.P., Kawczak, A.W., Garbark, D.B., inventors; assignee (2008) Furanic-modified amine-based curatives. US Patent 2008,020,847 A1.

109. Benecke, H.P., Kawczak, A.W., Garbark, D.B., inventors; assignees. 2010. Furanic-modified amine-based curatives. US Patent 2010,0280,186 A1.

110. Avantium (2012) Products and Applications. FDCA and Levulinics. Available at http://avantium.com/yxy.html (accessed 28 August 2014).

111. Quillerou, J., Belgacem, M.N., Gandini, A. *et al.* (1989) Urethanes and polyurethanes bearing furan moieties. 1. Synthesis and characterization of monourethanes. *Polymer Bulletin*, **21** (6), 555–562.

112. Belgacem, M.N., Quillerou, J., Gandini, A. *et al.* (1989) Urethanes and polyurethanes bearing furan moieties. 2. Kinetics and Mechanism of the Formation of Furanic and Other Monourethanes. *Comparative European Polymer Journal*, **25** (11), 1125–1130.

113. Belgacem, M.N., Quillerou, J. and Gandini, A. (1993) Urethanes and polyurethanes bearing furan moieties 3. Synthesis, characterization and comparative kinetics of the formation of diurethanes. *European Polymer Journal*, **29** (9), 1217–1224.

114. Boufi, S., Belgacem, M.N., Quillerou, J. and Gandini, A. (1993) Urethanes and polyurethanes bearing furan moieties 4. Synthesis, kinetics, and characterization of linear-polymers. *Macromolecules*, **26** (25), 6706–6717.

115. Boufi, S., Gandini, A. and Belgacem, M.N. (1995) Urethanes and polyurethanes bearing furan moieties 5. Thermoplastic Elastomers Based on Sequenced Structures. *Polymer*, **36** (8), 1689–1696.

116. Hui, Z. and Gandini, A. (1992) Polymeric Schiff bases bearing furan moieties. *European Polymer Journal*, **28** (12), 1461–1469.

117. Silvestre, A.J.D. and Gandini, A. (2008) Terpenes: major sources, properties and applications, in *Monomers, Polymers and Composites from Renewable Resources* (eds M.N. Belgacem and A. Gandini), Elsevier Ltd, London, pp. 17–34.

118. Wilbon, P.A., Chu, F. and Tang, C. (2013) Progress in renewable polymers from natural terpenes, terpenoids, and rosin. *Macromolecular Rapid Communications*, **34** (1), 8–37.

119. Corma, A., Iborra, S. and Velty, A. (2007) Chemical routes for the transformation of biomass into chemicals. *Chemical Reviews*, **107** (6), 2411–2502.

120. Albert, R.M. and Webb, R.L. (1989) *Fragrance and Flavour Chemicals* (eds Zinkel DF and Russel J). Pulp Chemical Association, New York, pp. 479–509.

121. Yang, M.C. and Perng, J.S. (2000) Camphene as a novel solvent for polypropylene: comparison study based on viscous behavior of solutions. *Journal of Applied Polymer Science*, **76** (14), 2068–2074.

122. McCoy, M. (2008) Biobased Isoprene. *Chemical & Engineering News*, **86** (38), 15.

123. Grau, E. and Mecking, S. (2013) Polyterpenes by ring opening metathesis polymerization of caryophyllene and humulene. *Green Chemistry*, **15** (5), 1112–1115.

124. Lowe, J.R., Martello, M.T., Tolman, W.B. and Hillmyer, M.A. (2011) Functional biorenewable polyesters from carvone-derived lactones. *Polymer Chemistry*, **2** (3), 702–708.

125. Firdaus, M., de Espinosa, L.M. and Meier, M.A.R. (2011) Terpene-based renewable monomers and polymers via thiol-ene additions. *Macromolecules*, **44** (18), 7253–7262.

126. Lu, J., Kamigaito, M., Sawamoto, M. *et al.* (1996) Cationic polymerization of beta-pinene with the AlC3/SbCl3 binary catalyst: Comparison with alpha-pinene polymerization. *Journal of Applied Polymer Science*, **61** (6), 1011–1016.

127. Roberts, W.J. and Day, A.R. (1950) A study of the polymerization of alpha-pinene and beta-pinene with Friedel crafts type catalysts. *Journal of the American Chemical Society*, **72** (3), 1226–1230.

128. Keszler, B. and Kennedy, J.P. (1992) Synthesis of high-molecular-weight poly (beta-pinene). *Advances in Polymer Science.*, **100**, 1–9.

129. Guine, R.P.F. and Castro, J. (2001) Polymerization of beta-pinene with ethylaluminum dichloride (C2H5AlCl2). *Journal of Applied Polymer Science*, **82** (10), 2558–2565.

130. Lu, J., Kamigaito, M., Sawamoto, M. *et al.* (1997) Living cationic isomerization polymerization of beta-pinene. 2. Synthesis of block and random copolymers with styrene or p-methylstyrene. *Macromolecules*, **30** (1), 27–31.

131. Lu, J., Kamigaito, M., Sawamoto, M. *et al.* (1997) Living cationic isomerization polymerization of beta-pinene. Initiation with HCl-2-chloroethyl vinyl ether adduct TiCl3(OiPr) in conjunction with nBu(4)NCl. *Macromolecules*, **30** (1), 22–26.

132. Snyder, C., McIver, W. and Sheffer, H. (1977) Cationic polymerization of beta-pinene and styrene. *Journal of Applied Polymer Science*, **21** (1), 131–139.

133. Kennedy, J.P. and Nakao, M. (1977) Co-polymerization of isobutylene with alpha-pinene. *Journal of Macromolecular Science, Part A—Pure and Applied Chemistry*, **A11** (9), 1621–1636.

134. Sheffer, H., Greco, G. and Paik, G. (1983) The characterization of styrene beta-pinene polymers. *Journal of Applied Polymer Science*, **28** (5), 1701–1705.

135. Khan, A.R., Yousufzai, A.H.K., Jeelani, H.A. and Akhter, T. (1985) Copolymers from alpha-pinene. 2. Cationic copolymerization of styrene and alpha-pinene. *Journal of Macromolecular Science—Chemistry*, **A22** (12), 1673–1678.

136. Li, A.L., Zhang, W., Liang, H. and Lu, J. (2004) Living cationic random copolymerization of beta-pinene and isobutylene with 1-phenylethyl chloride/TiCl4/Ti(OiPr)(4)/nBu(4)NCl. *Polymer*, **45** (19), 6533–6537.

137. Lu, J., Kamigaito, M., Sawamoto, M. *et al.* (1997) Living cationic isomerization polymerization of beta-pinene 3. Synthesis of end-functionalized polymers and graft copolymers. *Journal of Polymer Science Part A—Polymer Chemistry*, **35** (8), 1423–1430.

138. Kobayashi, S., Lu, C., Hoye, T.R. and Hillmyer, M.A. (2009) Controlled polymerization of a cyclic diene prepared from the ring-closing metathesis of a naturally occurring monoterpene. *Journal of the American Chemical Society*, **131** (23), 7960–7961.

139. Firdaus, M. and Meier, M.A.R. (2013) Renewable polyamides and polyurethanes derived from limonene. *Green Chemistry*, **15** (2), 370–380.

140. Lowe, J.R., Tolman, W.B. and Hillmyer, M.A. (2009) Oxidized dihydrocarvone as a renewable multifunctional monomer for the synthesis of shape memory polyesters. *Biomacromolecules*, **10** (7), 2003–2008.

141. Silvestre, A.J.D. and Gandini, A. (2008) Rosin: major sources, properties and applications. In *Monomers, Polymers and Composites from Renewable Resources* (eds Belgacem MN and Gandini A). Elsevier Ltd, London, 67–88.

142. Zinkel, D.E. and Russel, J. (1989) *Naval Stores: Production, Chemistry, Utilization*, Pulp Chemical Association, New York.

143. Maiti, S., Ray, S.S. and Kundu, A.K. (1989) Rosin: a renewable resource for polymers and polymer chemicals. *Progress in Polymer Science*, **14** (3), 297–338.

144. Zhang, J. (2012) *Rosin-based Chemicals and Polymers*, Shropshire, Smithers Rapra Technology Ltd.

145. Atta, A.M. (2012) Preparation and characterisation of epoxy binders based on rosin, in *Rosin-based Chemicals and Polymers* (ed J. Zhang), Smithers Rapra Technology Ltd, Shropshire, pp. 23–69.

146. Coppen, J.J.W. and Hone, G.A. (1995) *Gum Naval Stores: Turpentine and Rosin from Pine Resin*. Non-Wood Forest Products. Food and Agriculture Organization of the United Nations, Rome, p. 62.

147. Liu, X., Xin, W. and Zhang, J. (2009) Rosin-based acid anhydrides as alternatives to petrochemical curing agents. *Green Chemistry*, **11** (7), 1018–1025.

148. Wang, H., Liu, X., Liu, B. *et al.* (2009) Synthesis of rosin-based flexible anhydride-type curing agents and properties of the cured epoxy. *Polymer International*, **58** (12), 1435–1441.

149. Liu, X., Xin, W. and Zhang, J. (2010) Rosin-derived imide-diacids as epoxy curing agents for enhanced performance. *Bioresource Technology*, **101** (7), 2520–2524.

150. Wang, H., Wang, H. and Zhou, G. (2011) Synthesis of rosin-based imidoamine-type curing agents and curing behavior with epoxy resin. *Polymer International*, **60** (4), 557–563.

151. Zhang, J. (2012) Use of Rosin Acids as Rigid Building Blocks in the Synthesis of Curing Agents for Epoxies, in *Rosin-based Chemicals and Polymers* (ed J. Zhang), Smithers Rapra Technology Ltd, Shropshire, pp. 1–22.

152. Takeda, H., Schuller, W.H. and Lawrence, R.V. (1968) Thermal isomerization of abietic acid. *Journal of Organic Chemistry*, **33** (4), 1683–1684.

153. Atta, A.M., Mansour, R., Abdou, M.I. and Sayed, A.M. (2004) Epoxy resins from rosin acids: synthesis and characterization. *Polymers for Advanced Technologies*, **15** (9), 514–522.

154. Jin, J.F., Chen, Y.L., Wang, D.N. *et al.* (2002) Structures and physical properties of rigid polyurethane foam prepared with rosin-based polyol. *Journal of Applied Polymer Science*, **84** (3), 598–604.

155. Atta, A.M., El-Saeed, S.M. and Farag, R.K. (2006) New vinyl ester resins based on rosin for coating applications. *Reactive & Functional Polymers*, **66** (12), 1596–1608.

156. Atta, A.M., Elsaeed, A.M., Farag, R.K. and El-Saeed, S.M. (2007) Synthesis of unsaturated polyester resins based on rosin acrylic acid adduct for coating applications. *Reactive & Functional Polymers*, **67** (6), 549–563.

157. Wang, J., Wilbon, P.A., Yao, K. *et al.* (2012) *Rosin-derived Polymers and their Progress in Controlled Polymerisation*. Rosin-based Chemicals and Polymers. Smithers Rapra Technology Ltd, Shropshire. pp. 85–128.

158. Zheng, Y., Yao, K., Lee, J. *et al.* (2010) Well-defined renewable polymers derived from gum rosin. *Macromolecules*, **43** (14), 5922–5924.

159. Wang, J., Ya, K., Korich, A.L. *et al.* (2011) Combining renewable gum rosin and lignin: towards hydrophobic polymer composites by controlled polymerization. *Journal of Polymer Science Part A—Polymer Chemistry*, **49** (17), 3728–3738.

160. Wilbon, P.A., Zheng, Y., Yao, K. and Tang, C. (2010) Renewable rosin acid-degradable caprolactone block copolymers by atom transfer radical polymerization and ring-opening polymerization. *Macromolecules*, **43** (21), 8747–8754.

161. Wang, J., Chen, Y.P., Yao, K. *et al.* (2012) Robust antimicrobial compounds and polymers derived from natural resin acids. *Chemical Communications*, **48** (6), 916–918.

162. Yao, K., Wang, J., Zhang, W. *et al.* (2011) Degradable rosin-ester-caprolactone graft copolymers. *Biomacromolecules*, **12** (6), 2171–2177.

163. Pizzi, A. (2008) Tannins: major sources, properties and applications, in *Monomers, Polymers and Composites from Renewable Resources* (eds M.N. Belgacem and A. Gandini), Elsevier Ltd, London, pp. 179–200.

164. Pizzi, A. (1980) Tannin-based adhesives. *Journal of Macromolecular Science—Reviews in Macromolecular Chemistry and Physics*, **C18** (2), 247–315.

165. Pizzi, A. (1982) Condensed tannins for adhesives. *Industrial & Engineering Chemistry Product Research and Development*, **21** (3), 359–369.

166. Hergert, H.L. (1989) Condensed tannins in adhesives introduction and historical perspectives. *ACS Symposium Series.*, **385**, 155–171.

167. Simon, C. and Pizzi, A. (2002) Tannin-based adhesives. *Actualite Chimique*, **11–12**, 15–16.

168. Tondi, G., Pizzi, A. and Olives, R. (2008) Natural tannin-based rigid foams as insulation for doors and wall panels. *Maderas Ciencia y Tecnologia*, **10** (3), 219–227.

169. Meikleham, N.E. and Pizzi, A. (1994) Acid-catalyzed and alkali-catalyzed tannin-based rigid foams. *Journal of Applied Polymer Science*, **53** (11), 1547–1556.

170. Tondi, G. and Pizzi, A. (2009) Tannin-based rigid foams: Characterization and modification. *Industrial Crops and Products*, **29** (2–3), 356–363.

171. Tondi, G., Zhao, W., Pizzi, A. *et al.* (2009) Tannin-based rigid foams: a survey of chemical and physical properties. *Bioresource Technology*, **100** (21), 5162–5169.

172. Tondi, G., Fierro, V., Pizzi, A. and Celzard, A. (2009) Tannin-based carbon foams. *Carbon*, **47** (6), 1480–1492.

173. Celzard, A., Zhao, W., Pizzi, A. and Fierro, V. (2010) Mechanical properties of tannin-based rigid foams undergoing compression. *Materials Science and Engineering A—Structural Materials Properties Microstructure and Processing*, **527** (16–17), 4438–4446.

174. Li, X., Basso, M.C., Fierro, V. *et al.* (2012) chemical modification of tannin/furanic rigid foams by isocyanates and polyurethanes. *Maderas: Ciencia y Tecnologia*, **14** (3), 257–265.

175. Li X, Essawy HA, Pizzi A, *et al.*Modification of tannin based rigid foams using oligomers of a hyperbranched poly(amine-ester). *Journal of Polymer Research*. 2012;**19**(12), doi: 10.1007/s10965-012-0021-4.

176. Basso, M.C., Giovando, S., Pizzi, A. *et al.* (2013) Tannin/furanic foams without blowing agents and formaldehyde. *Industrial Crops and Products*, **49**, 17–22.

177. Lacoste, C., Basso, M.C., Pizzi, A. *et al.* (2013) Pine tannin-based rigid foams: mechanical and thermal properties. *Industrial Crops and Products*, **43**, 245–250.

178. Tondi, G., Pizzi, A., Delmotte, L. *et al.* (2010) Chemical activation of tannin-furanic carbon foams. *Industrial Crops and Products*, **31** (2), 327–334.

179. Tondi, G., OO, C.W., Pizzi, A. *et al.* (2009) Metal adsorption of tannin based rigid foams. *Industrial Crops and Products*, **29** (2–3), 336–340.

180. Abdullah, U.H.B. and Pizzi, A. (2013) Tannin-furfuryl alcohol wood panel adhesives without formaldehyde. *European Journal of Wood and Wood Products*, **71** (1), 131–132.

181. Li, X.J., Nicollin, A., Pizzi, A. *et al.* (2013) Natural tannin-furanic thermosetting moulding plastics. *RSC Advances*, **3** (39), 17732–17740.

182. Carothers, W.H., Dorough, G.L. and Natta, F.J.V. (1932) Studies of polymerization and ring formation. X. The reversible polymerization of six-membered cyclic esters. *Journal of the American Chemical Society*, **54** (2), 761–772.

183. NatureWorks LLC (2013) Ingenious materials made from plants not oil. Available at www.natureworksllc.com (accessed 28 August 2014).

184. European Bioplastics (2013) Driving the evolution of bioplastics. Available at en.european-bioplastics.org (accessed 28 August 2014).

185. Corbion Purac (2014) Shaping the future of biobased products. Available at www.purac.com/EN/Bioplastics/Home.aspx (accessed 28 August 2014).

186. Mitsui Chemicals (2013) PLGA. Available at www.mitsuichem.com/service/functional_chemicals/healthcare/plga/index.htm (accessed 28 August 2014).

187. Lasprilla, A.J.R., Martinez, G.A.R., Lunelli, B.H. *et al.* (2012) Poly-lactic acid synthesis for application in biomedical devices: a review. *Biotechnology Advances*, **30** (1), 321–328.

188. Liu, H.Z. and Zhang, J.W. (2011) Research progress in toughening modification of poly(lactic acid). *Journal of Polymer Science Part B—Polymer Physics*, **49** (15), 1051–1083.

189. Nampoothiri, K.M., Nair, N.R. and John, R.P. (2010) An overview of the recent developments in polylactide (PLA) research. *Bioresource Technology*, **101** (22), 8493–8501.

190. Averous L. (2008) Polylactic acid: synthesis, properties and applications. In *Monomers, Polymers and Composites from Renewable Resources* (eds Belgacem MN and Gandini A). Elsevier Ltd, London, 433–450.

191. Mehta, R., Kumar, V., Bhunia, H. and Upadhyay, S.N. (2005) Synthesis of poly(lactic acid): a review. *Journal of Macromolecular Science—Polymer Reviews*, **C45** (4), 325–349.

192. Auras, R., Harte, B. and Selke, S. (2004) An overview of polylactides as packaging materials. *Macromolecular Bioscience*, **4** (9), 835–864.

193. Duda, A. and Penczek, S. (2003) Polylactide poly(lactic acid): synthesis, properties and applications. *Polimery*, **48** (1), 16–27.

194. Sodergard, A. and Stolt, M. (2002) Properties of lactic acid based polymers and their correlation with composition. *Progress in Polymer Science*, **27** (6), 1123–1163.

195. Garlotta, D. (2001) A literature review of poly(lactic acid). *Journal of Polymers and the Environment*, **9** (2), 63–84.

196. Koivistoinen, O.M., Kuivanen, J., Barth, D. *et al.* (2013) Glycolic acid production in the engineered yeasts Saccharomyces cerevisiae and Kluyveromyces lactis. *Microbial Cell Factories*, **12**.

197. Lehermeier, H.J., Dorgan, J.R. and Way, J.D. (2001) Gas permeation properties of poly(lactic acid). *Journal of Membrane Science*, **190** (2), 243–251.

198. Martin, O. and Averous, L. (2001) Poly(lactic acid): plasticization and properties of biodegradable multiphase systems. *Polymer*, **42** (14), 6209–6219.

199. Edward, E.S., Rocco, A.P., inventors; American Cyanamid Co, assignee (1969) Polyglycolic acid prosthetic devices. US Patent 3,463,158.

200. Schneider, A.K., inventor; Ethicon Inc, assignee (1972) Polylactide sutures. US Patent 3,636,956 A.

201. Versfelt, C., Wasserman, D., inventors; Ethicon Inc, assignee (1974) Use of stannous octoate catalyst in the manufacture of l(–)lactide-glycolide copolymer sutures. US Patent 3,839,297 A.

202. Albertsson, A.C. and Varma, I.K. (2003) Recent developments in ring opening polymerization of lactones for biomedical applications. *Biomacromolecules*, **4** (6), 1466–1486.

203. Ramakrishna, S., Mayer, J., Wintermantel, E. and Leong, K.W. (2001) Biomedical applications of polymer–composite materials: a review. *Composites Science and Technology*, **61** (9), 1189–1224.

204. Sun, X. and Zhang, J.F. (2005) Poly(lactic acid) based bioplastics, in *Biodegradable Polymers for Industrial Applications* (ed R. Smith), CRC Press, Woodhead Publishing Limited, Cambridge, UK, pp. 251–288.

# 6

# Bio-Based Materials

## Antoine Rouilly[1] and Carlos Vaca-Garcia[1,2]

*[1]National Polytechnic Institute of Toulouse, France*
*[2]King Abdulaziz University, Center of Excellence for Advanced Materials Research,*
*Saudi Arabia*

## 6.1 Introduction

Man-made materials can be obtained from one of these three resources: minerals (silica, etc.), fossil resources (crude oil, natural gas, etc.) and vegetal resources (cotton, wood, etc.). The materials obtained from each of these are usually exclusive to their resource (e.g. glass can only be made from mineral resources and cellophane can only be obtained starting with vegetal resources). Sometimes, the same material can be obtained from fossil or vegetal resources if more than one transformation step is performed. For instance, polyethylene (PE) is the product of the polymerisation of ethylene. This gas is obtained in the petrochemical industry by steam cracking of light petrol molecules. It can also be obtained from the gasification of cellulose followed by the Fischer–Tropsch process, as described in Chapter 3. Consequently, the choice of both the source and the process depends on the availability and price of the resource and the technical feasibility and economics of the process.

A second approach to better understanding the development of new materials is the following. If a present material becomes scarce or too expensive, another material, more abundant and/or cheaper, can be used provided it fulfils the basic

*Introduction to Chemicals from Biomass*, Second Edition. Edited by James Clark
and Fabien Deswarte.
© 2015 John Wiley & Sons, Ltd. Published 2015 by John Wiley & Sons, Ltd.

requirements of performance. For instance, metal parts in cars were replaced by reinforced plastic during the period of a steel crisis because of their lower price and lower weight.

A third approach to be considered is the ecological and sustainable aspects of a material. For instance, the fossil resources and energy used for the production of plastics led to undesirable carbon dioxide emissions to the atmosphere increasing global warming. Moreover, the long persistence of some plastic materials in the environment causes visual pollution and suffocation of animals if their post-use disposal is not adequate.

The choice of a material and its source has therefore become a new challenge in our modern world. There is no reliable warranty on the supply of raw materials in our planet and we grow with more and more ecological consciousness. A replacement strategy has already begun to be implemented and this includes bio-materials. In this chapter we define biomaterials as those materials obtained from biomass (also called bio-based materials). A particular emphasis will be given to agricultural biomass.

Plant biomass sources can grow fast. Plants naturally convert the carbon dioxide from atmosphere into polymers (the so-called biopolymers) and other compounds such as sugars and lipids, easily convertible into materials. Biomaterials are there-fore the best choice to regulate the carbon cycle in the lithosphere provided that their life cycle analysis is positive, which is not a general rule.

## 6.2    Wood and Natural Fibres

Nature has always given to mankind a large palette of biomaterials from plants. The fact that they are natural does not mean that their performances are poor, as might be wrongly assumed. For instance, wood combines excellent elasticity, insulating and toughness properties. Linen provides a fresh and elegant textile in summer. Stubbles in thatched roofs are still used in rural houses or even in ancient Japanese temples; its strength and insulating properties account for this preference.

The common element in the cited examples is the mechanical resistance and the insulating properties. These advantages come not only from the macroscopic con-figuration of these materials (the hollow cylindrical structure of stubble, for instance) but mainly from their microscopic structure. Most vegetal fibres can be described through two models: wood fibres and cotton fibres. In order to better understand the mechanical properties of these fibres, let us first consider their molecular constitu-tion (Section 6.2.1) then their hierarchical structure (Section 6.2.2).

### 6.2.1    Molecular Constitution

All natural fibres essentially comprise three groups of components: polysaccha-rides, including cellulose, hemicelluloses (xylan, etc.), pectin, and so on; lignin; and water- and solvent-soluble compounds (waxes, minerals, etc.).

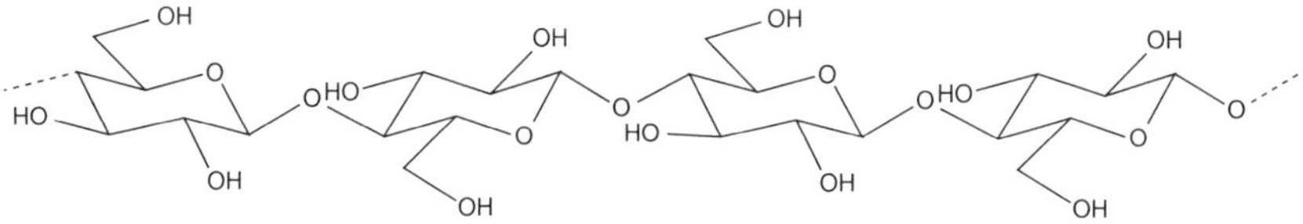

**Figure 6.1**    *Molecule of cellulose.*

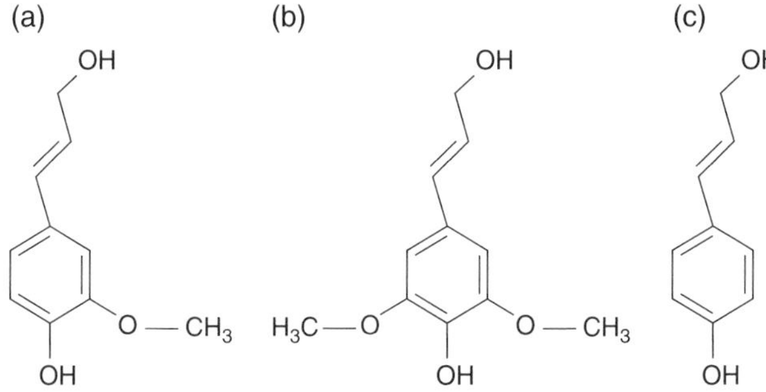

**Figure 6.2**    *Molecules from which the characteristic units of lignin are obtained: (a) trans-coniferyl alcohol; (b) trans-sinapyl alcohol; and (c) trans-p-cumaryl alcohol.*

*Cellulose* is by far the most abundant component, especially in cotton fibres. This biopolymer comprises cellobiose (repetitive unit), which is composed of two glucose building blocks in inversed position with respect to the plane of the cycle (Figure 6.1).

Such a configuration results from a β-1→4 glucosidic bond. Schematically speaking, cellulose is a 'flat' molecule with its hydroxyl groups pointing out of the 'ribbon'. The accessibility of OH groups favours the formation of intermolecular hydrogen bonds and the creation of crystalline zones. A cellulose molecule from a cotton fibre contains more than 7,000 glucose units. After extraction, the degree of polymerisation is reduced to about 1,000 units.

*Hemicelluloses* comprise different hexoses and pentoses (glucose, mannose, xylose, etc.). Since these heteropolysaccharides are often branched polymers, they cannot constitute crystalline structures. However, their function in the constitution of natural fibres is crucial. Together with lignin, they comprise the bonding matrix of the cellulose microfibres.

*Lignin* is an amorphous non-polar macromolecule, comprising phenyl-propane units. The structure of lignin depends on the source. Moreover, the extraction method modifies the structure of the lignin prior to analysis. Recent studies [1, 2] have shown fundamental differences with previously reported structures of lignin.

Lignin typically comprises three basic units (see Figure 6.2):

• the guaiacyl unit, derived from *trans*-coniferyl alcohol, which is abundantly present in softwoods;

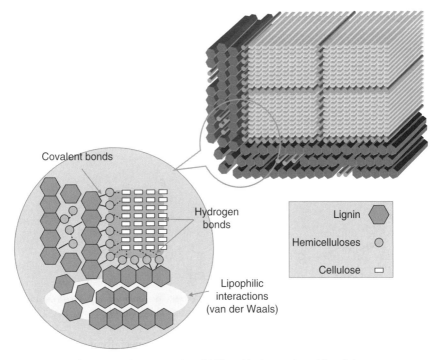

**Figure 6.3**   *Representation of LCC and its interaction with cellulose.*

- the syringyl unit, derived from *trans*-sinapyl alcohol, which is present with the guaiacyl unit in hardwoods; and
- the *p*-hydroxyphenyl unit, derived from the *trans-p*-cumaryl alcohol, which is characteristic of grasses.

Lignin is linked through covalent bonds (ester and ether) to hemicelluloses. The two macromolecules constitute the lignin-carbohydrate complex known as LCC. As the hemicelluloses can be linked to cellulose through hydrogen bonding, the LCC is capable of agglomerating around the cellulose microfibrils (Figure 6.3).

### 6.2.2   Hierarchical Structure of Wood and Timber

Employed for thousands of years as a structural material or a source of energy, wood is still widely used as a construction material because of its mechanical properties and its visual aspect. The diversity of wood species throughout the world and the elevated world production levels make it an abundant material available in almost all countries globally.

Wood is defined as the ligneous and compact material forming the branch, the trunk and the roots of trees. Timber is the term dedicated to sawn wood, usually

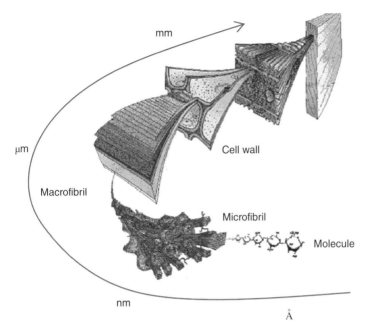

**Figure 6.4**   *Hierarchical structure of wood: from timber to cellulose.*

the trunk of the tree, used for construction purposes due to its strength. World production of sawlogs and veneer logs equates to 635 million m³. In Europe, Sweden, Finland and Germany account for practically half of the EU-28 production, which reaches 154.4 million m³ (FAOSTAT, http://faostat.fao.org/).

From the nano- to the macroscopic scale, wood is structured in a hierarchical way. The molecules of cellulose (40–50%) are arranged first in microfibrils then in macrofibrils, and they comprise the layers of the wall of every cell in the wood (Figure 6.4).

The cell wall (Figure 6.5) is composed of a primary and a secondary wall surrounding a void, the lumen (L). The primary wall (0.1–0.2 μm) presents no particular arrangement and comprises cellulose, hemicelluloses, pectin, proteins and lignin. The secondary wall consists of three layers, namely S1, S2 and S3, all of which are oriented layers of cellulose. In the S1 layer (0.2–0.3 μm) and in the S3 layer (0.1 μm), the microfibrils make an angle of 50–70° to the axis of the fibre. In the S2 layer (1–5 μm), which comprises 90% of the weight of the cell wall, the angle of microfibrils is between 10 and 30°. The shorter this angle, the higher the modulus and the strength of the fibre (and the lower the elongation at break).

Two neighbouring cells are separated by the middle lamella (M), which is composed of polysaccharides and lignin as depicted in Figure 6.3. Its thickness is between 0.2 and 1 μm.

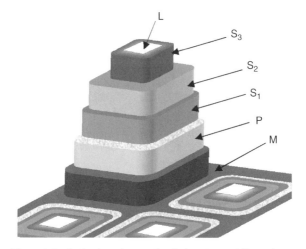

**Figure 6.5**   *Projection of a wood cell, showing its different layers.*

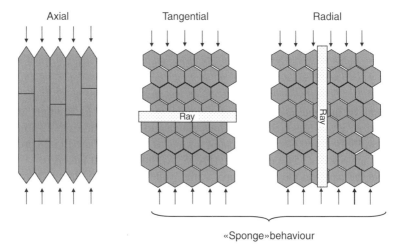

**Figure 6.6**   *Representation of the three different compression solicitations in wood, according to the position of the cells and the ray.*

At a mesoscopic level, the hollow cylindrical-shaped cells are orientated in the same direction of the trunk. Radial sheets of horizontal cells rich in lignin run from the centre of the trunk to the bark. They can be easily observed in a transversal cut of the trunk because of their darker colour. The orientation of both the cells and the rays causes anisotropy in wood. The mechanical properties of wood depend on the direction of the mechanical solicitation (Figure 6.6).

Let us consider a wooden pencil: it is quite easy to crush it with our teeth in a horizontal position, whereas it is very difficult to do the same if we place the

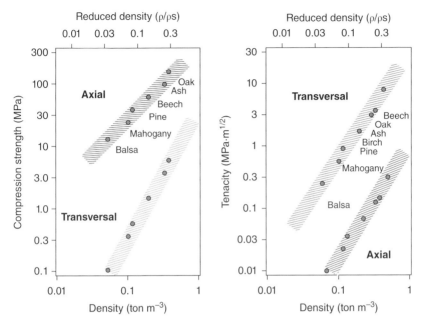

**Figure 6.7** *Mechanical properties of various wood species as a function of wood density.*

pencil in a vertical position (assuming a reasonably short pencil and good teeth!). A Canadian lumberjack always cuts the small wood rods with his axe in a vertical (i.e. axial) position. The explanation is found in Figure 6.7, where it can be seen that the pencil is between 30 to 100 times easier to crunch in a transversal position depending on the wood species. A similar analysis shows that the lumberjack requires about 30 times less effort to cut the rods in a vertical direction. Table 6.1 lists the main mechanical properties of different wood species.

The water content in wood influences its mechanical properties but also changes its dimensions. When completely dried wood is placed in a humid atmosphere, water uptake makes the cell walls swell in a linear relation to the water content of the wood. When the fibre reaches the saturation point (around 30% water, the exact amount dependent on the species), the water uptake can continue (especially if wood is in contact with liquid water) but water then starts to fill the lumen and the swelling of the cell wall stops. Swelling is different according to the section of the wood. Axial maximum swelling is limited to less than 1%. The transversal swelling is more significant (approximately 4% radial and 7% tangential). Swelling is a reverse phenomenon: humid wood shrinks during drying. For some applications, the dimensional instability of wood is a serious problem. Hydrophobic agents can be added, but chemical modification is also used. For instance, acylation of wood with aliphatic anhydrides can limit swelling by up to 75% (Figures 6.8 and 6.9) [4].

**Table 6.1** Mechanical properties of selected wood species with a water content of 12%. Sources: www.matweb.comand [3].

| | Wood species | Density (ton m$^{-3}$) | Fibre length (mm) | Strength at break (MPa) | | | Flexural modulus (GPa) |
| --- | --- | --- | --- | --- | --- | --- | --- |
| | | | | Compression | Tensile | Flexural | |
| Softwoods | Spruce, fir, red cedar | 0.3–0.9 | 3–7 | 35–45 axial<br>6–8 ⊥ | 90–100 axial<br>1.2 ⊥ | 50–70 | 8–10 |
| | Pines | 0.3–0.85 | 4 | 40–50 axial<br>7.5–8 ⊥ | 100–120 axial<br>1.8 ⊥ | 80–90 | 9–14 |
| Hardwoods | Poplar, mahogany, okoume | 0.3–0.5 | 1.1 | 30–40 axial<br>7.5–10 ⊥ | 80–100 axial<br>2 ⊥ | 65–85 | 9–11 |
| | Chestnut, soft oak, maple, ash, beech | 0.3–0.95 | 1.1–1.2 | 40–60 axial<br>12–15 ⊥ | 100–120 axial<br>3 ⊥ | 75–130 | 9–12.5 |
| | Hard oak, teck | 0.3–0.8 | 0.7–1.2 | 50–80 axial<br>18–20 ⊥ | 120–150 axial<br>4 ⊥ | 100–170 | 11.5–15 |
| | Azobe | 1.1 | | 90–100 axial<br>>20 ⊥ | 150–200 axial<br>5 ⊥ | 227 | |

**Figure 6.8**    *Wood acetylation reaction.*

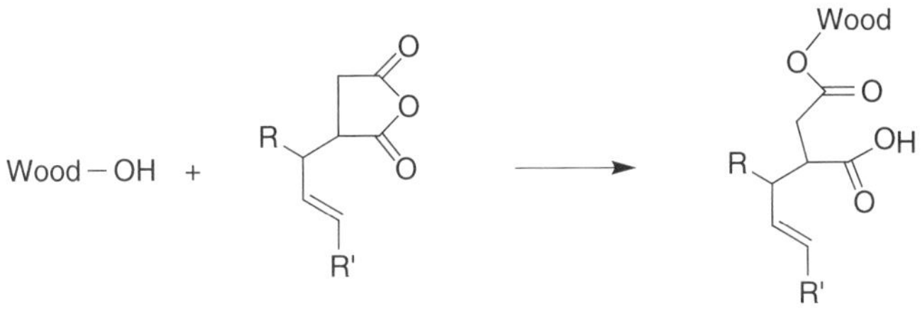

**Figure 6.9**    *Reaction of wood with alkenyl succinic anhydrides (as in Surfasam treatment [6]).*

Despite all the advantages of wood, its use as a material is affected by its durability. Softwoods have a limited durability and hardwoods are in general more resistant than softwoods; they both contain excellent nutrients for fungi and xylophagous insects (termites, longhorn beetle larvae, etc.) however. Consequently, the impregnation of wood with antifungal and insecticide compounds is a current practice. Among these, soluble salts of copper, chromium and arsenic (CCA) are the most used. However, human health concerns have forced wood treatment companies to eliminate the most toxic compounds on the market in certain countries. Products such as copper, chromium and boron (CCB) and copper and chromium without boron or arsenic (CCO) have therefore been recently introduced. Due to the Biocides 98/8/CE directive in Europe, wood treatment chemicals are now regulated to ensure that they do not present a significant danger to the environment, humans and animals.

In response to these ecological concerns, new industrial methods of preserving wood have been developed from intensive laboratory and pilot plant research, including the following.

- *Acetylation* (e.g. Accoya, http://www.accoya.com/acetylated-wood/): Wood is treated with acetic anhydride. Residual acetic acid from the esterification reaction is difficult to eliminate however, and a vinegary odour may remain.
- *Thermal treatment* (e.g. Retiwood, www.retiwood.com; WTT, www.wtt.dk): Wood undergoes partial pyrolysis at high temperatures in the absence of oxygen or in the presence of steam or reducing compounds. Readily biodegradable biomolecules are destroyed in the process.
- *Oleothermal treatment* (e.g. Oléobois [5]): Wood is 'fried' in hot vegetable oils, which create a barrier to aggressors, especially by the cross-linking of triglycerides. The use of siccative oils (linseed, sunflower) increases the efficacy of the treatment.

- *Oleochemical modification* (e.g. Surfasam [6], WoodProtect® [7]): Wood is reacted with fatty compounds such as alkenyl succinic anhydrides or fatty-acetic anhydrides. The alkenyl or fatty base comes from vegetable oils (fatty acid or fatty acid esters).

### 6.2.3 Plant Fibres

Plant fibres show variable properties according to their species, their age and their growing spot. They can be classified into categories according to their position in the plant (Figure 6.10).

By far, the most-produced fibre in the world is cotton. Its production is almost 30 times that of jute (Figure 6.11). The extensive industrial textile applications of cotton account for a large proportion of its use. As such, a few other fibres play a more important part than cotton in biomaterials production. It is interesting to note that most natural fibres (other than wood) used commercially come from emerging countries, especially from Asia.

The fibre characteristics depend on the source. Table 6.2 lists the main characteristics commercially-available vegetal fibres. A short description of each type of fibre and its application are given in the following sections.

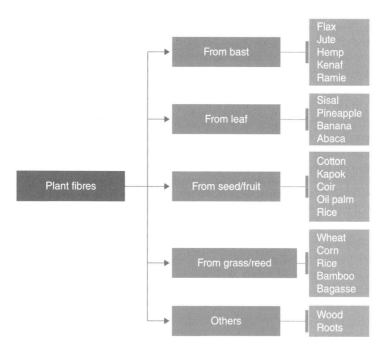

**Figure 6.10** *Plant fibres classified according to their position in the plant.*

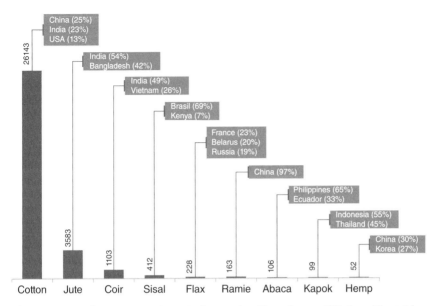

**Figure 6.11**   *Production (in metric tons) of some plant fibres. Source: FAO (http://faostat.fao. org/).*

**Table 6.2**   *Characteristics of the most common plant fibres*

|  | Linen | Ramie | Hemp | Jute | Sisal | Coconut | Cotton |
|---|---|---|---|---|---|---|---|
| Density (g cm⁻³) | 1.54 | 1.56 | 1.07 | 1.44 | 1.45 | 1.15 | 1.5–1.6 |
| Cellulose content (%) | 64–71 | 83 | 78 | 61–71 | 67–78 | 43 | 87–99 |
| Microfibril angle (°) | 10 | 7.5 | 6.2 | 8 | 20 | 45 |  |
| Diameter (µm) | 5–76 | 16–126 | 10–51 | 25–200 | 7–47 | 12–24 | 10–20 |
| Length (mm) | 4–77 | 40–250 | 5–55 |  | 0.8–8 | 0.3–1 | 10–55 |
| Shape factor (L/D) | 1700 | 3500 | 960 | 110 | 100 | 35 | 2000 |
| Modulus of elasticity (GPa) | 12–85 | 60–130 | 35 | 25–30 | 9–21 | 4–6 | 5–13 |
| Elongation at break (%) | 1–4 | 1.2–3.8 | 1.6 | 1.5–1.8 | 3–7 | 15–40 | 7–8 |
| Strength at break (MPa) | 600–2000 | 400–1000 | 390 | 390–770 | 350–700 | 130–175 | 290–600 |

### 6.2.3.1   *Cotton*

Cotton is a soft fibre that grows around the seeds of the cotton plant (*Gossypium* spp.) and practically all of the commercial cotton grown today worldwide comes from the American species *Gossypium hirsutum* and *Gossypium barbadense*.

The fibre is most often spun into thread and used to make textiles, making cotton the most widely used natural-fibre fabric in clothing today.

The structure of cotton fibres is similar to that of the secondary walls (S1, S2 and S3) of a wood cell. Cotton comprises 98% cellulose with almost no lignin (no middle lamella gathering all the fibres together). The natural wax on the surface of the fibres is very useful for threading as it lubricates their contact with the machines, minimising fibre damage. The long fibres are used for textile applications while the shorter fibres (linters) of cotton are used as a source of high-quality cellulose for biomaterials. Cotton is used for the chemical transformation of cellulose (regeneration, esters and ethers, see Section 6.3.1) for which high purity is required.

### 6.2.3.2    Linen or Flax Fibre

As used today, the word 'linen' is descriptive of a class of woven textiles used in homes. Linens were manufactured almost exclusively using fibres from the flax plant *Linum usitatisimum*. Today flax is a prestigious, expensive fibre and only produced in small quantities. Flax fibres can be identified by their typical nodes, which account for the flexibility and texture of the fabric. The cross-section of the fibre is made up of irregular polygonal shapes, which contribute to the coarse texture of the fabric. When adequately prepared, linen has the ability to absorb and lose water rapidly. It can gain up to 20% moisture without feeling damp.

Linen uses range from bed and bath fabrics, home and commercial furnishing items (wallpaper/wall coverings, upholstery, support for oil painting, etc), prestigious clothing, to industrial products (luggage, canvases, sewing thread, etc). Linen is preferred to cotton for its strength, durability and archival integrity.

### 6.2.3.3    Jute

Jute is a long, soft, shiny vegetal fibre that can be spun into coarse, strong threads. It is produced from plants in the genus *Corchorus*. Jute is one of the cheapest natural fibres, and is second only to cotton in amount produced and variety of uses. It can be grown in 4–6 months. The mechanical properties of jute hurd (inner woody core or parenchyma of the jute stem) are a cross between those of textile fibre and wood; it therefore has good potential to fight deforestation. Jute is the second most important vegetal fibre after cotton not only for cultivation but also for various other uses; for example, jute is used to make coarse cloth for wrapping bales and sacks. The fibres are also woven into curtains, chair coverings, carpets, area rugs and backing for linoleum.

While synthetic materials are replacing jute in many of these uses, synthetics would be unsuitable for some applications which depend upon the biodegradable nature of jute. Examples of such uses include: containers for planting young trees, which can be planted directly with the container without disturbing the roots; geotextiles, which is a lightly woven fabric made from natural fibres that is used for soil erosion control; seed protection; weed control; and many other agricultural and landscaping uses. Biodegradable jute geotextiles can be left to rot in the ground, keeping the ground cool and making the land more fertile.

### 6.2.3.4 Hemp

Hemp is the common name for *Cannabis sativa*, cultivated for industrial (non-drug) use. Licenses for hemp cultivation are issued in the European Union and Canada. Hemp grows quickly and produces strong fibres. In the past hemp was widely used for canvas (note the similar etymology) and other articles such as carpets and rope. The ultimate fibre is flatter, less regular and more lignified than linen. In Europe, the major application of hemp is in plastic-natural-fibre composites as filler, mainly for the automotive industry. The microscopic protuberances on the surface of the fibre represent the advantage of a stronger mechanical anchorage with the plastic matrix. Another recent application is the fabrication of insulating mats for housing.

### 6.2.3.5 Ramie

Ramie (*Boehmeria nivea*) is one of the oldest fibre crops, principally used for fabric production (including mummy cloths!) with antifungal and antibacterial properties. It is a bast fibre, and the part used is the bark (phloem) of the vegetative stalks. Unlike other bast crops, ramie requires chemical processing to de-gum the fibre (up to 25% mass loss).

Ramie is not as durable as other fibres, so is usually used as a blend with other fibres such as cotton or wool. It is known especially for its ability to hold shape, reducing wrinkling and shrinking, and introduces a silky lustre to the fabric appearance. It does not dye as well as cotton however, but its white colour is useful in textile industry. It is similar to flax in absorbency, density and microscopic appearance. Because of its high molecular crystallinity, ramie is strong but stiff and brittle and will break if folded repeatedly in the same place; it lacks resiliency and has a low elasticity and elongation potential. When wet, it exhibits greater strength. Spinning the fibre is difficult due to its brittle quality and low elasticity, and weaving is complicated by the hairy surface of the yarn, resulting from lack of cohesion between the fibres. A greater utilisation of ramie would depend upon the development of improved processing methods. Ramie is currently used to make products such as industrial sewing thread, packing materials, fishing nets and filter cloths. Shorter fibres and waste are used in paper manufacture.

### 6.2.3.6 Sisal

Sisal is an agave (*Agave sisalana*) that yields a stiff fibre used in making rope. Sisals are sterile hybrids of uncertain origin. Although originating from the port of Sisal in Yucatan, Mexico, they do not actually grow in Yucatan, which presently cultivates henequen (*Agave fourcroydes*). Sisal plants consist of a rosette of sword-shaped leaves about 1.5–2 m tall. The sisal plant has a 7–10 year lifespan and typically produces 200–250 commercially usable leaves. Each leaf contains around 1,000 packs of fibres. The fibre element, which accounts for only about 4% of the plant by weight, is extracted by decortication. Sisal is valued for cordage use because of its strength, durability, ability to stretch, affinity for certain dye-stuffs and resistance to deterioration in saltwater. Sisal is used by industry in three grades: lower-grade fibre is processed by the paper industry because of its high cellulose and hemicellulose content; medium-grade fibre is used in the cordage industry for making ropes; and, after treatment, higher-grade fibre is converted into yarns and used by the carpet industry.

Sisal products are undergoing rapid development, such as furniture and wall tiles made of resonated sisal. A recent development even expanded the range to interior car parts. Other products developed from sisal fibre include spa products, cat scratching posts, lumbar support belts, rugs, slippers and cloths. In recent years sisal has been utilised as a strengthening agent to replace asbestos and glass fibre as well as an environmentally friendly component in the automobile industry.

### 6.2.3.7    Coconut Fibres

The coconut palm (*Cocos nucifera*) is grown throughout the tropical world, for decoration as well as for its many culinary and non-culinary uses; virtually every part of the coconut palm has some human use. Coir (the fibre from the husk of the coconut) is used in ropes, rugs and mats, brushes and as stuffing fibre. It is also used extensively in horticulture for making potting compost.

### 6.2.3.8    Miscanthus

*Miscanthus giganteus* is a large, perennial grass imported from Asia. A perennial plant of the Poaceae family (also known as Gramineae or true grasses), Miscanthus is an interesting candidate for the production of biomass as its production poten-tial is particularly high for a very low amount of agricultural inputs. It can grow to heights of more than 3.5 m in 1 year and its production [8] can reach $30\,t\,ha^{-1}\,a^{-1}$.

While its fibres can be used for materials applications or as garden mulch, its high productivity makes Miscanthus the perfect energy crop. It is currently used in the European Union as a source of heat and electricity, especially in co-firing with coal as its mineral content is low, or is converted into biofuels such as ethanol.

### 6.2.3.9    Bamboo

From the same grass Poaceae family as Miscanthus, there are more than 1,450 species of bamboo across the world. Growing in non-continental climate condi-tions, bamboo is one of the fastest-growing plants on Earth, with reported growth rates of 100 cm in 24 hr.

A particular feature of bamboo is that its stem (culm) grows from the ground at its full diameter and to its full height in a single growing season of 3–4 months. Each new shoot grows vertically up to its mature height. Branches extend from the nodes and leaves then appear. During the following 2 years, the outer wall of the stem slowly hardens. Bamboo culms are considered mature after 3 years of growth.

While the shoots (new culms that come out of the ground) of some species of bamboo are edible (if properly prepared; it contains a toxin that has to be degraded before consumption), bamboo is mostly utilised for its natural resistance and fibres. Due to its high strength-to-weight ratio in its natural form, bamboo is traditionally used as construction material for houses or scaffolding; it also has applications in musical instruments, fishing rods, papermaking and textiles. For the latter application, because of the nodes fibres are too short to be processed directly and most bamboo textiles are made from viscose (see Section 6.3.1.2) obtained from bamboo cellulose; the fibres are dissolved in chemicals and then spun into new and highly resistant fibres.

## 6.3   Isolated and Modified Biopolymers as Biomaterials

Plants are wonderful chemical reactors that are able to make extremely complex macromolecules. These compounds are located in the cell wall (e.g. cellulose, lignin, hemicelluloses and pectin), constitute the plant's energy reserves (e.g. starch) or even have specific functions (e.g. proteins). Most of these biopolymers are useful to make industrial biomaterials and a significant amount of research has been and continues to be conducted on them [9].

In order to develop valuable applications, extraction methods for these biopolymers are necessary. The isolation of starch is one of the most simple, as it requires relatively straightforward physical methods for extraction and purification (wet milling being the most common). In contrast, the cell wall components require chemical treatment to break the covalent bonds of the LCC to liberate the cellulose fibrils. This can either be done with sulphuric acid containing solutions of hydrogen sulphites or with solutions of sodium hydroxide and sodium sulphate. These hot treatments reduce biomass chips to fibres by using a mild mechanical action. To further purify the raw pulp, a multi-stage refining procedure is needed in which alkali and oxidising agents are used to remove residual lignin. Finally, extractions with cold or hot alkali are used to remove pentanes and oligosaccharides. Pure cellulose (99%) can finally be obtained.

The characteristics of the isolated biopolymers depend on their structure. Cellulose and amylose are linear polymers whereas amylopectin, pectin and hemicelluloses are branched polymers. Pectin and amylopectin contain carboxylic groups that can significantly interact with water molecules. Amylose has a helix structure whereas cellulose molecule looks like a ribbon. The interactions with water and other neighbouring molecules are therefore different.

One of the most important aspects of these biopolymers is the fact that they cannot be used directly for thermoplastic applications. They are not reticulated polymers but, despite their linear or slightly branched structure, they have two disadvantages. First, their structure contains relatively fragile bonds or fragile configurations that break with high temperatures. This is the case for the glycosidic bond or the C–O–C bond in the saccharide cycle of the polysaccharides. The tertiary structure of proteins is also altered with heat. The second disadvantage is that they form numerous inter-molecular hydrogen bonds, resulting in some cases in very rigid structures. A lot of energy is necessary to break these bonds and make the macromolecules flow; the polysaccharides would decompose under the action of this energy before the poly-mer could attain their molten state. This is particularly true in the case of cellulose, and is the reason why paper burns and does not melt at high temperatures. In the case of starch, thermoplasticity occurs only with the help of an external plasticiser (water, glycerol, etc.) that disrupts the hydrogen bonding of the biopolymer.

In the following sections, we describe the technologies needed to overcome these and other disadvantages. Following treatment, natural polymer can be turned into profitable materials for industry. We have limited ourselves in this chapter to the main biopolymers currently used in commercial products.

### 6.3.1   Cellulose

The ribbon-like structure of the cellulose molecule (Figure 6.1) favours its organi-sation in oriented packs of about 50–100 molecules. If the organisation is com-pletely regular, crystallites are formed. In nature the crystallinity rate approximates 50% in most species (wood, cotton, etc.). The same molecule therefore has both crystalline and amorphous regions. The fibrous structure of cellulose is main-tained even after the chemical processes of pulping from wood. Indeed, pulping attacks and dissolves the LCC, that is, the middle lamella. The cell walls, which essentially comprise cellulose fibres, are therefore recovered without severe chemical degradation. Such a fibrous structure is useful for common bulk materi-als such as paper or absorbing cushions (the so-called non-wovens). Pure cellu-loses are also used in high-added-value applications such as hydrogels, stationary phases for chromatography or pharmaceutical formulations.

The potential of cellulose increases greatly after chemical modification. The esterification or the etherification of its hydroxyl groups leads to new biopolymers with very different properties, described in the following section.

We first consider a particular case of the modification of the cellulose molecule: regeneration. Two types of regeneration can be distinguished: with and without chemical modification.

#### *6.3.1.1   Regeneration of Cellulose with Chemical Modification*

During viscose processes, a reaction occurs between cellulose and $CS_2$ in an alka-line solution to form a viscous solution of cellulose xanthate. The resulting solu-tion is filtered to eliminate solid particles, before being extruded to form continuous

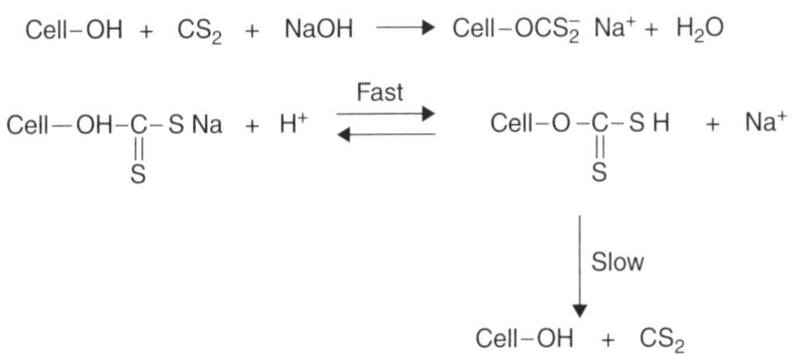

Cell–OH + CS$_2$ + NaOH $\longrightarrow$ Cell–OCS$_2^-$ Na$^+$ + H$_2$O

$$\text{Cell}-\text{OH}-\underset{\underset{S}{\|}}{C}-\text{S Na} + H^+ \;\underset{\xleftarrow{\hspace{2cm}}}{\overset{\text{Fast}}{\xrightarrow{\hspace{2cm}}}}\; \text{Cell}-\text{O}-\underset{\underset{S}{\|}}{C}-\text{S H} + Na^+$$

Slow

Cell–OH + CS$_2$

**Figure 6.12**   *Main reactions taking place in the viscose process for cellulose regeneration.*

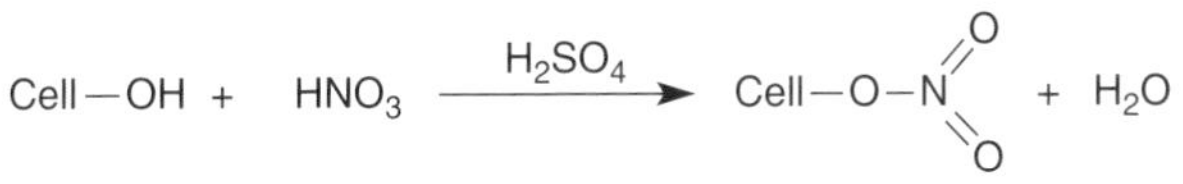

$$\text{Cell}-\text{OH} + HNO_3 \xrightarrow{\;H_2SO_4\;} \text{Cell}-\text{O}-N\underset{\diagdown O}{\overset{\diagup O}{}} + H_2O$$

**Figure 6.13**   *Synthesis of cellulose nitrate.*

sheets or threads or individual beads. On being plunged into a water solution containing sulphuric acid and other additives, cellulose xanthate is hydrolysed and converted into cellulose once more (Figure 6.12).

If the extrusion profile is a tiny cylinder, the obtained product is a continuous thread for textile applications. The industrial name of this product is Rayon®, or 'artificial silk', a brilliant fabric with good dying properties.

The viscose process has some variants. Depending on the quality of the cellulose and the composition of the regenerating bath, special high-added-value products can be obtained including the so-called *modal-polynosic* fibres, or the *modal-high wet modulus* fibres. If the extrusion profile is a sheet, the obtained product is Cellophane®. Again, some additives are added to the hydrolysing bath to obtain the transparency and plastic-like aspect of the sheet. Even though the tensile strength of this product is high, the shearing resistance is low which makes it an excellent film for packaging and conditioning of biscuits or candies.

The viscose process has been progressively abandoned (but not totally) due to environmental concerns as CS$_2$ is toxic and can easily cause an explosion. However, it remains historically important in the field of the chemistry of cellulose. The viscose process has set standards of variety, quality and cost that any new process must at least equal, or even surpass. If this is not possible, then it is only the safety and environmental restrictions which may cause the viscose process to be totally abandoned (Figure 6.13).

The cellulose carbamate process, which uses urea as modifying agent, represents a greener alternative to the xanthate process [10]. However, the process remains to be commercialised.

### 6.3.1.2    *Regeneration of Cellulose without Chemical Modification*

The processes and products described in this section use solvents to turn cellulose into a solution that can be filtered and extruded. No derivatisation of the cellulose hydroxyl groups occurs; precipitation therefore takes place instead of chemical regeneration.

Among the most ancient solvents for cellulose we find cuprammonium hydroxide and cupriethylenediamine, but only the former was used industrially to yield the cupro fibre. Later, other cellulose solvents were developed such as the *N*-methylmorpholine-*N*-oxide (NMMO) in water and the systems of lithium chloride with an aprotic polar solvent (dimethyl acetamide, dimethyl formamide and dimethyl sulfoxide). Only the former has been successful however, yielding the lyocell fibre. More recently, ionic liquids capable of dissolving cellulose have been developed. The most widely known is 1-*N*-butyl-3-methylimidazolium chloride (BMIMCl). However, no significant industrial development of yarn fabrication with this process has been reported so far. The reasons why some of these solvents have not found industrial outlets include poor mechanical properties of the fibre (too rigid or low elasticity) and difficulties in the control of all the process parameters and in the recycling of the solvent.

All the cellulose regeneration processes, with or without chemical modification, cause cellulose molecules to organise in a different crystalline form called *cellulose II* and sometimes *cellulose IV* (especially in high-wet-modulus- or HWM-modal fibres).

It is worth noting that the mercerisation process, introduced in the nineteenth century, also allows a *cellulose II* structure to be obtained, but without dissolution of the fibres and therefore with no reshaping. Cotton fibres are soaked in a concentrated (19%) NaOH solution then washed. Mercerised cotton is softer to touch and demonstrates more brilliance than natural cotton.

### 6.3.1.3    *Microcrystalline Cellulose*

Microcrystalline cellulose (MCC) is obtained by a controlled acid treatment intended to destroy the molecular bonding in the amorphous zones of cellulose. Typically, HCl or $H_2SO_4$ are used at 110°C for 15 min over native cellulose or regenerated cellulose. Colloidal gels are therefore obtained showing thixotropy. MCC is used in the preparation of pharmaceutical compressed tablets due to its binding and disintegration properties.

In the ice-cream industry, MCC is used to avoid the formation of large ice crystals, increasing smoothness. MCC is also used as a rheological modifier in water-based paintings and toothpaste. In chemistry laboratories, MCC is also used as a support for chromatography (column, thin layer, etc.).

### 6.3.1.4  Nanocellulose

The hierarchical structure of cellulose allows the extraction of nanoparticles by mechanical and chemical and/or enzymatic dissolution of its amorphous part [11]. These nanoparticles are called cellulose nanocrystals or microfibrillated cellulose (MFC) and can be obtained from various natural resources [12]. Defibrillation of individual microfibrils is obtained through mechanical shearing actions followed by a strong acid hydrolysis treatment to longitudinally cut the fibrils and change the size and aspect ratio of the cellulose-based nanoparticles (Table 6.3).

These nanoparticles have many advantages, including good mechanical properties (Young's modulus around 100 GPa), abundance, low density, high surface area (several hundred squared metres per gram) and biodegradability, and are today used to produce bionanocomposites [13]. Their main drawback however, which is related to their natural origin, is their hydrophilic nature, limiting the use of these natural nanofillers in classical thermoplastic and/or thermosets resins. As a result, many research programmes are now focusing on the surface modification of these nanoparticles [14].

### 6.3.1.5  Bacterial Cellulose

Bacterial cellulose (BC) is an ex-cellular product of various bacteria, but only the *Gluconacetobacter* produce enough cellulose to justify a commercial interest. While it has a gel-like appearance with a solid content of <1%, it is almost pure cellulose and contains no lignin and other extra compounds such as hemicelluloses and pectins. Apart from its purity, the second main advantage of BC is the possibility to grow it to any desired shape and structure, offering remarkable physical properties similar to those of nanocellulose particles from plant cellulose. Production yields are still relatively low (around $10$–$40\,g\,L^{-1}$) depending on the bacteria and substrate used, keeping its price high [15]. However, BC is already being used commercially in diet foods, filtration membranes, paper additives and biomedical products and devices [16].

**Table 6.3**  *Nanocellulose dimensions [13].*

| Cellulosic structure | Diameter (nm) | Length (nm) | Aspect ratio (L/D) |
|---|---|---|---|
| Microfibril | 2–10 | >10000 | >1000 |
| Microfibrillated cellulose | 10–40 | >1000 | 100–150 |
| Cellulose whisker | 2–20 | 100–600 | 10–100 |
| MCC | >1000 | >1000 | ~1 |

### 6.3.2    Cellulose Derivatives

#### *6.3.2.1    Cellulose Inorganic Esters*

Cellulose nitrates are the most important inorganic esters of cellulose. Depending on the degree of substitution (DS), that is, the average number of hydroxyl groups modified in a unitary glucose unit, the cellulose nitrates go from a resistant but inflammable polymer for film and photographic applications (celluloid grade, DS = 1.9) through a polymer for lacquers (DS = 2.05–2.35) to a powerful explosive that burns spontaneously in air (gun cotton, DS = 2.7).

The industrial production of vegetal parchment usually sold as paper for the bakery industry involves the formation of another cellulose inorganic ester: cellulose sulphates. As part of this process, a continuous sheet of paper is immersed for a few seconds in a concentrated sulphuric (65–75%) acid bath maintained at low temperature. Sulphates are formed and acid hydrolysis of the cellulose fibres starts. The sheet is then passed through several rinsing baths to hydrolyse the sulphates and to eliminate the sulphuric acid (Figure 6.14). A continuous matrix of gelatinised cellulose is formed on both sides of the paper, which protects the inner mat of cellulose fibres. The latter ensures rigidity and mechanical resistance, whereas the continuous matrix shows extraordinary hydrophobic and lipophobic properties as well as high resistance to temperature. As a result, kitchen parchment is resistant to humidity and to greasy food and it is perfectly adapted for cooking.

#### *6.3.2.2    Cellulose Organic Esters*

The synthesis of cellulose organic esters can be accomplished in many ways. The acylation of the hydroxyl groups of cellulose require strong agents such as acid chlorides (Figure 6.15) or acid anhydrides. The former are preferred for long fatty chains but it is necessary to use a strong base such as pyridine to neutralise the formed HCl, which can cause extensive degradation of the biopolymer. Other systems without pyridine have been proposed to limit the degradation of cellulose

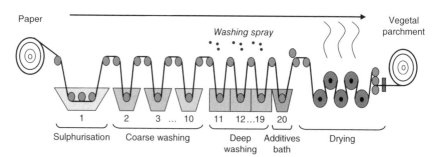

**Figure 6.14**    *Process for the production of vegetal parchment.*

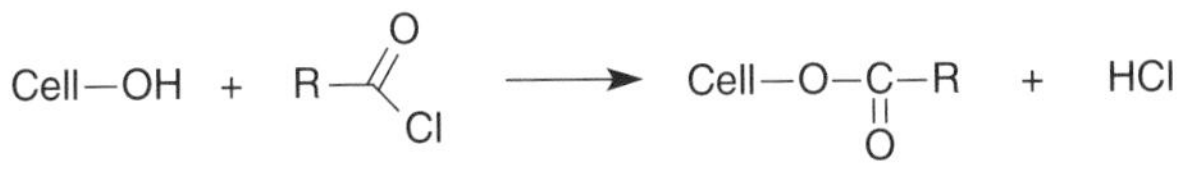

$$\text{Cell}-\text{OH} \;+\; \text{R}-\overset{\text{O}}{\underset{\text{Cl}}{\diagup\!\!\diagup}} \;\longrightarrow\; \text{Cell}-\text{O}-\overset{}{\underset{\text{O}}{\text{C}}}-\text{R} \;+\; \text{HCl}$$

***Figure 6.15*** *Synthesis of cellulose organic esters by reaction with fatty acid chlorides.*

and include the use of partial vacuum [17] or dry nitrogen gas flow [18] to take HCl out of the reactor.

Nevertheless, the use of fatty acids (>C6) is still possible but only with the use of a co-reagent that forms *in situ* new stronger entities. Among others, we can cite trifluoroacetic anhydride [19] and *N,N*-dicyclohexylcarbodiimide [20].

*Cellulose acetates* are by far the most important organic ester. Diacetate (DS = 2.4) is produced either in filament form for fibres or in powder for melt processing. Diacetate filaments are obtained by dissolution in acetone, extrusion through a spin and then evaporation of the solvent. The obtained fibres are used in textiles (called simply 'acetate') and in cigarette filters (tow). Triacetate (DS = 2.9) finds application in brilliant easy-to-dye textiles.

The industrial production of cellulose diacetate utilises acetic anhydride and sulphuric acid as catalyst. Reaction is conducted at a low temperature and cellulose starts to dissolve in the acetylation bath as the reaction progresses. The reaction is conducted until practically complete acetylation. The homogeneous solution obtained is then hydrolysed to reduce DS to a value lower than 2.4. Precipitation in diluted acetic acid, washing with water and final drying yields cellulose acetate flakes.

The dramatic reduction of hydrogen bonding renders cellulose acetates thermoplastic. However, their softening points are very close to their decomposition temperature (around 300°C). As a result, external plasticisers (e.g. phthalates) are often used to process cellulose diacetate in plastic applications (eyewear, screwdriver handles, etc).

*Cellulose mixed esters* (acetate-butyrate, acetate propionate) are used in the same plastic applications. The irregularity introduced by two different substituents diminishes the glass transition to around 120°C. Moreover, the combination of a short side-chain such as acetate and a long chain from a fatty acid (in a 2.4:1 ratio respectively) results in very interesting biomaterials; they are as hydrophobic as simple fatty esters but with better mechanical properties (Figure 6.16) and higher glass transition (Figure 6.17). Their biodegradability is also increased due to changes in crystallinity.

Cellulose esters of long-chain carboxylic acids (up to $C_{20}$) are interesting thermoplastic materials. An exhaustive description of all the synthesis methods has recently been reported [21]. These derivatives present softening points between 70 and 250°C depending on the DS and the number of carbon atoms in the acyl chain [22]. The higher the DS value, the lower the softening point. Nevertheless, the difficulty in obtaining high DS values without using toxic chemicals is a

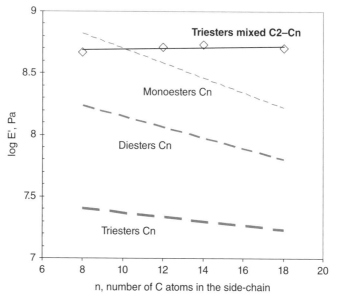

**Figure 6.16**    *Modulus of elasticity (E') of simple fatty esters and mixed acetic-fatty triesters of cellulose (DMA measurements).*

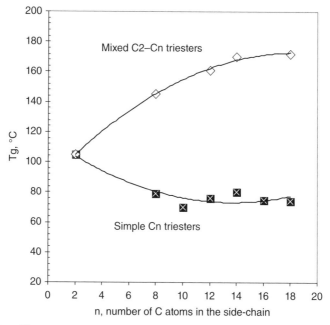

**Figure 6.17**    *Glass transition temperatures of simple fatty triesters and mixed acetic-fatty triesters of cellulose.*

**Table 6.4** *Main changes observed in cellulosic fibres after esterification of the hydroxyl groups with organic acids.*

|  | Cellulose fibres | Esterified cellulose fibres | Reference |
|---|---|---|---|
| Thermoplasticity | Do not melt, thermal degradation above 200°C | Show apparent melting point | [23] |
| Hydrophobicity | Hydrophilic; swell in water | Water resistance; dimensional stability | [24] |
| Biological resistance | Easily attacked by insects and fungi | Resistance to termites and fungi | [25] |
| Biodegradability | Readily biodegradable | Retarded biodegradation | [26] |
| Solubility | Insoluble in ordinary organic solvents | Soluble in various organic solvents |  |
| Inflammability | Burn easily | Do not burn readily | [25] |
| Weathering resistance | Fast degradation due to water absorption (rotting and biological attack) | Increased UV and water resistance | [25] |

limiting point for the commercialisation of long-chain esters of cellulose in the plastic industry.

Nevertheless, cellulose fatty esters with low DS values show other qualities such as a high hydrophobicity. The development of water-repellent cellulosic materials (i.e. cotton, wood), has led to interesting applications in the textile and wood industries. For instance, the direct esterification of timber with fatty acids (and their derivatives) has resulted in materials with extraordinary outdoor durability and resistance to biological attack (e.g. rotting, termites). Industrial exploitation of this technology has recently started in France (WoodProtect [7]). In this case, the water-repellent properties conferred to wood and the lack of recognition from predators' enzymes account for these properties.

In general terms, esterification with organic acids lead to significant changes in the properties of cellulosic fibres, as described in Table 6.4.

### 6.3.2.3    *Cellulose Ethers*

Cellulose ethers are the most widely produced cellulose derivatives. Depending on the nature of the substituent that replaces the hydroxyl function, the cellulose ether can be soluble in water or in organic solvents, yielding viscous solutions. When added to water at concentrations of around 1–2%, viscosity can rise up to 50 Pa s (50,000 times the viscosity of pure water). They are therefore used as rheological modifiers in numerous formulations and are good substitutes for xanthan gum. The raw material can either be wood pulp or cotton linters. The latter, which have a higher degree of polymerisation, are used for the production of cellulose ethers with high viscosity.

*Carboxymethyl cellulose* (CMC) is a white solid, without odour and harmless. Its sodium salt is more common, obtained by reaction of alkali cellulose with sodium chloroacetate instead of chloroacetic acid. The modification of the rheological properties of sodium CMC depends on the DS and on the degree of polymerisation. Commercial sodium CMC has DS values between 0.6 and 0.9 and is water-soluble. Sodium CMC is widely used in many applications, not as a biomaterial itself but as a component of many materials. These include additive in textiles (e.g. ironing, printing, anti-shrinking), paper (e.g. retention of fillers, coating, glue for wallpaper) and cement (e.g. retarder agent).

*Methyl cellulose* (MC) is a yellowish or white solid, with no odour or flavour. It is obtained by reaction of alkali cellulose with methylene chloride. It is soluble in water with DS values of between 1.4 and 2. MC and its derivatives (mixed ethers with hydroxyethyl, hydroxypropyl and hydroxybutyl) form gels when the water solutions are heated. The temperature of gelatinisation depends on the DS, concentration, degree of polymerisation and the presence of salts or organic solvents. MC is present in the formulation of different materials and formulations in a similar manner to those of CMC.

*Ethyl cellulose* (EC) is a water-insoluble cellulose ether. It is produced by the reaction of alkali cellulose with ethylene chloride and has film-forming and thermoplastic properties. As a plastic, it can be processed by extrusion and injection. It is hard, stiff and with good resistance to impact. It is soluble in the molten state with other thermoplastics. As for its film-forming properties, it is used in the formulation of varnishes, inks and glues. It can also form removable coatings.

### 6.3.3    Starch

Starch is the main energy reserve of superior vegetal plants. It is found in large quantities in wheat, potato, corn and manioc. Starch is a homopolymer (99%) of D-anhydroglucopyranose units. Two different configurations exist: amylose and amylopectin (Figure 6.18). The branched polymer is present in the proportion 70–80% [27]. Native starch is present under the form of partially crystalline (25–40%) granules (up to 100 µm diameter), demonstrating a complex structure which has been the subject of thousands of scientific papers.

Starch grains are insoluble in water at room temperature. At 50–60°C, starch absorbs water reversibly and hydrogen bonding is reduced. Above 60°C (the exact temperature depends on the native source) the structure of starch is modified irreversibly, crystallinity disappears and gelatinisation occurs.

The paper industry is the main non-food outlet for starch and consumes 17% of the European starch production. Starch–cellulose–starch bonds are created and contribute to the internal cohesion of paper sheets.

Thermoplastic materials may be obtained by extrusion in the presence of water at temperatures of around 160–200°C. If an external plasticiser (i.e. glycerol or sorbitol) is added [28], the glass transition temperature decreases. Even though these materials are fully biodegradable, their affinity with water is a huge inconvenience when considered to replace traditional plastic materials. Blends [29] and

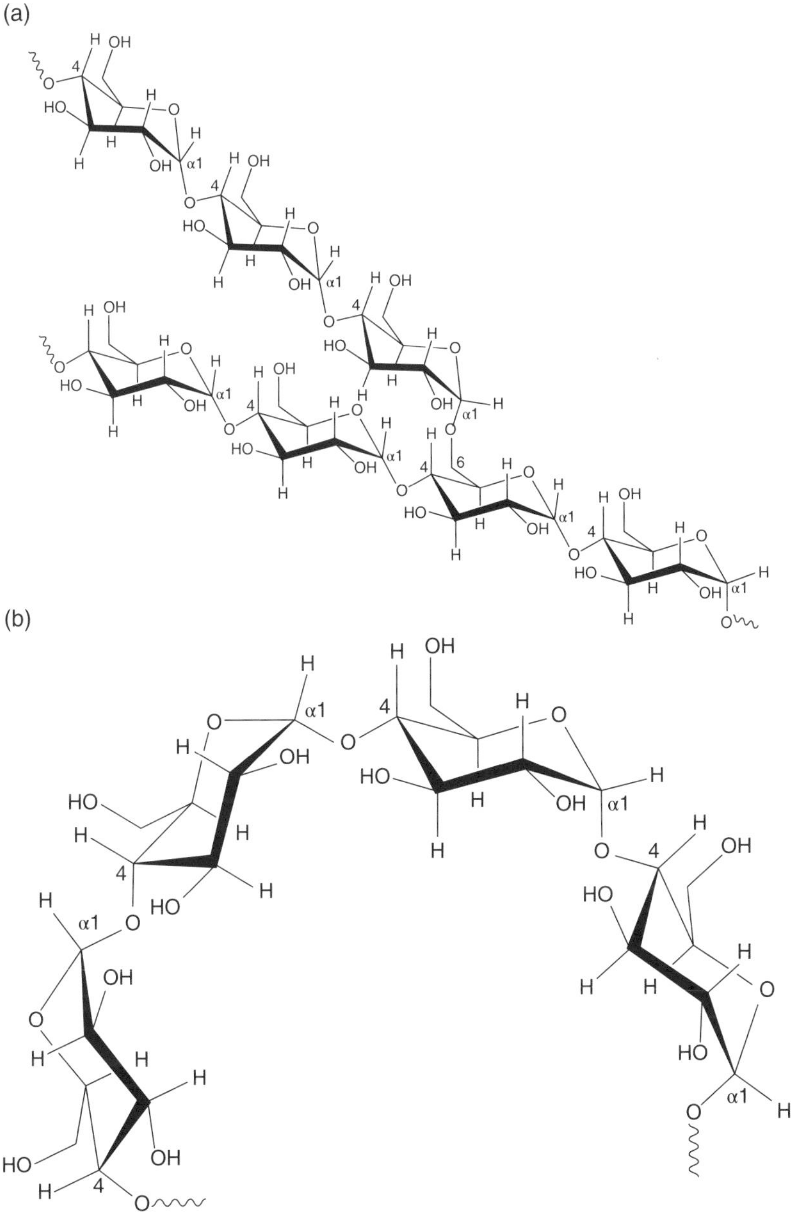

(a)

(b)

*Figure 6.18*   *Molecules of (a) amylopectin and (b) amylase.*

chemical modifications are required to overcome this issue and turn starch into a material that can be perfectly thermally processed [30].

### 6.3.4    Starch Derivatives

The *oxidation* of starch leads to the formation of carbonyl and carboxyl groups in the polymer chain. Depolymerisation also occurs and starch turns yellowish. Advantageously, the viscosity and the gelatinisation temperature of oxidised starches are lower than that of native starch. They are used in the paper industry for coating and glues.

*Non-ionic starch ethers* are used in the food industry to avoid the water release from frozen food and in the paper industry as a coating agent.

*Cationic starch ethers* are used in the paper industry to increase the cohesion and the rigidity of cellulose fibres and as a flocculant for the selective separation of negatively charged particles.

*Reticulated starch*, obtained by reaction with bifunctional reagents, increases the water retention capacity and diminishes the swelling of starch grains, thus increasing their mechanical and thermal resistance. This is particularly important for the production of highly absorbent diapers.

*Starch esters* are thermoplastic materials with properties that are similar but not exactly the same as those of cellulose esters. In particular, starch acetate (DS < 0.2) is used as a coating agent in the paper industry and as a food or detergent additive.

### 6.3.5    Chitin and Chitosan

*Chitin* is a polysaccharide comprising *N*-acetylglucosamine that forms a hard, semi-transparent biomaterial found throughout the natural world. Chitin is the main component of the exoskeletons of crab, lobster and shrimp. Chitin is also found in insects (e.g. ants, beetles, and butterflies), cephalopods (e.g. squids and octopus) and even in fungi. Nevertheless, the main industrial source of chitin by far is crustaceans.

Because of its similarity to cellulose in terms of structure, chitin may be described as cellulose with one hydroxyl group on each monomer replaced by an acetylamine group. This allows for increased hydrogen bonding between adjacent polymers, giving the polymer improved strength.

The properties of chitin as a tough and strong material make it an interesting material for surgical thread. Additionally, its biodegradability means it wears away with time as the wound heals. Moreover, chitin has some unusual properties that accelerate healing of wounds in humans. Chitin has even been used as a stand-alone wound-healing agent.

Industrial separation membranes and ion-exchange resins can also be made from chitin, especially for water purification. Chitin is also used industrially as an

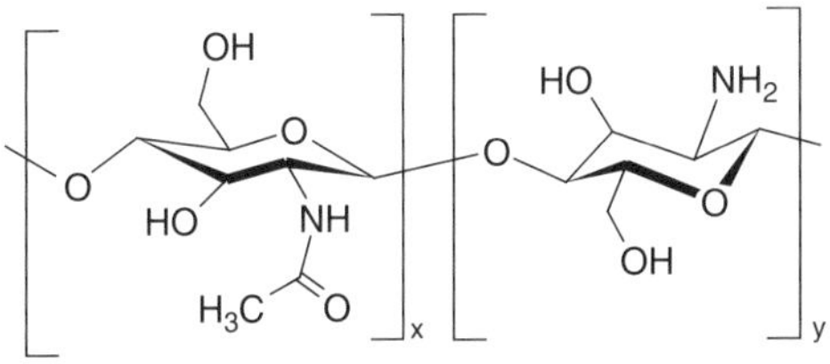

**Figure 6.19**    *General formula for chitosan and chitin (in the case of chitin y = 0).*

additive to thicken and stabilise foods and pharmaceuticals. Since it can be shaped into fibres the textile industry also utilises chitin, especially for socks as it is claimed that chitin fabrics are naturally anti-bacterial and anti-odour (http://www.swicofil.com/products/055chitosan.html). Chitin can also be used as a binder in dyes, fabrics and adhesives and, in some processes, to size and strengthen paper.

*Chitosan* is produced commercially by deacetylation of chitin. It is a linear polysaccharide composed of randomly distributed β-(1-4)-linked D-glucosamine (deacetylated unit) and *N*-acetyl-D-glucosamine (acetylated unit). The degree of deacetylation in commercial chitosans is in the range 60–100% (Figure 6.19).

The amino group in chitosan has a p$K_a$ value of about 6.5. This means that chitosan is positively charged and soluble in acidic to neutral solutions with a charge density dependent on pH and the deacetylation extent. In other words, chitosan readily binds to negatively charged surfaces such as mucosal membranes. Chitosan can also enhance the transport of polar drugs across epithelial surfaces, and is biocompatible and biodegradable. Purified qualities of chitosan are available for biomedical applications.

Chitosan possesses flocculating properties, which are used in water processing engineering as a part of the filtration process. It can remove phosphorus, heavy minerals and oils from the water. In the same manner, it is used to clarify wine (as a substitute for egg albumin) and beer.

### 6.3.6    Proteins

#### 6.3.6.1    *Molecular Structure*

Proteins are much more complex polymers than carbohydrates; there are 20 common monomers (in fact 22 proteinogenic, but only 20 are encoded by the universal genetic code) with polymeric chains composed of amino acids (Figure 6.20, left). From these monomers all kinds of secondary interactions can arise: Van der Walls, ionic, hydrogen bonding and hydrophobic interactions and, through a redox mechanism involving the amino acids bearing sulphur, a reversible covalent disulphide bonding is also possible.

As a result of this complicated molecular structure, proteins are organised in an intricate macromolecular organisation with a primary structure (the amino acid

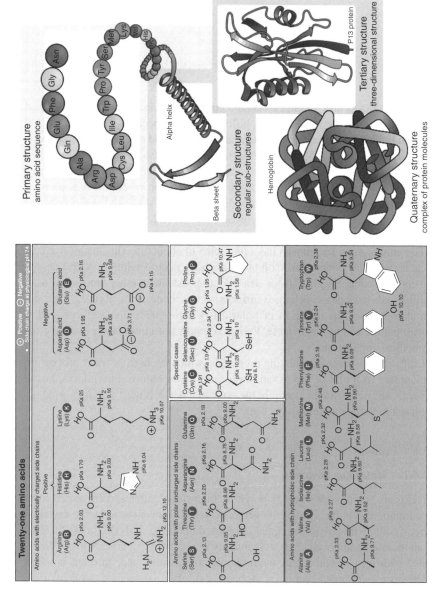

**Figure 6.20** *The 21 amino acids found in eukaryotes, grouped according to their side-chains (left). Macromolecular organisation of proteins (right).*

sequence), a secondary structure (α-helix and β-sheet substructures), a tertiary structure (3D structure) and finally a quaternary structure (complex of protein molecules) (Figure 6.20, right). Their use in forming biomaterials and their thermomechanical transformation is therefore more complicated, and their macromolecular structure modification has to be considered. However, there has been significant research and development during the last decades aimed at texturing and transforming proteins for food and non-food applications [31]. They can afford materials with specific properties and, in particular, an increased moisture resistance when compared with carbohydrate-based materials.

### 6.3.6.2    *Oilseed Proteins*

Main developments concerning protein-based materials originated from the beginning of the century with Ford's research into producing lighter cars using soybean proteins. The same kinds of operations and materials can be reproduced with other sources of protein, such as sunflower or rapeseed [32].

The texturation of oilseed proteins is essentially governed through the control of their denaturation, which is commonly defined as any non-covalent change in their structure. This loss of structure is relatively similar to the fusion of crystallites in semi-crystalline polymers and may alter the secondary, tertiary or quaternary structure of the molecules. Denaturation can be caused by many factors including heat, pH, dielectric constant and ionic strength.

Once thermal denaturation is achieved, the phenomenon being extremely dependent on moisture content, polypeptides behave like classical amorphous polymers undergoing a glass transition at a temperature between 0 and 190°C for a moisture content between 0 and around 30% (Figure 6.21). Proteins can then be shaped by hot-pressing, extrusion or injection-moulding. The Iowa State University research group headed by Jay-Lin Jane has, for example, characterised many different materials from soybean proteins in the 1990s. The Laboratory of Agro-Industrial Chemistry has also has characterised all kind of materials from sunflower proteins in the early 2000s.

### 6.3.6.3    *Cereal Proteins*

Cereal proteins such as wheat gluten or corn zein are also interesting candidates for the production of biomaterials. They are abundant as co-products of the starch industry; they are well-defined but unfortunately relatively expensive because of their many possible uses as food additives or encapsulating materials.

Pure zein is odourless, tasteless, hard and water insoluble. Since it is edible, it has application in processed foods and pharmaceuticals, in competition with chitosan and chitin. Historically it has been used in the manufacture of a wide variety of commercial products, including coatings for paper cups, soda bottle cap linings, clothing fabric, buttons, adhesives and binders. It is now used for the

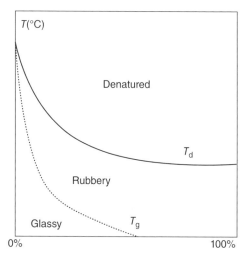

**Figure 6.21** *Dependence of denaturation temperature (T$_d$, solid line) and glass transition temperature (T$_g$, dashed line) of proteins on water content.*

encapsulation of foods and drugs. The main barrier to greater commercial success has been its historic high cost until recently. It is argued that one solution might be to extract zein as a by-product of bioethanol production. From a high-value-added perspective, Chinese researchers have conducted intensive research on the use of zein as biomedical material [33].

The main properties of wheat grain, especially in pastries, reside primarily in the storage protein in its endosperm, gluten. It has many specific properties that are used in various food products but also some interesting viscoelastic properties and low water solubility that can be used for non-food applications [34]. Two technological approaches are usually employed – wet and dry processes – with the latter taking most advantage of the specific properties of these proteins. They possess thermoplastic properties under low moisture conditions, allowing their formation through extrusion and injection-moulding, but are also capable of cross-linking through sulphydryl–disulphide interchange, giving them elasto-meric-like properties when highly plasticised. The company Syral in France commercialises Meriplast®, a gluten-based material that is transformed and behaves like an elastomer (low-temperature extrusion section and high-temperature mould section), but is biodegradable and entirely bio-based.

### 6.3.6.4 *Animal Proteins*

The animal kingdom also produces many different kinds of proteins which can be used for various non-food applications and, in particular, in the production of materials.

**Figure 6.22**    *Twin-screw extruder, pellets and film blowing of caseinates-based plastic.  Courtesy of F. Prochazka.*

Casein is the name of a family of proteins that are commonly found in mammalian milk. It was used to produce galalithe by reaction with formaldehyde, one of the first declared plastics. Nowadays new processes exist to produce biodegradable, edible and water-soluble packaging [35]. The process involves sodium caseinates, obtained from the extraction of cow milk, and consists of a classical double-step operation: compounding of glycerol-plasticised transparent pellets and film blowing (Figure 6.22).

Other types of proteins such as keratin, for example from feathers or extracted from blood, have also been tested to produce biodegradable biomaterials in the last decades.

### 6.3.6.5   *Animal Fibres*

Animal fibres, for example silk or wool, consist largely of proteins. As for textile applications, these fibres can be used like vegetal fibres to produce various kinds of materials.

Applications for keratin, the fibrous structural protein constituting human hair and wool, include composites [36], biocomposites or biomedical applications [37], but also in the construction sector.

Silk is also a fibre of major interest. Mimicking spider silk biosynthesis is still a dream of many researchers [38] and classical silk finds many applications in composites and biomedical devices such as tissue scaffolds [39].

### 6.3.6.6    *Lignin Derivatives*

A portfolio of biomaterials can be obtained from lignin derivatives. First the lignophenol derivative, which contains a diphenylpropane unit formed by binding a carbon atom at an ortho-position of a phenol derivative to a carbon atom at a benzyl-position of a phenylpropane fundamental unit of lignin, and binding an oxygen atom of the hydroxyl group and a β-positional carbon atom under alkali conditions to obtain an arylcoumaran derivative [40]. The latter can be reticulated to form polymers shaped under hot-moulding.

A similar strategy involves liquefying biomass with phenol under acidic conditions to obtain phenolic monomers. The type of monomers obtained varies largely [41] and these monomers can reticulate with the help of increased temperatures or with formaldehyde to obtain novolac-type or resol-type resins.

Two commercial materials made from or including lignin are the blend XyloBag™ marketed by the American company CycleWood Solutions Inc. and Arboform® from German company Tecnaro.

## 6.4    Agromaterials, Blends and Composites

### 6.4.1    Agromaterials

The food-processing industry produces a large amount of waste and co-products rich in fibres. The nature of these fibres and the types of biopolymers contained within vary largely. The three main food industries (vegetable oil, starch and sugar) produce millions of tons of oilcakes, stalks, pulp and bagasse. Most of these residues are often used for animal feed. However, their high cellulose content and their low price make them the source of choice for the fabrication of materials known as 'agromaterials' [42].

The availability of these residues depends on regional agricultural productions and on the commitment of agricultural cooperatives in favour of agromaterials. In France for instance, there are currently corn crops dedicated exclusively to the production of thermoplastic agromaterials.

In this case the biopolymers are directly plasticised by thermomechanical means and transformed through the classical forming technologies of the plastic industry: injection-moulding, extrusion and thermo-forming. These agromaterials (e.g. wood) maintain a natural aspect and are sensitive to atmospheric conditions, but have no shape restriction (Figure 6.23).

Agromaterials demonstrate that it is possible to profitably transform raw agricultural products without separation (Table 6.5) and that almost all non-cellulosic

**Figure 6.23** *Examples of thermoplastic agromaterials obtained from sunflower oilcake and whole corn plant.*

**Table 6.5** *Average composition (%, dry matter basis) of the three most common feedstocks used for the production of agromaterials.*

|  | Sunflower oilcake | Sugar beet pulp | Whole corn plant |
| --- | --- | --- | --- |
| Lipids | 1 | 0 | 3 |
| Sugars | 6 | 0 | 7 |
| Ash | 7 | 5 | 5 |
| Cellulose | 12 | 23 | 21 |
| Hemicelluloses and pectins | 17 | 48 | 24 |
| Lignin and phenolics | 13 | 2 | 3 |
| Starch | 1 | 0 | 30 |
| Protein | 34 | 7 | 7 |

biopolymers can be plasticised *in situ* to constitute a natural continuous matrix for cellulosic fibres.

Destructuring the native organisation of the raw agricultural products is possible with the combination of thermal, mechanical and chemical effects in an extruder (Figure 6.24). In the first constraint zone, the matter is roughly crushed. In the second zone, a compression (up to 20 bars) in the reverse screw induces a non-degrading break of the structure when the moisture content and the temperature are 20–30% and 110–130°C, respectively. Water plays the roles of plasticiser and lubricant, avoiding degradation under high shear. The same phenomenon occurs with a higher compression and a lower shear in the die to complete the transformation.

Starch plasticisation is obtained by gelatinisation of the grains in low-moisture conditions leading to the 'melting' of the starch grains, the key phenomenon in the transformation of the whole corn plant. The plasticised starch forms a continuous matrix in which the defibrated fibres are embedded, as showed in Figure 6.25.

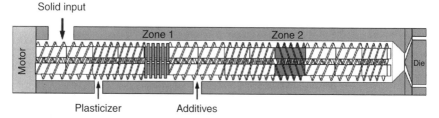

**Figure 6.24** *Twin-screw extruder configuration for the transformation of raw agricultural products.*

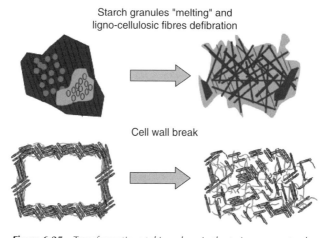

**Figure 6.25** *Transformations taking place in the twin-screw extruder.*

The transformation of sunflower oilcake is similar. The globulin corpuscles are denatured in low moisture conditions to form a continuous matrix. The real difference lies in the structure of the peptide chains compared to starch. Many different interactions take place between the proteins, and their texturation does not result in a simple 'thermoplastic' flow as occurs for starch.

Sugar beet pulp is made of primary cell walls. The breakage of the structure has to be performed at a smaller scale, and greater energy is needed to break some of the covalent bonds between the cell wall polysaccharides. The use of the die is therefore necessary to increase the residency time in the extruder. The final structure consists of cellulose micro-fibres embedded in a pectin and hemicelluloses matrix.

Injection moulding is the first targeted forming technology for agromaterials. Without any complex technological modification, injection-moulded objects can be obtained quickly and easily. At the end of the twin-screw extrusion process, the obtained granulates are stable and mouldable by injection. The French company Vegeplast is the world leader of this technology. Typically, the shaped materials are dense (density c. 1.4 kg m$^{-3}$), relatively stiff and brittle (flexural modulus c. 4 GPa; strength at break c. 17 MPa), and water sensitive. The latter is a profitable

**Figure 6.26**   *Film obtained by extrusion of sugar beet pulp.*

characteristic for the biodegradation of the agromaterials, which reaches 100% in a very short time. A particular application is to assist in the pricking out of seedlings. A pot made of any of these agromaterials is strong enough to ensure the mechanical support of growing small plants. When they are big enough, the plants can be transplanted to the soil without removing the pots. The latter will biodegrade and provide nutrients to the plant, which has been demonstrated to grow faster and to a larger size.

In the particular case of protein-rich granulates (oilcakes), thermo-moulding is often more suitable because it takes advantage of their cross-linking abilities [43]. The mechanical properties of these materials are lower, but the advantage is that oilcake can be used as it is and the materials produced are more resistant to water.

Regarding its microstructure, sugar beet pulp is the best candidate to extrude films (thickness >200 µm) [44]. Figure 6.26 depicts a film produced using sugar beet pulp and glycerol as the external plasticiser.

### 6.4.2   Blends of Synthetic Polymers and Starch

Blends of starch and a synthetic polymer (usually polyethylene) are products of commercial importance. Two families of blends are currently available: those using dehydrated starch pellets and those using gelatinised or thermoplastic starch. In both cases, the mixture is produced with the synthetic polymer by extrusion. Further processing by moulding or blowing is also possible, depending on the kind of starch used.

*Dehydrated starch* acts as a charge and diminishes the mechanics properties of the polymer. It is not possible to add more than 20%; in practice, the rate oscillates between 6 and 10%. In order to obtain a better compatibility between starch and the polymer, starch might be rendered hydrophobic by chemical treatment as described in Section 6.3.4.

One of the principal objectives in these systems is biodegradability. In this regard, auto-oxidation agents and organometallic catalysts are often added to the blend. They are intended to break the synthetic polymer chains to the point where they became digestable by micro-organisms, thus making starch accessible. Degradation begins at low rate and is accelerated when exposed to degradation conditions (compost, etc.). In most cases, the material is biofragmented but not fully biodegraded or bio-assimilated when exposed to degradation. In order to reduce water absorption of starch blends, and therefore their mechanical properties, the addition of zein seems to be efficient [45].

Examples of commercial starch-containing blends include Ecopolym (Polychim) and Ecostar (St Lawrence Starch Company). In the former, the synthetic polymer is polyethylene and starch is present at 10%. This is associated to one catalyst that promotes the decomposition by oxidation and by cleavage of chains due to the produced radicals. The latter product associates PE with a mixture of starch and auto-oxidant unsaturated fatty acids. The global content of starch is between 6 and 15%. The degradation process then follows two mechanisms: the starch is fragmented then assimilated by micro-organisms; and the interaction between the auto-oxidants and the metallic complexes from soil or water yields peroxides that attack the synthetic polymer chains. These types of products are normally used in mulch films, bags and packing.

Other systems have been investigated. The combination of starch and polyester has been claimed to be fully biodegradable [46]. Others are partially biodegradable such as the starch/polyethylene/poly-ε-caprolactone blends [47] and their derivatives or the combinations of starch and modified polyesters.

The product Mater-Bi, produced by Novamont who have revolutionised the starch-based biomaterials sector over the last two decades, deserves a particular mention. The commercial success of this biodegradable and biocompostable plastic relies on two main factors: the economy of scale that allows costs to be reduced, and the diversity of formulations to develop different end-products (plastic bags, tableware, toys, etc.). More than 210 references in Chemical Abstracts are available on this (registered) keyword, and the number of patents related to different formulations and developments is also impressive. Mater-Bi can essentially be described as a blend of starch with a small amount of other biodegradable polymers and additives. The actual compositions are only known by a very few people.

### 6.4.3   Composites with Natural Fibres

#### 6.4.3.1   *Wood–Plastic Composites*

Vegetal fibres (including wood fibres) represent a good replacement solution for glass and carbon fibres in the reinforcement of composites based on a thermoplastic matrix. The advantages of vegetal fibres are both economical and ecological:

- they are inexpensive and less abrasive for the processing equipment;
- they are of low density, lightening the whole composite material; and
- they have a limited environmental impact.

In 75% of cases, wood fibres are the preferred filler for thermoplastic matrices (their high availability may account for this). The resulting composite material is internationally known as WPC, which stands for wood–plastic composites. They have been produced industrially since the 1980s and the market has increased in the last decade, especially in the United States, reaching 700,000 metric tons with an 11% increase rate per annum. In Europe, the WPC market is considered as emerging. The best estimations indicate that it reached 100,000 metric tons in 2005 [48]. In America, more than 50% of WPC is used for parquets and decking made with polyethylene. In Europe, the automobile sector is the dominant consumer, using polypropylene-based composites. For these two types of WPC, the most studied parameters are: fibre/matrix adhesion; filler content; granulometry of the fibres; extrusion parameters; and durability of the WPC when exposed to water, sunlight, fungi and insects.

The adhesion between the fibre and the matrix is by far the crucial parameter of the composite materials. The mechanical properties of the WPC depend greatly on it and on the compatibility of the filler and the matrix. The only practical way to improve the adhesion is by adding a coupling agent to the formulation, that is, a molecule capable of establishing bonding between the filler and the matrix. Three kinds of bonding are involved: covalent, hydrogen and non-polar interactions. They all contribute to the reduction of the natural incompatibility between the hydrophilic fibres and the hydrophobic matrix. The most used coupling agent for polyolefins-based composites is maleated polypropylene (MAPP). It is prepared from polypropylene (PP) and maleic anhydride (Figure 6.27). The anhydride function reacts with wood fibres and the attached PP moiety is fully compatible with free PP.

The addition of MAPP at around 1–2% ensures the perfect covering of the wood fibres by polypropylene or polyethylene. The micrographs of Figure 6.28 clearly show the lack of adhesion between the fibres (in dark colour) and the PP matrix (in light colour). The mechanical properties are increased by at least 30% when a coupling agent is used.

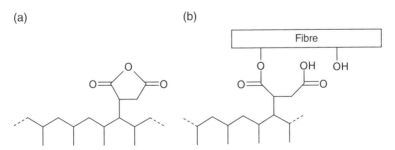

*Figure 6.27*  (a) Maleated polypropylene and (b) the adduct formed with vegetable fibres by covalent bonding.

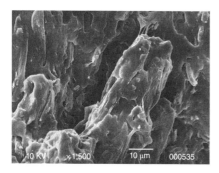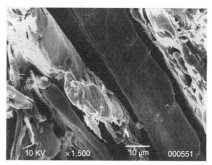

*Figure 6.28*   *Scanning electron micrographs of fracture surfaces of polypropylene/wood fibres composites. Left: with maleated PP; right: without MAPP.*

For PVC composites, used in housing and in the automobile industry, no coupling agent is used because the PVC is rather polar and compatible with wood fibres. Some research has been performed and the positive effect of aminosilanes [49] and poly(methylene(phenyl isocyanate)) [50] has been demonstrated.

Note that most of the WPCs are not biodegradable materials. The synthetic matrix remains a fossil resource with poor biodegradability properties, even if accompanied by natural fibres. Nevertheless, these materials also show some advantages. Indeed, polypropylene needs to be reinforced with a filler when recycled in order to maintain its mechanical properties. WPCs are therefore a perfect solution for prolonging the life of olefins. Regarding the recycling of the WPC itself, only PVC/wood composites have been demonstrated not to lose their properties when thermomechanical recycling is applied [51].

### 6.4.3.2    Biocomposites (with Biodegradable Thermoplastic Matrix)

The use of natural fibres in biodegradable plastics is in the same way as their use in polyolefins (WPC), but is documented differently because of specific applications. The main goal is identical: reduce the amount of matrix (because it is very expensive) without degradation of the mechanical properties. All the issues are similar: improvement of the fibre wetting and the influence of the thermomechanical processing on the fibre size and repartition [52].

### 6.4.3.3    Natural Fibres in Thermosets

The development of plant fibre composites for structural applications is growing, especially in the automotive and aerospace industries. When compared to glass or carbon fibres, natural fibres present the advantage of being cheaper, non-toxic upon inhalation, renewable, biodegradable and recyclable and also consume less energy during their production. On the other hand, their mechanical properties are not as good and their processing requires some adjustments.

According to Shah [53], the key aspects in developing this kind of materials include: fibre type (nature, extraction process, surface modification); fibre volume fraction; fibre/filler geometry and interfacial properties; reinforcement packing arrangement and orientation; and matrix type and composite manufacturing technique. From his literature review, the author concluded that it was better to use bast fibres (mainly from flax, hemp or jute), prepared especially for composite applications (as opposed to textile processes) with high fibre content (up to 60% in volume), using pre-impregnation techniques followed by compression-moulding or infusion. Concerning the matrix, thermoset matrices (especially epoxy-based) are more suitable due to better compatibility, lower viscosity and processing temperature, and better mechanical properties.

### 6.4.4 Wood-Based Boards

When wood sheets or particles of wood are bound together to create a stiff board, many applications are possible. These applications depend on the mechanical properties and the granulometry of the wood employed and on the density of the board.

In a first classification, we can distinguish between boards made with (1) an external synthetic binder, such as urea-formaldehyde or phenol-formaldehyde resins for thermosets, and (2) internal natural binders. The products described in the following sections do not constitute an exhaustive list.

#### 6.4.4.1 *Wood Boards with Synthetic Binder*

Different commercial products are classified according to the size or the form of the raw material (wood). All of them require the binder (around 10%) to be sprayed on the wood material before undergoing hot pressing at temperatures and pressures of 150–200°C and 20–25 bar.

Wood sheets, obtained by 'unrolling' steamed wood rods with a blade, can be bound together to form thicker boards. The main characteristic of these 'sandwiches' is the alternation of the fibre orientation by 90° in every layer. The anisotropy of the whole board is therefore limited in the length and in the width. Three, five or seven layers are usually used in these panels, referred to as *plywood*.

The inner layers of plywood are often replaced by oriented strand board (OSB), obtained from wood chips, to reduce costs; the two outer layers of wood sheets ensure the esthetical appearance of the panel. The orientation is desired to simulate the characteristics of a wood panel, with limited swelling and higher resistance in the fibre direction. Chips are orientated by air with a blowing machine. These panels are commonly used in building sites and often comprise part of the walls in particular houses, especially in the United States. If no orientation is given, the panel is called *waferboard*.

When sawdust is used, particleboards are obtained. The bigger granulometries are used for *fibreboard*, which is the material mostly used in the fabrication of affordable furniture and kitchens. The smallest particle sizes, that is, wood flour, are used for medium-density fibreboard (MDF), which has the advantage of showing a smooth surface after cutting, avoiding the need for a plastic covering on the edges (as particleboard does).

Urea-formaldehyde resins are used for applications in which the panel is not in contact with water. For applications with a high level of humidity, phenol-formaldehyde resins are required. In all cases, free formaldehyde constitutes a dangerous pollutant that is slowly released and can be particularly toxic in confined environments. Recent research has been carried out into the possibility of substituting this type of resin by natural binders such as cross-linkable proteins [54, 55].

### 6.4.4.2   Wood Boards without External Binder

In this case, water is sprayed on sawdust instead of a synthetic binder. Hot-pressing of the material is carried out to partially hydrolyse some macromolecules contained in the wood, and thermally degrade some of the free sugars and other small molecules. The degradation products contribute to the cohesion of the particles.

Depending on the pressure applied, more or less rigid boards are obtained. The boards of lowest density ($0.15–0.5\,\mathrm{kg\,m^{-3}}$), processed with the lowest pressure, are called insulation board. As the name indicates, they are used in housing for thermal and phonic insulation of the walls and ceilings. In practice, they are used alongside more rigid materials to obtain a stronger product.

If the pressure used in the fabrication is increased, the density of the product is higher ($0.5–1.45\,\mathrm{kg\,m^{-3}}$); *hardboard* is then obtained. This kind of board is used in furniture parts in which the mechanical resistance required is not very high, such as the back panel of inexpensive bookcases for example.

### 6.4.5   Materials for Construction

Apart from wood- and fibreboards, many agricultural products, by-products and waste are now used in construction materials; these applications concern more industrial applications, however. Although the scientific literature on this subject is sparse, some studies and reviews have recently been published including a paper [56] from the Visvesvaraya National Institute of Technology (India)

The types of material used in this domain include the following.

- *Insulation materials:* as a substitution for glass wool and polyurethane, which are both of fossil origin and bear a high amount of grey energy, many natural fibres are now used in commercial products (e.g. paper waste, hemp fibres, wood fibres, palm leave fibres, etc.). Their use is mainly in fibre panels obtained after carding and bounding with thermoplastic fibres but other methods are possible, including wet projection or waste paper blowing.

- *Clay bricks:* to decrease weight, enhance mechanical properties and decrease heat conductivity, agricultural waste is now used in masonry composites. For crude earth bricks, improved properties are obtained from the reinforcement of the clay-based materials; for classical fired bricks, the use of organic substances creates porous materials from the degradation of these fillers [57].
- *Cementitious materials:* the concrete industry faces huge energy issues and the use of natural fibres could be one option to improve the life-cycle performance of cementitious materials. Lignocellulosic fibres are used as reinforcement, creating two different problems: the relative incompatibility between the matrix and these natural fillers; and the influence of these organic and hydrophilic materials on the water transfer during the drying [58].

## 6.5 Conclusion

The world of biomaterials has been defined in this chapter from the point of view of technical feasibility. However, it can only be fully apprehended through a deep knowledge across numerous disciplines including agronomy, chemistry, biochemistry, economy, sociology, medicine, engineering and ecology. Moreover, the development of these biomaterials will depend not only on technical aspects, but also on political decisions and trends. The determination of industrialised societies to progress is expected to be the main driver for the promotion of ecologically friendly materials.

## References

1. Banoub, J.H., Benjelloun-Mlayah, B., Ziarelli, F., *et al.* (2007) Elucidation of the complex molecular structure of wheat straw lignin polymer by atmospheric pressure photoionization quadrupole time-of-flight tandem mass spectrometry. *Rapid Communications in Mass Spectrometry*, **21**, 2867–2888.
2. Forss, K. and Fremer, K-E. (2000) The nature of lignin: a different view. In *Lignin: Historical, Biological, and Materials Perspectives* (eds W.G. Glasser, R.A. Northey, and T.P. Schultz). American Chemical Society, Symposium Series 742, pp. 100–116.
3. Dulbecco, P. and Luro, D. (2001) *L'essentiel sur le bois*, Centre Technique du Bois et de l'Ameublement, Paris, p. 184.
4. Hill, C.A.S. and Jones, D. (1996) The dimensional stabilization of Corsican pine sapwood due to chemical modifications with linear chain anhydrides. *Holzforschung*, **50**, 457–462.
5. Dumonceaud, O., Thomas, R. (2004) Process and apparatus for wood impregnation. French patent FR 2870773.
6. Morard, M., Vaca-Garcia, C., Stevens, M. *et al.* (2007) Durability improvement of wood by treatment with methyl alkenoate succinic anhydrides (M-ASA) of vegetable origin. *International Biodeterioration & Biodegradation*, **59**, 103–110.
7. Magne, M., El Kasmi, S., Dupire, M., *et al.* (2003) Procédé de traitement de matières lignocellulosiques, notamment du bois ainsi qu'un matériau obtenu par ce procédé. French patent FR 2838369.

8. Zub, H.W. and Brancourt-Hulmel, M. (2010) Agronomic and physiological performances of different species of Miscanthus, a major energy crop. A review. *Agronomy for Sustainable Development*, **30**, 201–214.

9. Gandini A. (2011) The irruption of polymers from renewable resources on the scene of macromolecular science and technology. *Green Chemistry* **13**, 1061–1083.

10. Ekman, K., Eklund, V., Fors, J. *et al.* (1984) Regenerated cellulose fibers from cellulose carbamate solutions. *Lezinger Berichte*, **57**, 38–40.

11. Dufresne, A. (2013) Nanocellulose: a new ageless bionanomaterial. *Materials Today*, **16**, 220–227.

12. Siró, I. and Plackett, D. (2010) Microfibrillated cellulose and new nanocomposite materials: a review. *Cellulose*, **17**, 459–494.

13. Siqueira, G., Bras, J. and Dufresne, A. (2010) Cellulosic bionanocomposites: a review of preparation, properties and applications. *Polymers*, **2**, 728–765.

14. Missoum, K., Belgacem, M.N. and Bras, J. (2013) Nanofibrillated cellulose surface modification: a review. *Materials*, **6**, 1745–1766.

15. Lin, S.-P., Calvar, I.L., Catchmark, J.M. *et al.* (2013) Biosynthesis, production and applications of bacterial cellulose. *Cellulose*, **20**, 2191–2219.

16. Freire, C.S.R., Fernandes, S.C.M., Silvestre, A.J.D. and Neto, C.P. (2013) Novel cellulose-based composites based on nanofibrillated plant and bacterial cellulose: recent advances at the University of Aveiro – a review. *Holzforschung*, **67**, 603–612.

17. Kwatra, H., Caruthers, J. and Tao, B. (1992) Synthesis of long chain fatty acids esterified onto cellulose via the vacuum-acid chloride process. *Industrial & Engineering Chemistry Research*, **31**, 2647–2651.

18. Thiebaud, S. and Borredon, M.E. (1995) Solvent-free wood esterification with fatty acid chlorides. *Bioresource Technology*, **52**, 169–173.

19. Hamalainen, C., Wade, R. and Buras, E.M., Jr (1957) Fibrous cellulose esters by trifluoroacetic anhydride method. *Textile Research Journal*, **27**, 168–168.

20. Samaranayake, G. and Glasser, W. (1993) Cellulose derivatives with low DSI. A novel acylation system. *Carbohydrate Polymers*, **22**, 1–7.

21. El Seoud O., Heinze T. (2005) Organic esters of cellulose: new perspectives for old polymers. *Advances in Polymer Science* **186**:103–149.

22. Vaca-Garcia, C., Gozzelino, G., Glasser, W.G. and Borredon, M.E. (2003) DMTA transitions of partially and fully substituted cellulose fatty esters. *Journal of Polymer Science. Part B.*, **41**, 281–288.

23. Shiraishi N., Aoki T., Norimoto M., Okumara M. (1983) Make cellulosics thermoplastic. *Chemtech* (**June**):366–373.

24. Rowell, R.M. and Keany, F.M. (1991) Fiberboards made from acetylated bagasse fiber. *Wood and Fiber Science*, **23**, 15–22.

25. Rowell R.M. (1997) Chemical modification of agro-resources for property enhancement. In *Paper and Composites from Agro-Based Resources*. R.M. Rowell, R.A. Young and J.K. Rowell (eds.). Lewis Publishers CRC, Boca Raton, pp. 351–375.

26. Glasser, W., McCartney, B. and Samaranayake, G. (1994) Cellulose derivatives with low degree of substitution. 3. The biodegradability of cellulose esters using a simple enzyme assay. *Biotechnology Progress*, **10**, 214–219.

27. Park, I.M., Ibáñez, A.M. and Shoemaker, C.F. (2007) Rice starch molecular size and its relationship with amylose content. *Starch/Stärke*, **59**, 69–77.

28. Averous, L. (2004) Biodegradable multiphase systems based on plasticized starch: a review. *Journal of Macromolecular Science, Polymer Reviews*, **44**, 231–274.

29. Wang, X.-L., Yang, K.-K. and Wang, Y.-Z. (2003) Properties of starch blends with biodegradable polymers. *Journal of Macromolecular Science, Part C: Polymer Reviews*, **43**, 385–409.

30. Liu, H., Xie, F., Yu, L. *et al.* (2009) Thermal processing of starch-based polymers. *Progress in Polymer Science*, **34**, 1348–1368.

31. Verbeek, C.J.R. and van den Berg, L.E. (2010) Extrusion processing and properties of protein-based thermoplastics. *Macromolecular Materials and Engineering*, **295**, 10–21.

32. Rouilly, A. and Vaca-Garcia, C. (2013) Industrial use of oil cakes for material applications, in *Economic Utilisation of Food Co-products* (eds A. Kazmi and P. Shuttleworth), Royal Society of Chemistry, Cambridge, pp. 185–214.

33. Dong, J., Sun, Q. and Wang, J.-Y. (2004) Basic study of corn protein, zein, as a biomaterial in tissue engineering, surface morphology and biocompatibility. *Biomaterials*, **25**, 4691–4697.

34. Lagrain, B., Goderis, B., Brijs, K. and Delcour, J. (2010) Molecular basis of processing wheat gluten toward biobased materials. *Biomacromolecules*, **11**, 533–541.

35. Belyamani, I., Prochazka, F. and Assezat, G. (2014) Production and characterization of sodium caseinate edible films made by blown-film extrusion. *Journal of Food Engineering*, **121**, 39–47.

36. Conzatti, L., Giunco, F., Stagnaro, P. *et al.* (2013) Composites based on polypropylene and short wool fibres. *Composites Part A: Applied Science and Manufacturing*, **47**, 165–171.

37. Vasconcelos, A. and Cavaco-Paulo, A. (2013) The use of keratin in biomedical applications. *Current Drug Targets*, **14**, 612–619.

38. Rising, A., Widhe, M., Johansson, J. and Hedhammar, M. (2011) Spider silk proteins: recent advances in recombinant production, structure–function relationships and biomedical applications. *Cellular and Molecular Life Sciences*, **68**, 169–184.

39. Hardy, J.G. and Scheibel, T.R. (2010) Composite materials based on silk proteins. *Progress in Polymer Science*, **35**, 1093–1115.

40. Funaoka, M. (2005) Novel lignin derivatives, molded products using the same and processes for making the same. US Patent 154,194.

41. Lin, L., Nakagame, S., Yao, Y. *et al.* (2001) Liquefaction mechanism of β-O-4 lignin model compound in the presence of phenol under acid catalysis. Part 2. Reaction behaviour and pathways. *Holzforschung*, **55**, 625–630.

42. Rouilly, A. and Rigal, L. (2002) Agro-materials: a bibliographic review. *Journal of Macromolecular Science, Part C: Polymer Reviews*, **42**, 441–479.

43. Rouilly, A., Orliac, O., Silvestre, F. and Rigal, L. (2005) New natural injection-moldable composite material from sunflower oilcake. *Bioresource Technology*, **97**, 553–561.

44. Rouilly, A., Jorda, J. and Rigal, L. (2006) Thermo-mechanical processing of sugar beet pulp. I. Twin-screw extrusion process. *Carbohydrate Polymers*, **66**, 81–87.

45. Gaspar, M., Benko, Z.S., Dogossy, G. *et al.* (2005) Reducing water absorption in compostable starch-based plastics. *Polymer Degradation and Stability*, **90**, 563–569.

46. Tokiwa, Y., Takagi, S., Koyama, M. (1993) Starch-containing biodegradable plastic and method of producing same. US Patent 5,256,711.

47. SK Corp. (2001) Biodegradable linear low-density polyethylene composition and film thereof. Korean patent KR20010084444.

48. Anonymous (2006) Wood-plastic composite growth taking off in Europe... while strong WPC growth continues in the USA. *Additives for Polymers*, **5**, 9–11.
49. Kokta, B.V., Maldas, D., Daneault, C. and Beland, P. (1990) Composites of poly(vinyl chloride)—wood fibers. III: Effect of silane as coupling agent. *Journal of Vinyl Technology*, **12**, 146–53.
50. Kokta, B.V., Maldas, D., Daneault, C. and Beland, P. (1990) Composites of poly(vinyl chloride)-wood fibers. I. Effect of isocyanate as a bonding agent. *Polymer-Plastics Technology and Engineering*, **29**, 87–118.
51. Augier, L., Sperone, G., Vaca-Garcia, C. and Borredon, E. (2007) Influence of the wood fibre filler on the internal recycling of PVC-based composites. *Polymer Degradation and Stability*, **92**, 1169–1176.
52. Faruk, O., Bledzki, A.K., Fink, H.-P. and Sain, M. (2012) Biocomposites reinforced with natural fibers: 2000–2010. *Progress in Polymer Science*, **37**, 1552–1596.
53. Shah, D.U. (2013) Developing plant fibre composites for structural applications by optimising composite parameters: a critical review. *Journal of Materials Science*, **48**, 6083–6107.
54. Silvestre, F., Rigal, L., Leyris, J., Gaset, A. (2000) Aqueous adhesive based on a vegetable protein extract and its preparation. European patent EP997513.
55. Yang, I., Kuo, M., Myers, D. and Pu, A. (2006) Comparison of protein-based adhesive resins for wood composites. *Journal of Wood Science*, **52**, 503–508.
56. Madurwar, M.V., Ralegaonkar, R.V. and Mandavgane, S.A. (2013) Application of agrowaste for sustainable construction materials: a review. *Construction and Building Materials*, **38**, 872–878.
57. Gorhan, G. and Simsek, O. (2013) Porous clay bricks manufactured with rice husks. *Construction and Building Materials*, **40**, 390–396.
58. Pacheco-Torgal, F. and Jalali, S. (2011) Cementitious building materials reinforced with vegetable fibres: a review. *Construction and Building Materials*, **25**, 575–581.

# 7

# Biomass-Based Energy Production

## Mehrdad Arshadi[1] and Anita Sellstedt[2]

[1]*Department of Forest Biomaterials and Technology, Swedish University of Agricultural Sciences (SLU), Sweden*

[2]*Department of Plant Physiology, UPSC, Umeå University, Sweden*

## 7.1   Introduction

The transport sector generates substantial airborne emissions of different gases and particles that are produced by the combustion of various fossil fuels in vehicle engines. These emissions have adverse effects on human health and are also known to cause vast environmental changes. Several of the emitted gases are greenhouse gases such as $CO_2$, $CH_4$, and $N_2O$. There is very strong evidence of a correlation between greenhouse gas emissions and global climate changes that are causing global warming. Together with the world's growing energy demand, the depletion of global fossil fuel reserves and the increasing importance of long-distance transport provide a strong impetus for the development of new energy sources.

Biomass is a renewable energy source comprising a wide range of biological materials that can be used as fuels and as precursors for the production of other industrial chemicals. Energy generated from biomass is usually referred to as bioenergy. Today, bioenergy is a practical and increasingly widely available

*Introduction to Chemicals from Biomass*, Second Edition. Edited by James Clark and Fabien Deswarte.
© 2015 John Wiley & Sons, Ltd. Published 2015 by John Wiley & Sons, Ltd.

option for heating that is being adopted by many industries and households that are looking to use more sustainable energy sources.

In the European Union (EU) a set of binding laws called the Climate and Energy Package were introduced to ensure that the EU would meet its ambitious climate and energy targets by 2020. These targets, known as the "20-20-20" targets, are expressed in three key objectives:

1. to reduce EU-wide greenhouse gas emissions by 20% relative to their 1990 levels;
2. to increase the share of EU energy consumption generated from renewable resources to 20%; and
3. to increase EU-wide energy efficiency by 20%.

Meeting these targets will require action on several fronts, including the development of various different bioenergy products that are tailored to match the resources and facilities available in different regions. In addition, it will be important to identify pre-existing infrastructure that could be used to facilitate the transition to a more bioenergy-focused system. Finally, economic, environmental, national security, and technological factors must be considered for each of the affected countries.

In general, biomass-derived fuels may be solids (e.g. chips, pellets), liquids (e.g. ethanol, biodiesel), or gases (e.g. biogas, hydrogen). They can be classified based on the processes used in their production, which include physical upgrading (Section 7.2), microbiological processes (Section 7.3), thermochemical processes (Section 7.4), and chemical processes (Section 7.5), as described in Figure 7.1.

## 7.2    Physical Upgrading Processes

### 7.2.1    Refinement of Biomass into Solid Fuels

The production of solid fuels from renewable resources (e.g. biomass) has become more important due to the growing global demand for energy and environmental concerns. Upgraded or refined solid biofuels include powdered fuels and densified fuels such as briquettes and pellets. Some of the most common raw materials for pellet production are shavings and sawdust from sawmills. However, there is also an increasing interest in alternative biological raw materials such as grasses and agricultural residues including corn stover, olive seeds, wheat straw, and peat [1].

### 7.2.2    Wood Powder

Wood powder is an upgraded fuel that is burned in large-scale combustion plants for heat production. However, it is also suitable for power generation. It is made from sawdust, shavings, and bark. The raw material is crushed, dried, and milled into fine particles in order to obtain a fuel with optimal properties. Many different grades of wood powder can be produced that differ in terms of physical properties

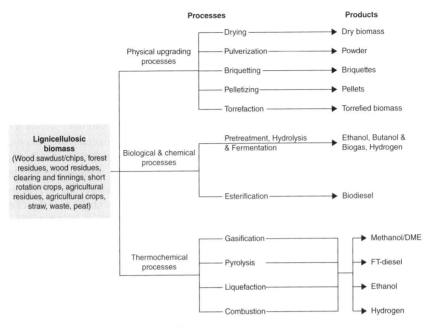

**Figure 7.1**    *Different types of bioenergy production processes from biomass.*

such as particle size distribution, particle shape, and moisture content. The powder is usually handled in closed systems from milling to its storage in silos to avoid the risk of dust explosions. The properties of the powder are determined by the raw materials and type of mill used in its production [2].

### 7.2.3    Briquette Production

Briquettes are another type of upgraded biofuel. Agricultural and forestry residues and other biological materials are often difficult to use as biofuels because they are heterogeneous, uneven and bulky in nature, and awkward to handle. These problems can be eliminated by densifying the material into compact, regular shapes. This is done by forcing the biomass through a die at high pressure using a screw or piston-press. The friction between the raw material and equipment increases the temperature inside the press, softening the lignin present within the materials, which then acts as a binder. Typical briquettes have diameters of 35–90 mm. The process has many advantages as follows.

- It increases the net calorific content of the material per unit volume. Briquettes typically have densities of 800–1300 kg m$^{-3}$ whereas the bulk density of loose biomass is 10–20 kg m$^{-3}$ and that of wood chips is about 300 kg m$^{-3}$.
- It generates a product of uniform size with a well-defined quality and moisture content.

- It generates a product that is easy and cheap to handle, transport, and store.
- Briquettes facilitate the optimization of combustion processes, resulting in higher efficiency, lower emissions, and reduced ash production.
- Briquette combustion typically requires considerably less investment in furnaces and purification equipment than some alternatives.

Biomass briquetting can be performed using an exogenous binder or by direct compacting with no added binder. One example from a pilot plant (Biofuel Technology Centre, Sweden) is presented below. In this case, bales of raw material are put on a conveyor and transported to a shredder. The shredded material is then stored in a silo. The shredder cuts the material into particles with diameters ranging from the microscopic to 15 mm. Briquette production is initiated by transporting the material from the silos through a mixer and a separator to a buffer silo above the briquette press. At the bottom of the buffer silo, an auger then transfers the material into the press. The briquette press has a high production capacity of 450–500 kg h$^{-1}$. The humidity of the raw material prior to pressing can be determined using a Haymatic moisture meter, which measures the material's electrical conductivity and uses that to roughly estimate the moisture content at the sample surface.

Several parameters affect the mechanical strength of the briquettes, including the moisture content of the raw material and the pressure and temperature at which they are formed. A wide range of raw materials have been used in briquette production, including dry household waste fractions in combination with cellulose-rich materials such as straw or reed canary grass. In general, the resulting briquettes have favorable combustion properties and the incorporation of waste into the biofuel does not increase the emissions of organic compounds formed during combustion and heat production [3].

### 7.2.4    Pellet Production

Fuel pellets made from sawdust represent a renewable energy source for heat production that has become increasingly popular in recent years. About 2.2 million tons of wood pellets were manufactured in Sweden in 2010, making this country the biggest fuel pellet producer in Europe. Moist solid biomass, such as sawdust and other feedstocks used in pellet production, normally has a moisture content of more than 50 wt%. It is often stored outdoors before being pelletized, in order to ensure an adequate supply of fuel to sustain pellet production during winter time.

A wood pellet can be regarded as a small, round, hard unit of stored bioenergy. They are typically cylindrical with diameters of 6–8 mm and variable lengths. Over the last decade, softwood pellets have emerged as a leading source of renewable energy that is primarily used for heat and power production and which has the potential to replace fossil fuels such as oil and natural gas in many applications. The energy value of 1 ton of pellets is about 5.0 MWh, which is equivalent to that of 0.5 m$^3$ of oil.

Because pellets are a renewable fuel and have a closed $CO_2$ emission cycle, their usage represents a step towards a sustainable energy system. Wood pellets have been used for heat and power generation in both the household and industrial sectors. They are currently produced in several countries, some of which export them overseas. For example, Canada ships significant quantities of pellets to Europe every year [4]. The raw materials used in pellet production are often forest by-products such as sawdust, mostly from pine and spruce, as well as cutter shavings and, in some cases, the bark of the same species. However, there is considerable interest in identifying new raw materials for pellet production because lignocellulose-rich material from softwoods is also used in the pulp and paper industry. Moreover, cellulose could also be used in ethanol production. Other biomass sources include straw, various grasses, corn stover, and other agricultural products [5]. It is possible to use mixtures of different lignocellulose-containing materials in pellet production and there is ongoing research in Sweden and China on pelletizing various mixtures including corn stover, rape seed cakes, eucalyptus leaves, bark, peat, hemp, cassava stems and cotton.

The first-generation pellet presses were developed for feed production. However, experiments conducted over several years have resulted in the development of modified presses that have been optimized for use with woody raw materials. In traditional wood pellet production processes the raw material is dried to a moisture content of 5–15%, which requires significant amounts of energy. The dried material is then ground into a fine powder consisting of particles that are less than 3 mm in diameter. This powder is then pressed through cylindrical holes in a pellet matrix (die) to form short sticks called pellets. The pressure inside the pellet press is very high (about 70–100 MPa). Production capacity can be increased by treating the powder with steam before pressing to adjust the moisture content of the raw material and soften its lignin/hemicelluloses to improve particle binding. During pelletizing, the temperature of the material rises to more than 120°C. This further softens wood components such as lignin and therefore helps to bind the particles together, eliminating the need for an exogenous binder in many cases. In general, the use of binders increases production costs and has an adverse effect on the fuel's combustion properties. The newly formed pellets must be cooled to ambient temperature relatively quickly to prevent the absorption of moisture from the surrounding air.

One drawback of the pelletizing process is its high consumption of electricity. Another is the relatively large quantity of fine particles and dust that is generated during pellet production and storage. This increases the cost of pellet production, because the dust and fines must be removed (to avoid dust explosions, among other things) before the pellets can be packaged and delivered. It is also important to remove any potential metal, stone, or sand contaminants that may have become mixed with the sawdust before it is pelletized. Figure 7.2 outlines the key processes that occur in a pilot pellet manufacturing plant.

The interactions between particles during pellet production are governed by a complex set of forces [5]. Once manufactured, the pellets are much more

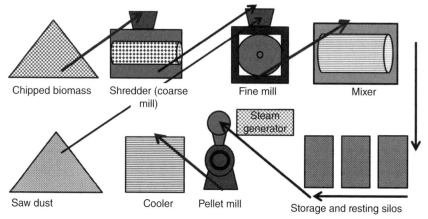

**Figure 7.2**   *Key processes in pilot pellet manufacturing plant Biofuel Technology Center at SLU in Umeå. Reproduced with permission from Håkan Örberg of Biofuel Technology Center at SLU.*

amenable to extended storage and transportation without loss of quality than sawdust, which has a moisture content of 45–55% and requires large volumes of storage space. In addition, unprocessed sawdust is much more sensitive to microbial activity and can undergo various chemical changes during storage.

When used as fuels, pellets have several advantages over the raw materials used in their production (primarily sawdust):

- a lower moisture content (about 40–45% less water content);
- more homogeneity and so less variation in moisture content and particle size (pellets therefore have superior combustion properties);
- more even combustion and lower levels of undesirable emissions;
- higher density and lower transportation costs;
- higher energy density and much easier feeding of burners;
- lower storage costs; and
- the ability to be stored for extended periods of time without risk of mold or other microbiological degradation.

Pellets are usually stored in closed warehouses or silos to ensure that they do not absorb water. During prolonged storage or transportation, wood pellets may emit some volatile organic compounds (VOCs) such as aldehydes and low-molecular-weight carboxylic acids into the storage room or container. These compounds may have unpleasant smells and can potentially represent health hazards. They are formed by the oxidation of fatty and resin acids present in the wood pellets [6].

The pellets are delivered to customers in bulk by lorry, train, or boat, in large or small sacks. They are burned in pellet furnaces at small-, medium-, and large-scale facilities to generate heat. The quality of the pellets can be evaluated using a few key parameters, which is important because their properties may vary from batch to batch. The most important physical parameters are bulk density, pellet density, and durability (which can be measured by determining the fine particle content of the raw material). Key physical and chemical parameters include the pellets' moisture content and their calorific heat value (as an indicator of their energy content), contents of elements such as K, Mg, Ca, N, S, Cl, ash content, and ash melting behavior (which determines the degree of ash agglomeration that will occur in the furnace). The net calorific heat value of typical pellets is about $19\,MJ\,kg^{-1}$, which is less than half the value for the same quantity of oil. Pellets contain water, flammable components, and ash. The amount of ash generated during pellet combustion is typically less than 0.5% of the pellets' gross weight. Ash contains several elements such as phosphorus, potassium, calcium, magnesium, and silicon. The presence of large quantities of fines (notably, sawdust released from pellets during transportation or otherwise) in pellets increases ash formation and causes other problems during combustion. In the worst-case scenario, ash may melt in the burner and damage the furnace. If the pellet combustion process is not optimized, various harmful emissions will be generated including CO (due to an inadequate oxygen supply or an insufficiently high combustion temperature), $NO_x$ (due to an overly high combustion temperature that allows its formation from N in the pellets and/or the air), $CO_2$, hydrocarbons, and particulate matter. The chlorine and sulfur contents of the pellets are also important because they affect the production of HCl, dioxins, and $SO_x$ during combustion.

### 7.2.5   Storage of Solid Biomass

Solid biomass is often stored for extended periods of time to facilitate logistics and to maintain high productivity at pelletization plants on a year-round basis. In addition, the prolonged storage of sawdust may increase the quality of the resulting pellets. The stored biomass often changes during storage. For example, the storage of pine and spruce sawdust in large quantities causes its fatty and resin acid content to decline gradually over the first 12 weeks (relative to the dry mass of the raw material). However, storage for longer periods do not cause any further reductions, indicating that the sawdust has become mature (i.e. its fatty and resin acid content stabilized) after 12 weeks of storage [7]. These compounds have significant effects on pellet production, and pellets made from mature sawdust often exhibit superior durability [8].

In Sweden, wood pellets are primarily manufactured from spruce and pine sawdust and shavings. Because these raw materials contain significant quantities of readily degraded extractives (notably, fatty and resin acids), it is expected that the chemical composition of the raw material will change during pelletization and subsequent storage. This is reflected in the self-heating of pellets and the occurrence of odour problems [6, 9]. It is also well known that dry lignocellulosic material can generate heat due to moisture uptake, which presents a risk of spontaneous combustion.

Under certain conditions, wood pellets can emit appreciable quantities of gaseous VOCs, for example aldehydes and ketones, and non-condensable gases (this latter process is known as off-gassing), primarily CO, $CO_2$, and $CH_4$. The emissions of condensable gases often give the pellets a pungent smell.

### 7.2.6    Torrefaction Technology

Torrefaction is a technology for the pretreatment of solid biomass [10]. It is primarily of interest because it generates a product with superior properties to untreated biomass. In particular, torrefied biomass has a greater energy density per unit volume and has the added benefit of being hydrophobic.

Broadly speaking, there are three types of thermal treatment that can be used to torrify woody raw materials:

1. *conventional drying processes*: involves heating at 100–130°C to remove moisture;
2. *torrefaction*: anoxic thermal treatment at 200–350°C with low particle heating rates and a long residence time (>1 min); and
3. pyrolysis: anoxic treatment at 400–800°C.

The torrefaction of lignocellulosic biomass is an upgrading process for biomass drying that removes some VOCs and decomposes the reactive hemicellulose fraction. The moisture content of torrefied biomass is very low. Torrefaction is a mild pyrolytic process that improves certain properties of the raw material, generating a product with a higher carbon content and net calorific value than untreated wood. It involves heating the woody material to 200–350°C for several minutes in the absence of oxygen. Torrefaction at different temperatures yields products with different properties and chemical compositions. Notably, the partial removal of VOCs (around 20% of the VOC content of the raw material is lost) dramatically changes the material's properties.

Torrefaction is likely to become more important in the future as a method for the pretreatment of biomass prior to gasification [11]. Torrefied biomass can be compressed into fuel pellets of high physical and energy density to reduce the amount of space required for its storage and the cost of transporting it. Torrefied pellets are less prone to moisture absorption than those made from untreated sawdust, and the tendency to absorb moisture diminishes as the severity of the torrefaction process increases [12].

The pelletization of torrefied Norway spruce was shown to yield pellets with around 80–90% of the durability of pellets made from untorrefied material [13]. The lack of moisture in the torrefied material increases the glass transition temperature of the remaining carbohydrate polymers and limits the scope for the formation of robust bridges between particles. However, there are several ways of overcoming this lack of bonding. For example, an additive with a high bonding capacity could be introduced after the torrefaction process to compensate for the loss of hydrogen bonding sites due to torrefaction [14].

## 7.3    Microbiological Processes

### 7.3.1    Organisms and Processes

Various microbial processes can be exploited to utilize energy that has been stored in biomass by photosynthesis. These processes can generate useful biofuels such as hydrogen, butanol, and biogas. Moreover, biomass can be converted into ethanol; this is commonly done using fungi but can also be achieved with bacteria. Finally, certain algae can be used to produce biodiesel, as described in the following section.

### 7.3.2    Hydrogen Production

Hydrogen is the most abundant element in the universe and is also a very efficient energy carrier. This extremely small molecule is metabolized by a variety of organisms including bacteria, archaea, cyanobacteria, and green algae. The simple conversion of protons into hydrogen gas is a reversible reaction in most organisms that can perform it, and is catalyzed by a group of enzymes known as hydrogenases that have evolved a wide range of subunits to solve this problem under different circumstances. Thirteen different hydrogenases have been identified to date [15], most of which perform some energy-related function. There are excellent reviews on the biodiversity and functions of hydrogen-producing organisms (e.g., [15, 16]) and on the structure and function of the three known hydrogenase classes: [Ni-Fe]-hydrogenases, [Fe-Fe]-hydrogenases, and [Fe]-hydrogenases [17].

The hydrogenase reaction has the following stoichiometric formula:

$$2H^+ \leftrightarrow H_2 + 2e^-$$

As indicated above, the process is reversible so hydrogen may either be produced or consumed. Four categories of hydrogen-related processes and organisms that perform them have been delineated as listed below:

1. photoautotrophic hydrogen production;
2. photoheterotrophic hydrogen production;
3. heterotrophic hydrogen production; and
4. heterotrophic hydrogen production coupled to photo-production.

Microorganisms that are capable of both hydrogen production and photosynthesis are said to be phototrophic while those that metabolize hydrogen but cannot perform photosynthesis are said to be heterotrophic. Phototrophic organisms can be divided into photoautotrophs and photo-heterotrophs. The phototropic organisms are represented mainly by microalgae and cyanobacteria and are capable of using light as an energy source and carbon dioxide as a carbon source. The photoheterotrophs are cyanobacteria that, despite being able to use both light energy and carbon dioxide, also require organic carbon to perform nitrogen fixation. The heterotrophic bacteria utilize carbon compounds for their hydrogen production and are therefore excellent catalysts for the decomposition of organic waste materials.

### 7.3.3    Classification of Hydrogen-Forming Processes

#### 7.3.3.1    *Photoautotrophic Hydrogen Production*

Representative photoautotrophic organisms include algae such as *Scenedesmus* sp. and *Chlamydomonas reinhardtii* [18]. Interestingly, Melis *et al.* [18] developed a two-stage photosynthesis and hydrogen production system based on the green alga *C. reinhardtii*. It was shown that lack of sulfur caused a specific but reversible decline in the rate of the oxygen-producing step of photosynthesis. In enclosed cultures, this imbalance in the photosynthesis-respiration relationship due to S-deficiency resulted in the net consumption of oxygen, leading to anaerobiosis in the culture and increased hydrogen production. The two-step system allows the culture to grow and maintain an active photosynthesis apparatus in a complete growth medium. However, if the medium is made S-deficient, hydrogen is evolved. Recently, Melnicki *et al.* [19] showed that the poly-β-hydroxybutyrate content of the algae increased within 24 h of S-starvation. This implies that this process can be used to produce both hydrogen and biologically important compounds.

#### 7.3.3.2    *Photoheterotrophic Hydrogen Production*

The photoheterotrophic microorganisms are represented by cyanobacteria such as *Nostoc* sp. and *Anabaena variabilis*. They are capable of photosynthesis using sunlight and carbon dioxide in a similar manner to the photoautotrophs. In addition, they are also capable of nitrogen fixation mediated by a nitrogenase enzyme. Organic carbon is required to enable this process, as shown by the stoichiometric formula:

$$N_2 + 8H^+ + 8e^- + 16ATP \rightarrow 2NH_3 + H_2 + 16ADP + 16Pi$$

Hydrogen production requires energy that is obtained from light according to the following expression:

$$C_2H_4O_2 + 2H_2O + light\,energy \rightarrow 2CO_2\ gas + 4H_2\ gas$$

These organisms therefore have two enzymes that produce hydrogen: a nitrogenase and a hydrogenase. There is a very wide range of photoheterotrophs that could be investigated to determine their capacity for hydrogen production as well as their potential for genetic modification and tailoring for biofuel production.

#### 7.3.3.3    *Heterotrophic Production of Hydrogen*

Heterotrophic microorganisms generate hydrogen by metabolizing organic compounds via processes such as that represented by the stoichiometric equation below:

$$C_6H_{12}O_6 + 4H_2O \rightarrow 2CH_3COO^- + 2HCO_3^- + 4H_2$$

This equation is based on the use of a 6-carbon compound, but a wide range of compounds can be used under natural conditions. Heterotrophic bacteria such as *Clostridium* sp., *Enterobacter* sp., *Ralstonia* sp., *Rhodobacter* sp., and *Frankia* sp. cannot perform photosynthesis and therefore require a supply of carbohydrates to produce hydrogen. Organisms of this sort occur naturally in ecosystems all over the world. Various industrial processes have been developed that use these natural systems and attempt to mimic their capacity for hydrogen production. One strain of *Frankia*, designated R43, was demonstrated to produce large quantities of hydrogen under anaerobic conditions using the carbon source propionate [20]. It was subsequently proven that its hydrogen-producing enzyme is present in both its hyphae and its vesicles [20].

There is a wide range of organic waste products generated by various industries that could potentially be used as carbon sources for large-scale heterotrophic hydrogen production. The abundance of different types of waste were recently reviewed by Redondas *et al.* [21]. The preferred substrates for such processes are carbohydrates because their conversion is thermodynamically superior to that of alternative substrates. Glucose is a carbohydrate that is present in most industrial flows. The bioconversion of 1 mol of glucose will theoretically yield 12 mol of hydrogen gas. After accounting for the reaction's stoichiometry, the bioconversion of 1 mol of glucose into acetate would produce 4 mol of hydrogen gas per mole of glucose. It has also been demonstrated that sulfate-reducing bacteria can produce large amounts of hydrogen, with 100% efficiency using formate as their substrate [22].

### 7.3.3.4   *Coupling Heterotrophic Hydrogen Production to Photoproduction*

The bioconversion of biomass by heterotrophic bacteria typically yields hydrogen gas and acetate because the organic material is not completely oxidized. Dark fermentation therefore produces both hydrogen and other carbohydrates that must be removed because they have adverse effects on the energy balance of the hydrogen-producing organism. This can be achieved by coupling one of the products of dark fermentation, acetate, to an additional container in which a second fermentation occurs. This second step is a photoheterotrophic process in which the organic acids produced in the first reaction are converted into hydrogen and carbon dioxide [23]. It has been shown that if dark and photofermentation processes are performed equally in this way, the theoretical yield of 12 mol $H_2$ per mole of glucose can be achieved [24].

### 7.3.4   Butanol Production Using Bacteria as Biocatalysts

Butanol is an advanced biofuel as well as an important industrial solvent. Butanol production occurs naturally in several species of *Clostridium* sp. that are capable of synthesizing it from carbohydrates. Interest in biological butanol production increased in the early 1980s due to the increase in oil prices during the 1970s, and the process garnered a renewed interest over the last 10 years [25]. All of the bacteria that catalyze this process are heterotrophic, that is, they utilize pre-made carbohydrates as substrates. The production of butanol starts with the conversion of glucose into pyruvate via glycolysis. The pyruvate is then transformed into butanol via a five-step process involving acetyl-Coenzyme A [25]. The *Clostridium* species that catalyze this process go through three distinct growth phases: an exponential phase during which acids are produced; a stationary phase during which the acids are transformed into solvents; and third, a phase in which spores are formed. Two of the stoichiometric reactions catalyzed by *C. acetobutylicum* are presented below [25]:

1. $1 \text{ glucose} \rightarrow 1 \text{ butanol} + 2 \text{ CO}_2 + 2 \text{ ATP}$
2. $1 \text{ glucose} \rightarrow 0.3 \text{ acetone} + 0.6 \text{ butanol} + 0.2 \text{ ethanol} + 2.3 \text{ CO}_2 + 1.2 \text{ H}_2$

Reaction (1) above is the simpler, but reaction (2) occurs more often in bacteria. Recently, a novel pathway for the production of butanol and isobutanol was identified in *Saccharomyces cerevisiae* [26]. The substrate in this case was glycerol and it was therefore suggested that mixtures of amino acids from side-stream processes could be used to produce this biofuel. Interestingly, the bacterium *Escherichia coli* has also been considered a candidate for this purpose and Atsumi *et al.* [27] have shown it to be capable of butanol production after genetic modification.

### 7.3.5   Microbiological Ethanol Production

At present, the vast majority of feedstocks for ethanol production comes from grain or agricultural sources. However, it has been suggested that it may be unethical to use feedstocks that could be used for food production to generate biofuels. There is therefore great interest in developing processes for ethanol production that use lignocellulosic biomass.

#### 7.3.5.1   *Lignocellulosic Feedstocks for Microbiological Ethanol Production*

Lignocellulosic or woody biomass consists of carbohydrate polymers such as cellulose, hemicellulose, lignin and, to a much lesser extent, extractives, acids, salts, and minerals.

Cellulose and hemicellulose can be hydrolyzed to sugars and fermented to ethanol. Cellulose is a polymer of glucose–glucose dimers, and the linkages between these dimers as well as the hydrogen bonds between polymeric glucose strands make it very difficult to break up. The process of saccharification involves the addition of water to break these bonds and liberate glucose for ethanol production.

Hemicellulose consists of short, highly branched chains of different sugars such as 5-carbon sugars xylose and arabinose and 6-carbon sugars mannose, glucose, and galactose. Because of its branched structure, it is relatively easy to hydrolyze hemicellulose for fermentation processes.

Lignin is one of the most resistant materials on Earth and cannot be used in fermentation processes. It is present in all biomass to a variable extent, depending on its origin.

Grain and woody materials are examples of biomass with somewhat variable compositions. For example, switchgrass contains 33% cellulose/glucan, 22% xylan, and 18% lignin while aspen contains 53% cellulose/glucan, 19% hemicellulose, and 19% lignin [28]. Aspen biomass has a slightly higher content of sugars, making it valuable for ethanol production. Glucose and xylose are the main fermentable sugars in all lignocellusosic materials, from rice to hardwoods. The only exception is softwood, which is also relatively rich in 6-carbon sugar mannose. In addition, pine has a high content of lignin, which can be used for pellet production. Grasses meanwhile are relatively poor in lignin [29]. These differences in composition all impose different requirements on any potential ethanol production process. It was shown by Huang *et al.* [28] that aspen could sustain a higher production of ethanol than switchgrass or hybrid poplar. It was also pointed out that aspen could contribute to wood chip production [30]. Poplar has also recently become a feedstock of interest because its genome has been sequenced and it can be genetically engineered to have a sugar content that is more favorable for ethanol production.

### 7.3.5.2   *Microbiological Ethanol Process: Pretreatment*

The production of ethanol requires the pretreatment of biomass before fermentation. This may involve physical, chemical, or biological processing. Physical pretreatments include commSunion (dry, wet, or vibratory ball milling), irradiation (electron beam irradiation or microwave heating), steaming, or hydrothermolysis [28]. Chemical pretreatments include treatment with acids, alkaline substances, solvents, ammonia, sulfur dioxide, or other chemicals. Biological pretreatments involve the addition of decomposing microorganisms or enzymes. This makes them environmentally friendlier than chemical or physical alternatives. However, to date they have also proven to be slower and thus less economically viable. Interestingly, however, the pretreatment of cellulose with certain enzymes can give relatively high yields of ethanol. Depending on the source of the biomass to be used for ethanol production, there is a wide range of different enzymes that can be used for such pretreatments. In addition, the biomass must be prepared before it can be pretreated with enzymes. This first step of pretreatment is often physical, such as milling. This exposes the cellulose to the enzymes that will degrade it. Cellulose is the main constituent of biomass from forests. Targeted enzymes are selective for the conversion of cellulose to glucose, and therefore produce no degradation by-products that may be encountered when using acid conversion technology. For example, the Canadian company Iogen Ltd. initially subjects its woody raw materials to a modified steam explosion process and then treats them with highly efficient cellulases that were developed in-house. However, different enzymes are required for other biomass types.

### 7.3.5.3    *Fermentation Process*

The production of ethanol undergoes glycolysis, a common biochemical pathway that is present in animals as well as yeast and bacteria. The overall reaction scheme for the conversion of 6-carbon sugars into ethanol is:

$$C_6H_{12}O_6 \rightarrow 2C_2H_5OH + 2CO_2$$

Similarly, the stoichiometric equation for the reaction with a 5-carbon sugar is:

$$3C_5H_{10}O_5 \rightarrow 5C_2H_5OH + 5CO_2$$

There are at least three different processes that can be used to convert cellulosic biomass into ethanol. In separate hydrolysis and fermentation (SHF), the pre-treated biomass is treated with a cellulase that hydrolyzes the cellulose to glucose at 50°C and pH 4.8. The resulting material is then fermented. In simultaneous fermentation and saccharification (SSF), the hydrolysis and fermentation occur in the same bioreactor. Third, in direct microbial conversion (DMC), the microorganisms that produce the cellulase also perform the fermentation.

Commercial ethanol production is primarily done by microorganisms such as *S. cerevisiae*, *Zymomonas* sp., and *Candida* sp. Of these, the well-known fungi *S. cerevisiae* (also known as baker's yeast) is the most widely used. Natural *S. cerevisiae* may be used, but genetically modified or "recombinant" strains are also available. Interestingly, several authors have described the genetic modification of *S. cerevisiae* using different techniques [31]. It has been demonstrated that the efficiency of fermentation by *S. cerevisiae* can be dramatically increased by cloning three exogenous genes into it [31].

In addition, a very interesting patent recently described a novel mix of fungi that can ferment both 6- and 5-carbon sugars into ethanol. This patent also states that the mixture of fungi enhances ethanol production by *S. cerevisiae* and can therefore be used in existing industrial set-ups [32, 33].

### 7.3.6    Production of Biodiesel from Plants and Algae

Biodiesel is a fuel that is derived from vegetable oils or algae. It is a biodegradable and non-toxic diesel equivalent. Diverse feedstocks can be used for biodiesel production, including rapeseed and soybean oils. In addition, animal fats including grease and chicken fat and sewage waste can be used as substrates for algae that produce biodiesel (Table 7.1).

Biodiesel is made from vegetable oils consisting of glycerol esters of fatty acids. A process of transesterification is used to exchange the glycerol components

**Table 7.1** *Comparison of some sources of biodiesel. Adapted from Cristi [34].*

| Crop (L ha[-1]) | Oil yield (Mha)[a] | Land area required | Percentage of existing US cropping area[a] |
|---|---|---|---|
| Corn | 172 | 1540 | 846 |
| Soybean | 446 | 594 | 326 |
| Microalgae[b] | 136,900 | 2 | 1.1 |
| Microalgae[c] | 58,700 | 4.5 | 2.5 |

[a] Required to meet 50% of all USA demand for transport fuel.
[b] Assuming 70% oil (by wt) in biomass.
[c] Assuming 30% oil (by wt) in biomass.

of the triglyceride molecules for methanol. This produces fatty acid methyl esters with straight saturated and unsaturated hydrocarbon chains, as described in the section on chemical processes (Section 7.5). The biological processes involved in biodiesel production are illustrated in the following reaction formula:

$$\text{Sunlight} + \text{carbon dioxide} \rightarrow \text{algal biomass} + \text{biodiesel}$$

Two key characteristics must be considered when selecting algal species for biodiesel production: the rate of cellular growth and the lipid content. For example, certain *Chlorella* species grow rapidly and produce large quantities of oil [35]. The various existing processes for producing biodiesel fuels have been reviewed by Perego and Ricci [36].

### 7.3.7 Biogas Production

Naturally occurring ecosystems such as water-logged swamps, bogs and marshes are inhabited by microorganisms that have evolved for millions of years under oxygen-limiting or oxygen-free conditions. These microorganisms are capable of utilizing organic and inorganic substrates to fuel their own metabolic activity. Biogas production, that is, the production of methane and carbon dioxide with trace amounts of hydrogen gas, nitrogen gas, and hydrogen sulfide, is a naturally occurring process that has attracted increasing levels of interest. The process is outlined by the following reaction formula:

$$\text{Complex organic materials} \rightarrow \text{monomeric organic compounds} \rightarrow \text{intermediate products} \rightarrow H_2 + CO_2 + \text{acetate} \rightarrow CH_4 + CO_2.$$

A more detailed description is presented in the flowchart in Figure 7.3. The first part of the process is the degradation of the complex organic compounds (lipids, proteins, carbohydrates) into mono- and oligomers (fatty acids, amino acids, monomers of carbohydrates) via a process of hydrolysis. This is typically the rate-limiting step in the anaerobic production of methane. The resulting species are

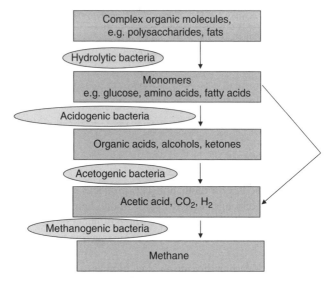

**Figure 7.3**    *Flowchart outlining the process of biogas production.*

fermented, producing intermediary compounds such as alcohols, long-chain fatty acids, and acetate. These undergo anaerobic oxidation that yields hydrogen gas, carbon dioxide, and acetate. The final step, methanogenesis, involves the conversion of the hydrogen, acetate, and carbon dioxide into various products that are dominated by methane and carbon dioxide. However, by-products such as hydrogen and nitrogen gas are also formed.

### 7.3.7.1    *Microorganisms Involved in Biogas Production*

A wide range of morphologically diverse methanogenic bacteria has been isolated, including rods, cocci, plate-shaped, and filamentous types. Molecular techniques such as the analysis of 16S rRNA sequences have revealed the true identities of the methanogens. To date, seven different methanogenic groups have been described, including species of *Methanobacterium, Methanococcus, Methanosarcina* and *Methanopyrus.*

The reactions that lead to methane formation involve microorganisms in the anaerobic chain that generally act cooperatively to generate the final product. The microorganisms at the end of the reaction chain are strongly dependent on those at the beginning. In some cases, the bacteria involved have adapted completely to a collaborative lifestyle. Such bacteria are typified by the hydrogenotrophic methanogens such as *Methanobacteriaceae* sp., which perform the final step of methanogenesis. Interestingly, another group of bacteria, the acetotrophic methanogens (e.g., *Methanosarcina* sp.) are also involved in the last step, as shown in Figure 7.3. It is noteworthy that hydrogen is produced during several steps of the methane pathway, indicating that totally hydrogen and methane formation coexists in nature [15].

### 7.3.7.2   Methods for Enhancing Biogas Production

Scientists have proposed a variety of ways to enhance biogas production as reviewed by Yadvika *et al.* [37]. These include the use of additives, gas enhancement by the recycling of digested slurry, variation of operational parameters, and the use of fixed films. For the sake of brevity, we only discuss additives and the variation of process conditions (e.g. the temperature and C/N ratio) in this chapter. Additives are primarily used to promote microbial activity and to maintain favorable conditions in terms of pH in order to minimize the inhibition of acetogenesis and methanogenesis. The parameter that has the strongest impact on biogas production is the temperature. Biogas can be produced at temperatures ranging from <30°C (psychrophilic) to 30–40°C (mesophilic) or even 40–50°C (thermophilic). Anaerobic bacteria typically function best in the mesophilic and thermophilic temperature ranges, so these temperatures yield more biogas. The reaction can also occur at psychrophilic temperatures, but it proceeds much more slowly under such conditions. It is also important to maintain an appropriate substrate composition in terms of the C/N ratio. The anaerobic bacteria active in biogas production use carbon 25–30 faster than nitrogen [37], so the substrate C/N ratio should be between 20:1 and 30:1 in order to establish optimal conditions for the bacteria.

## 7.4   Thermochemical Processes

Thermochemical processes for biomass conversion are known to be more efficient than the alternatives in terms of reaction times (which range from a few seconds to minutes) as well as the percent of utilization of different parts and degradation of plant biomass including lignins. They can be used to convert biomass into energy products such as heat and electricity, transport fuels such as methanol and dimethyl ether (DME), and other chemicals. Thermochemical processes can be divided into four main groups, as described in Chapter 3:

1. *gasification* of biomass, that is, its conversion into gaseous hydrocarbons such as methane;
2. *pyrolysis*, that is, the heating of biomass in the absence of oxygen to produce various energy products;
3. *direct liquefaction* of biomass by high-temperature pyrolysis or by high-pressure-to-liquid products; and
4. *combustion* in the presence of air to convert chemical energy into heat or mechanical power.

There are several factors that influence the choice of conversion process including the desired form of bioenergy, the type of and quantity of the available biomass feedstocks, and environmental and economical constraints. Biomass materials such as agricultural and forestry residues are often difficult to use directly in thermochemical conversion processes for the production of certain energy products due to their heterogeneous and bulky nature. They therefore need to undergo some kind of pretreatment or physical upgrading process prior to use, as discussed in Section 7.2. Biomass may also has a high water content, which must often be reduced prior to thermochemical treatment. For all these reasons, biomass is often crushed, dried, and milled before thermochemical processing. As discussed in Section 7.2.6, torrefaction is another pretreatment process that can be used to improve the thermochemical conversion of biomass by increasing gasification efficiency and reducing smoke production during gasification.

### 7.4.1    Thermal Processing Equipment

There are four main types of thermal processing equipment that are widely used for biomass gasification, including: fixed bed gasifiers (and moving bed systems); fluidized bed reactors (gasifiers); circulating fluidized bed (CFB) reactors; and entrained flow systems, as highlighted in Figure 7.4. Passing gas over the biomass at a high velocity in thermal processing systems generally promotes mixing and enhances combustion efficiency.

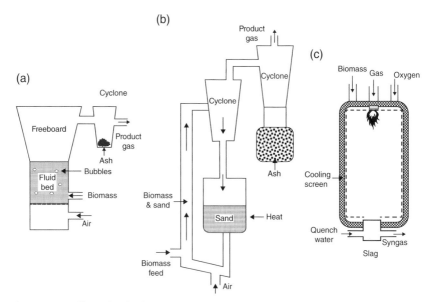

**Figure 7.4**    *Different kinds of reactors: (a) fluid bed reactor; (b) CFB; and (c) entrained flow reactor. Reproduced with permission from Rashmi Kataria, Department of Forest Biomaterials and Technology at SLU.*

Fixed bed reactors are the simplest type of gasification equipment. They have operating temperatures of 850–1,400°C [37, 38] and have been used for many years. They consist of an upright cylindrical container that has an inlet and an outlet for the gases into which the feedstock is fed from above; ashes are removed at the base of the container. There are several different fixed bed reactor designs including the updraft (counter current) gasifier, the downdraft (co-current) gasifier, and the cross-flow (crossdraft) gasifier, as shown in Figure 7.5.

In updraft reactors, the biomass is fed in at the top of the reactor and moves towards the bottom while gas flows in the opposite direction. The biomass is passed through drying, distillation, and reduction zones in which it is converted to gas and ash is removed. Air enters the reactor from the bottom and the syngas outlet is situated at the top of reactor. This design therefore allows vapours from liquid compounds and tars to distil (flow) over. Conversely, in downdraft gasifiers, such materials can only exit via the highest temperature zone, where they are destroyed. In these designs, the incoming air is introduced in the middle of the reactor. This minimizes tar formation and may obviate the need for further purification of the resulting syngas. In the cross-flow gasifier (reactor) design, the syngas is drawn off from the side opposite to that on which the incoming gas enters, preventing any contact between the reaction products and the fresh feedstock. Syngas generated in this way is invariably highly contaminated with tar. There are also some other gasification reactor designs available such as the entrained bed, twin fluid bed, bubbling fluid bed, and circulating fluid bed reactors. These are not discussed in this work, but are described at length by Bridgwater [39]. Several pilot plants for biomass gasification using different reactor technologies have been established, including entrained flow facilities in Germany (Freiberg) and Sweden (Piteå) as well as a circulating fluidized bed facility in Austria (Güssing).

Another type of gasifier is the fluidized bed reactor, which is based on counter-current flows of gas and a particulate material (the fluidizing medium) which is usually a kind of quartz sand bed with a controlled particle size (i.e. a fixed particle size distribution). This method maintains a uniform temperature throughout the bed, enhancing the overall yield of gasification at a lower temperature than is required in fixed bed reactors. One drawback of this method is the high loss of heat in the syngas and its residual fuel and ash content. However, a cyclone can be used to separate the fly ash and particulate matter from the syngas and there are several other methods that can be used to overcome these problems. The process of biomass gasification in a fluidized bed reactor is outlined in the following:

- biomass is fed into the fluidized bed;
- biomass is heated with hot sand (to around 1,000°C), forming syngas and char;
- the syngas and char are removed using a cyclone;
- the syngas is purified further before being used as a raw material for fuel production; and
- the flue gas (the excess heat gas) may be used in steam production and power generation.

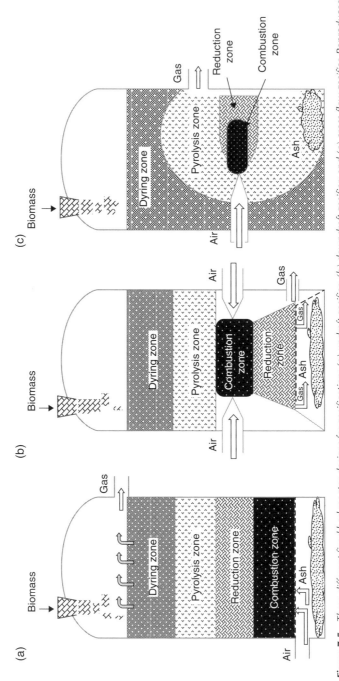

**Figure 7.5** Three different fixed bed reactor designs for gasification: (a) updraft gasifier; (b) downdraft gasifier; and (c) cross-flow gasifier. Reproduced with permission from Rashmi Kataria, Department of Forest Biomaterials and Technology at SLU.

Ash-related problems such as agglomeration may occur and exchange of the sand bed may be required after some time [39]. One way to overcome some problems with agglomeration is to use a mixture of biomass and peat in order to increase the melting point of the raw material in the furnace. It is also possible to use lime (which contains CaO and other Ca salts) instead of quartz in the fluidized bed reactor to increase the agglomeration temperature of some biomass types under the gasification conditions [40].

In CFB systems, biomass particles are placed on a bed of hot sand and all of the materials including the gas, biomass particles, and sand move together. In CFB reactors, char is ignited externally and separated using cyclones, and its residence time is almost the same as that of gas. In addition, heated solids are returned to the gasifier after the removal of flue gasses [41]. The main advantages of CFB reactors over alternative designs are their flexibility with respect to feed biomass, high biomass flow rate, scalability, and low amount of tar production.

Entrained flow reactors are based on a simple technology whereby fine solid particles are injected into a high-velocity oxygen stream and undergo rapid gasification at high temperature (up to 1,400°C). These reactors have not been very successful due to their poor rates of heat transfer between gases and solids. In addition, they suffer from high sample preparation costs due to the very small particle sizes required (100 μm to 1 mm) and their need for substrates with very low moisture contents [41].

### 7.4.2    Gasification

Gasification involves the conversion of carbon from biomass into gases (CO, $CO_2$) in the presence of controlled quantities of air at high temperatures (above 800°C). The resulting $CO_2$ reacts with hydrogen and is directly converted into methane. Certain other chemicals can also be formed in the presence of appropriate catalysts including diesels and 1-alkenes [42]. Notably, such processes are used in the production of Fischer–Tropsch (FT) diesel.

#### 7.4.2.1    Fischer–Tropsch Diesel

In 1923, two German scientists, Franz Fischer and Hans Tropsch, investigated the conversion of syngas (CO, $CO_2$, and $H_2$) into a number of different useful organic compounds. FT diesel can be produced by the gasification of coal or biomass. Coal or carbohydrates of different molecular size from biomass react with oxygen and steam to generate CO and $H_2$ via the Lurgi process. The resulting mixture is called synthetic gas (or syngas or synthesis gas) and can be directly used in gas turbines for power generation. Alternatively, in the presence of different catalysts and under different process conditions, it can be converted into different fuels such as methanol, DME, methane, or FT diesel. The composition of the products is determined by process parameters such as the temperature, pressure, and residence time as well as the catalyst that is used in the process. One of the advantages of syngas production is the wide range of feedstocks or raw materials that it can accommodate. When syngas is produced from a biomass precursor it is called bio-syngas in order to distinguish it from syngas made from fossil raw materials. The raw syngas may be contaminated with sulfur compounds (e.g. $H_2S$), nitrogen compounds (e.g. HCN, $NH_3$), halides (e.g. HCl), and heavy organic compounds that are known collectively as "tar." Unpurified gas produced by gasification may also contain benzene, toluene, and xylenes (mixture of these three aromatic hydrocarbons known as BTX) [43]. These contaminants may inactivate the catalysts. The raw syngas must therefore often be subjected to a gas cleaning step prior to conversion into bio-fuels such as DME or methanol. Tar separation is a very complex and expensive process involving a series of absorption steps, but it can remove some contaminants. There are also several other methods for tar removal such as thermal cracking, catalytic cracking and scrubbing methods as discussed by Hamelinck *et al.* [43].

Diesel is a mixture of aliphatic hydrocarbons (alkanes) with carbon chains of different lengths ($C_{12}$–$C_{22}$) that contains no aromatic hydrocarbons. A generalized illustration of the process by which carbon in biomass is converted into fuels such as methane and other alkanes by the FT method is shown in the following:

$$\text{Lurgi process}: C + H_2O\left(\text{Steam}\right) \xrightarrow{\text{Heat}} \underset{\text{Syngas}}{CO + H_2}$$

$$\text{FT synthesis}: CO + H_2 \xrightarrow[\text{Heat, Pressure}]{H_2,\,Fe} \text{Alkane} + H_2O$$

or

$$CO + H_2 \xrightarrow[\text{Heat, Pressure}]{2H_2,\,Ni} CH_4 + H_2O$$

The process requires high temperatures (200–350°C) and pressures (10–40 bar). The carbon source for FT diesel has varied throughout the years; during World War II, the Germans used coal to produce liquid fuel. During the apartheid era, South Africa was boycotted by most of the oil-producing countries and had to produce fuels and chemicals from coal by FT synthesis for several years. South Africa also produces various compounds such as ethane and propene for the manufacture of polyethylene (PE), polyvinyl chloride (PVC), polypropylene (PP), and acrylonitrile by FT processes [42]. Nowadays, natural gas is used to produce synthetic gas. Biomass has not yet been used commercially as a feedstock for the production of synthetic gas; it is an appealing feedstock however because it is $CO_2$-neutral source whereas fossil fuel precursors are not. A range of biomass materials have been used in the laboratory-scale production of FT diesel but several limitations remain to be addressed before these processes will be suitable for commercial use, including the need for gas cleaning steps and the development of scale-up technologies. FT diesel production can be combined with electricity production to increase energy efficiency and reduce production costs. FT diesel made from biomass is free from sulfur and nitrogen and contains no (or very few) aromatic compounds, which makes it more environmentally friendly than diesel produced from fossil resources.

### 7.4.3    Pyrolysis

By definition, the pyrolysis of carbon-containing compounds entails their incomplete thermal degradation in the absence of oxygen at moderate temperatures (350–700°C) under pressure. It produces various organic liquids and gases along with char. The liquid product of biomass pyrolysis is known as bio-oil or pyrolysis oil (and by several other names, including biofuel-oil, wood liquid, and wood oil). Bio-oils are a mixture of different molecules (alcohols, aldehydes, ketones, esters, and phenolic compounds) derived from the fragmentation of lignin, cellulose, hemicellulose, and extractives. It is much easier to handle and to transport bio-oil than solid biofuel. However, the composition of bio-oil can change during storage if the products in the condensate (bio-oil) did not reach thermodynamic equilibrium during pyrolysis. The moisture content of bio-oil is typically about 15–30 wt% of that in the original feedstock used in the pyrolysis process [44].

Biomass pyrolysis processes can be divided into three categories: slow pyrolysis (carbonization); fast pyrolysis; and flash pyrolysis. Slow pyrolysis is a conventional method involving a slow heating rate of $0.1–1°C s^{-1}$. This gives a high yield of char relative to liquid and gaseous products and has been utilized for thousands of years in the production of charcoal. Slow pyrolysis (or carbonization) requires low temperatures and very long residence times, with typical vapour residence times of 5–30 min.

Fast pyrolysis uses much higher heating rates $(10–200°C s^{-1})$ and high temperatures (500–700°C) for short periods of time, and is considered to yield better results than slow pyrolysis. For example, the pyrolysis of pine wood samples at 550°C releases large quantities of aldehydes, ketones, and methoxylated phenols. The most important products derived from pine by pyrolysis are turpentine and pine oil (sometimes called tall oil). Biomass should normally be dried to a moisture content of around 10% before fast pyrolysis [45]. Recently, biomass has been converted to bio-oil and then into hydrogen by catalytic steam reforming. However, the yield of this process is relatively low. The main aim of such processes is to achieve a high yield of liquid or gaseous biofuels and other chemicals [46].

Flash pyrolysis of biomass usually occurs at 500–700°C and involves very high heating rates (e.g. $300°C min^{-1}$) over very short periods of only a few seconds. It therefore requires feedstocks with very small particle sizes (105–250 μm) [47].

The gaseous products of fast pyrolysis require rapid cooling or quenching to minimize secondary reactions of the intermediate products (radical components).

These radicals are very reactive and can undergo secondary reactions such as cracking and carbon deposition. They can however be stabilized by the addition of hydrogen (quenching). There are a range of chemicals that can be produced from bio-oil including food flavouring agents and phenols. These compounds are obtained by extraction or by performing subsequent reactions. Many examples of power generation from biomass liquids produced by fast pyrolysis processes have been reported [48].

Several parameters affect the yield and the composition of the volatile material generated during biomass pyrolysis, including the origin of the biomass, its chemical and structural composition, the pyrolysis temperature, and the particle size, to name but a few [49].

Overall, fast/flash pyrolysis is a promising technique for the production of liquid fuels from biomass in high yields. Such materials could potentially serve as renewable alternatives to fossil fuel precursors for various chemicals and fuels [44].

### 7.4.4   Liquefaction

Liquefaction is a thermochemical conversion process in which liquid fuels are generated from biomass. It is sometimes called the biomass to liquid (BTL) process. It involves the treatment of biomass at low temperatures (300–350°C) and high pressures (5–20 MPa) with a residence time of about 30 min, and is often performed using a catalyst in the presence of hydrogen (in which case the process is called catalytic liquefaction). However, non-catalytic aqueous liquefaction of biomass is also possible; this process is known as direct liquefaction. Under these conditions, water remains in the liquid state and has a range of interesting properties, including low viscosity and high capacity for dissolving inorganic compounds. In liquefaction processes, biomass is directly converted into products without drying. This is important because drying is relatively energy intensive [50]. The yields and composition of the resulting bio-oil and char differ depending on whether the catalytic or direct liquefaction process is used. Liquefaction and pyrolysis are often confused, but there are several differences between them. For instance, pyrolysis usually occurs at higher temperatures and lower pressures than liquefaction and there is no need to dry the biomass before liquefaction; on the other hand, pyrolysis often requires a pre-drying step. There is less interest in liquefaction than pyrolysis because the latter requires more expensive reactors and fuel feeding systems. Liquefaction generates a mixture of gas, liquid (bio-oil), and solid products in varying proportions depending on the reaction conditions [51, 52].

In addition to hydrothermal liquefaction, biomass conversion can be performed by means of supercritical water oxidation (SCWO) and supercritical water gasification (SCWG). SCWO processes are primarily used to degrade industrial waste at temperatures above the critical temperature of water (374°C). This causes the oxidation of biomass to produce thermal energy and a gaseous stream of $CO_2$.

However, SCWO processes can also produce gases, including $H_2$ and $CH_4$. Because the temperature is maintained at 500°C or below, no oxidation takes place and catalysts are generally required for substrate conversion [50].

### 7.4.5   Combustion

Direct biomass combustion (i.e. the burning of biomass in air) is another way of converting its chemical energy into heat and electricity. It is the most common process for biomass-to-energy conversion worldwide, accounting for around 97% of all biomass energy production. It is an economical technology that is generally better understood than its alternatives, and is extensively used on a commercial scale [53]. Combustion is usually performed in furnaces, boilers, steam turbines, and turbo-generators. The combustion produces hot gases at temperatures of 800–1,000°C or more. The heat produced must be used immediately for heat and/or power generation because storage is not viable. Biomass combustion involves two key steps: devolatilization to char and volatiles, and combustion of the volatiles and char. Several variables affect the combustion process, including the volatiles, tar, and ash content of the raw material [54]. It is therefore essential to properly characterize the various potential biomass feedstocks. Biomass is characterized with respect to both its chemical and physical properties. Chemical characteristics of interest include moisture content, ash content, abundance of ash-forming elements, concentration of other inorganic elements, and ash melting behaviour during combustion. These depend on the origin of the biomass and the agricultural conditions under which it was produced. In some cases, the biomass must be pre-dried to reduce its moisture content to <50% before combustion. Relevant physical characteristics of biomass feedstocks include particle size, particle shape, and bulk density. There are several advantages to upgrading biomass before combustion, as discussed in Section 7.2.

Combustion may be performed on a small scale (for household heat production), an intermediate scale (in hospitals, government office buildings, etc) or in large-scale combustions plants for industrial purposes [54]. In the US, rice husk conversion has been established as a particularly useful commercial option for large-scale combustion [55].

The effective combustion of biomass normally yields nothing more than $CO_2$ and $H_2O$. However, if the biomass has been contaminated (treated) with chlorine, as in the case of PVC-coated wood, dioxins may also be formed during combustion (primarily during the cooling of the flue gas). Dioxins are a group of organic compounds that may have some carcinogenic properties [3].

Co-firing is a process in which biomass and coal are burned together, either directly or indirectly. This can increase the cost-effectiveness of the overall process and increase the consumption of biomass for energy production, while reducing greenhouse gas emissions. There are three types of co-firing processes depending on the way in which the coal and biomass are combined. In the first type, biomass and coal are mixed together and then burned in a boiler. In the second type, biomass is processed separately and then introduced into the boiler through a separate ingress point to that used for coal (both fuels are still burned simultaneously). In the final category, which is also known as indirect co-firing, biomass is gasified before the co-firing process is initiated by the addition of coal [53].

## 7.5    Chemical Processes

Many existing energy products that are used as fuels in the transport sector are made from either petroleum-based materials or from biomass by some kind of chemical process. The most common biofuels used in transport today are DME, methanol, ethanol, butanol, and biodiesel.

### 7.5.1    Dimethyl Ether (DME)

Dimethyl ether (DME) is the simplest ether that remains in the gas phase at typical room temperatures and pressures. DME is non-carcinogenic but highly flammable, and has the chemical formula $CH_3OCH_3$. It can be used as a fuel for diesel engines, either by itself or as a blend with diesel because of its limited tendency to self-ignite. DME has a high cetane number of 55–60 (diesel has a cetane number of 40–55) and therefore has a shorter ignition time and better combustion properties than diesel. DME produced for use as a fuel is usually handled under pressure in liquid form. Because it can be produced from synthesis gas, it can ultimately be generated from many raw materials (including biomass) via gasification.

However, most of the DME produced today is derived from fossil raw material. One common method involves the dehydration of methanol to DME. A drawback of DME is that its use as a fuel requires the installation of pressure vessels and a new distribution system; this would significantly increase fuel distribution costs if adopted on a large scale.

DME has also been used as an additive for methanol to reduce its ignition time. Moreover, it has found applications as a residential fuel for cooking and heating, power generation, and hydrogen-rich fuel cells [56].

Other ethers such as methyl-tert-butyl ether (MTBE) and ethyl-tert-butyl-ether (ETBE) have been used as fuel additives that can replace fossil-derived aromatic compounds to increase the fuel's octane number. One of the advantages of MTBE and ETBE is that they can also be used as fuels themselves in automotive motors.

### 7.5.2    Biodiesel

Biodiesel is a fuel made from natural (biological) renewable resources that can be used directly in conventional diesel motors (engines). Biodiesel has several advantages compared to diesel produced from fossil precursors. It is degradable, non-toxic, contains no sulfur, and releases fewer emissions during combustion [57]. Biodiesel is often referred to as fatty acid methyl esters (FAME) and can be produced from a number of different raw materials including palm oil, soybean oil, rapeseed oil, sunflower oil, and several other vegetable oils via a straightforward process of chemical modification known as transesterification.

One negative aspect of biodiesel is that its quality may change during storage due to oxidative and hydrolytic reactions. Another important limitation relates to the availability of raw materials for biodiesel production. One of the most common kinds of biodiesel is made from rapeseed oil, rapeseed methyl ester (RME). RME has the advantage of being renewable (unlike fossil diesel), non-toxic, and less flammable than many other fuels such as ethanol. RME has the same cetane number, viscosity, and density as diesel but contains no aromatic compounds and is biologically degradable, causing only minor soil contamination if spilled.

Up to 3 tons of rapeseed can be produced from 1 ha of land. The main fatty acids in rapeseed oil are oleic acid, linoleic acid, and linolenic acid. The oil is pressed from the plant and, after some purification, allowed to react with methanol in the presence of potassium hydroxide as a catalyst to produce fatty acid methyl esters, as shown in Figure 7.6.

The transesterification reaction generates two byproducts. One is glycerol, which can be used in the cosmetics industry or as a raw material for producing other industrial chemicals such as 1,3-propanol or 1,2-propane-diol (propylene glycol). The other is a protein-rich cake that is left over after pressing the oil from the seeds, and can be used as animal feed [58]. RME can be blended with diesel to produce a mixed fuel, and can be handled using existing diesel distribution systems without modification.

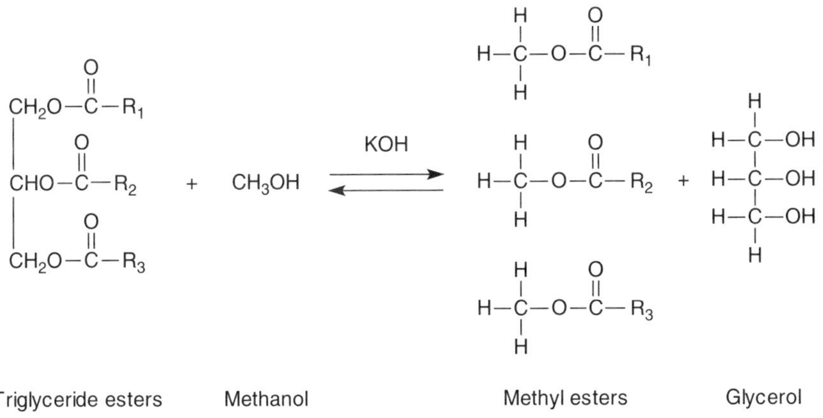

**Figure 7.6** *Transesterification of triglycerides to methyl esters and glycerol. R1, R2, and R3 are the alkyl chains of various fatty acids found in rapeseed oil, for example, oleic acid.*

## 7.6    Primary Alcohols

Methanol, ethanol, and butanol are liquid biofuels that can be synthesized from biomass and used in both four-stroke gasoline (Otto engines) and diesel engines. These alcohols can be prepared from sugarcane, sugar beet, wheat, barley, corn, switch grass, agricultural residues, wood, and many other industrial wastes. The most important characteristic of alcohols that makes them suitable as fuels for Otto engines is their high octane number. Fuels with high octane numbers can be burned at high compression ratios in engine cylinders, increasing the efficiency of the combustion process and reducing fuel consumption without presenting any risk of uncontrolled self-ignition. However, ethanol and methanol have low cetane numbers (8 and 5 respectively), which means that they can only be used in conjunction with an ignition improver of some kind (e.g. di-tert-butyl-peroxides). Such additives are usually costly. Fuels with excessively low cetane number ignite slowly and reduce engine performance.

Conventional gasoline engines can operate using gasoline–ethanol blends containing up to 5% ethanol without requiring any modification. Because ethanol is a renewable fuel, this can significantly reduce the net emissions of $CO_2$ into the atmosphere. An unfortunate drawback of ethanol and methanol is that they are highly flammable and burn with flames that can be difficult to see (i.e. low levels of low smoke).

As described in Chapter 4, ethanol can also be used to produce other related chemicals such as ethyl acetate, acetic acid, and acetaldehyde via chemical processes such as oxidation and esterification.

### 7.6.1    Methanol

Methanol is a colourless liquid with no particular smell at room temperature. It can be used as a fuel in both Otto and diesel engines. Methanol is highly toxic, corrosive, and flammable and, because it burns with a flame that is hard to see, methanol fires are not always immediately apparent to the eye. However, spillages of methanol into bodies of water or the soil degrade relatively quickly and do comparatively little environmental harm.

Methanol can be produced from fossil fuels and from biomass. It is usually produced from natural gas but can also be produced through gasification of biomass followed by conversion of the synthetic gas produced to biomethanol in the presence of catalysts at high pressures and temperatures (e.g. 220–275°C, 50–100 bar, and Cu/Zn/Al as catalysts). The product is generally contaminated with DME and water, which are removed by distillation.

Methanol can be blended with gasoline in small quantities to act as an oxygenate (i.e. a high-octane oxygen-containing compound). In addition, various methanol-heavy blends have been used as vehicle fuels, the most common of which is known as M85 (85% methanol and 15% gasoline). Neat methanol (M100) can also be used as a vehicle fuel.

The energy content of methanol on a per-liter basis is almost half that of petrol but its octane number is much higher, making it an energy-efficient fuel. The octane number of gasoline can be increased by adding aromatic compounds such as benzene or toluene. However, these compounds are carcinogens. They can be replaced with methanol, which has the added benefit of increasing the octane number of gasoline while presenting such health hazards.

Methanol is a bulk chemical that is produced around the world and is used extensively in industry as an important intermediate in the production of other chemicals such as formaldehyde, dimethyl ether, methyl tert-butyl ether, and acetic acid.

### 7.6.2 Ethanol

Ethanol is a colorless liquid that has many applications in the medical and food industries. It is soluble in water and can be used in pure form as a fuel or in blends with gasoline or diesel. It is also less toxic than gasoline, diesel, or methanol. Ethanol spills can be broken down by bacteria to carbon dioxide and water.

Blends of ethanol with gasoline containing up to 25% ethanol can be used in unmodified Otto engines; a 75:25 blend of gasoline and ethanol has been widely used in Brazil since 2002 and is known as gasohol. Diesel and ethanol are difficult to blend and an emulsifying agent must be added to enable the formation of a stable, homogenous solution. Pure ethanol has also been used in heavy vehicles with diesel engines, and blends containing up to 85% ethanol with 15% gasoline (E85) have been used in flexible fuel vehicles (FFV) such as the Ford Taurus.

Ethanol combustion produces much fewer emissions than that of fossil fuels. For example, the emissions of particulate matter, $NO_x$, CO, and other organic compounds from ethanol-burning engines are much lower than those produced by diesel combustion.

Ethanol can be produced by various processes and also from different raw materials including cellulose, waste from paper mills, excess wine in Europe, and other biological materials. Its production has increased in several countries in recent years including the USA, China, Brazil, India, and some European countries.

In contrast to methanol, ethanol can also be produced biochemically by the fermentation of carbohydrates (sugar) obtained from many different raw materials such as sugarcane, sugar beet, and corn. The three feedstock types used in ethanol production are: sugar feedstocks such as sugarcane; starch feedstocks such as cereal grains and potatoes; and cellulose feedstocks such as forest products and agricultural residues. More details on the three feedstock types are provided in the following sections.

### *7.6.2.1 From Sugar Feedstock*

Ethanol production from sugarcane (*Saccharum* sp.) is one of the easiest and most efficient fermentation processes because sugarcane contains about 15% sucrose. The glycosidic bond in the disaccharide can be broken down to two sugar units which are free and already available for fermentation. The sucrose is separated from the sugarcane by pressing chopped and shredded material. The residual solid left over after pressing is fibrous and is usually used as fuel in the sugar mill. Several steps are involved in the isolation of pure solid sugar, including several crystallizations steps. Sugarcane production requires a tropical climate and is very popular in Brazil, which has the largest area of land devoted to sugarcane cultivation and was the first and biggest producer of bio-ethanol in the world. Brazil also produces the cheapest bio-ethanol from sugarcane in the world today; the second cheapest is made from corn in the USA.

Sugar beet (*Beta vulgaris* L.) is a plant whose roots contain large amounts of sucrose. Sucrose consists of one glucose and one fructose unit that are linked by a glycosidic bond, that is, it is a disaccharide. The bulk of the world's sugar beet is produced in Europe, Russia, and the US. The 10 biggest beet-producing countries in Europe produced 242 million metric tons of sugar beet in 2005.

### 7.6.2.2   *From Starchy Feedstock*

Ethanol is much easier to produce from cereal grains such as barley, wheat, and corn than from cellulose-rich material. The process involves several steps: milling of grains; hydrolysis of starch to sugar units; fermentation by yeast; distillation; and removal of water from ethanol. After grinding the raw material, it is mixed with water and enzymes to break down the starch into sugar units. The free sugar can be fermented by yeast or bacteria, which convert it into ethanol and carbon dioxide. As the concentration of ethanol increases to around 15%, the rate of fermentation falls because high alcohol concentrations kill the yeast or bacteria. It is then necessary to separate the ethanol from the other material in the fermentation tanks by distillation. Distillation can increase the ethanol concentration of the mixture to around 95%. In order to remove the remaining water from the ethanol solution, it can be dried using different agents to a concentration of 99.9% ethanol, yielding so-called absolute ethanol.

The distillation and purification steps require significant amounts of energy and absolute ethanol will reabsorb water from the air over time. In Sweden, an ethanol company has been producing $50,000 \, m^3 \, a^{-1}$ ethanol from grain since 2000. One litre of absolute ethanol can be obtained from around 3 kg of wheat.

### 7.6.2.3   *From Cellulose Feedstock*

Ethanol production from cellulose feedstocks, which are often called "next-generation feedstocks", is depicted in Figure 7.7. It can be performed using cellulose-rich biomass from many different sources including wood, fast-growing plants such as switch grass or reed canary grass, and crop residues from food production such as corn stover. Ethanol production from lignocellulose material involves two processes: (1) hydrolysis of cellulose and hemicellulose into different

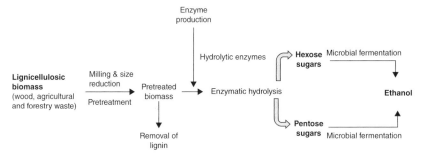

**Figure 7.7** *Ethanol production from lignocellulosic biomass.*

sugars; and (2) fermentation of the released sugars by yeast or bacteria. Ethanol can be produced from lignocellulosic material by chemical and microbiological hydrolysis processes. There are two main types of chemical hydrolysis process, one that uses high acid concentrations (the concentrated hydrochloride acid process, or CHAP) and another that was developed through a collaborative endeavor between Canada, America, and Sweden (the CASH process) and uses dilute acid. In the microbiological process, cellulose is broken down to sugar units by cellulase enzymes. One big challenge in this ethanol production process is to avoid degradation of sugar into other organic compounds such as furfural or 2-methyhydroxyfurfural in order to maintain high ethanol yields [59]. These compounds act as inhibitors in the fermentation step.

A unique pilot plant for ethanol production from lignocellulosic feedstocks was established at Ö-vik, Sweden in May 2004. The aim of the pilot plant was to develop efficient continuous technologies for the various process steps in ethanol production from forestry raw materials and other lignocellulosic feedstocks. Different raw materials require different conditions during the production process, which also needs to be optimized for every raw material. It was also considered important to demonstrate that large-scale lignocellulosic ethanol production was possible and to show that the production costs for cellulose-based ethanol could be decreased. The plant is fitted with several pieces of equipment that make it possible to perform two-step dilute acid and/or enzymatic hydrolysis on site. In 2005, ethanol from wood chips (softwood, i.e. spruce) was produced in the plant for the first time. The plant can process up to 2 tons of biomass per day, producing about 400 L of ethanol.

The lignin by-product is used for other purposes rather than just heat production and power generation, and there is ongoing research into its potential application as a precursor for the synthesis of other more valuable chemicals. In addition, the lignin can be pelletized together with other biomass. For example, at the Biofuel Technology Centre in Umeå, Sweden, lignin has been mixed with other feedstocks to produce pellets that enable its efficient transportation and use as a solid biofuel. The cost of ethanol production will decrease dramatically if methods that enable the conversion of lignin into valuable products can be developed.

### 7.6.3    Butanol

Biobutanol has many similarities to bio-ethanol and also some comparative advantages. It can be blended with gasoline in relatively large quantities without damaging unmodified engines, has a roughly 30% higher energy content, is better able to tolerate water contamination, and can be used together with ethanol and blended with gasoline. Its low vapour pressure reduces the vapour pressure of the ethanol/gasoline blend and makes it easier to blend ethanol with gasoline. It can also be used in existing vehicles without any modification to their engines, does not require a new distribution system, and can be delivered using the existing fuel supply infrastructure. It can be produced by the fermentation of biomass from the same feedstocks as bio-ethanol (e.g. sugarcane, sugar beet, corn, wheat) and is currently attracting significant commercial interest. A number of companies across the word are developing processes to turn biomass into biobutanol at a pilot scale.

## 7.7    Conclusions

Biomass is a renewable carbon-neutral energy source that reduces $CO_2$ concentration in the atmosphere as well as the use of fossil fuels. The collected biomasses are often bulky and wet and impractical in terms of transportation and storage requirements. Physical upgrading (pretreatment) such as drying and also densification to bales, briquettes, and pellets are therefore essential in many situations. There has recently been more focus on fractionation of different parts of certain types of biomass, since they contain varying levels of water and have different chemical content. For example, heartwood and sapwood have different amounts of fatty/resin acids; fractionation of heartwood and sapwood before biodiesel production may therefore improve the yield and thereby the cost-effectiveness of the process.

Microorganisms are vital in converting biomass to bioenergy. During biogas production there is a variety of bacteria involved ranging from aerobic to anaerobic. For ethanol production there are several kinds of biomass (e.g. sugarcane, cereal grain, and lignocellulose) with various accessibility to sugars; different processes must be applied for separation/isolation of the sugars. Most sugars can be fermented to either ethanol or butanol, while hydrogen can be produced by both dark fermentation and photoautotrophy.

To conclude, there are no simple concepts for replacing fossil fuels as energy sources with biomass due to several factors such as resources, demands, storage, transportation, and other economic aspects. Fortunately, different processes are already available for conversion of various biomasses to gaseous, liquid, and solid energy carrier. There are also a variety of ways to utilize the renewable biomass, not only as an energy source but also for bulk and fine chemicals production.

# References

1. Olsson, M. (2006) Wheat straw and peat for fuel pellets—organic compounds from combustion. *Biomass and Bioenergy*, **30**, 555–564.
2. Paulrud, S., Mattsson, J.E. and Nilsson, C. (2002) Particle and handling characteristics of wood fuel powder: effect of different mills. *Fuel Processing Technology*, **76**, 23–39.
3. Hedman, B., Burvall, J., Nillson, C. and Marklund, S. (2005) Emissions from small-scale energy production using co-combustion of biofuel and the dry fraction of household waste. *Waste Management*, **25**, 311–321.
4. Magelli, F., Boucher, K., Bi, H.T. *et al.* (2009) An environmental impact assessment of exported wood pellets from Canada to Europe. *Biomass Bioenergy*, **33**, 434–441.
5. Mani, S., Tabil, L.G. and Sokhansanj, S. (2006) Effect of compressive force, particle size and moisture content on mechanical properties of biomass pellets from grasses. *Biomass and Bioenergy*, **30**, 648–654.
6. Arshadi, M. and Gref, R. (2005) Emission of volatile organic compounds from softwood pellets during storage. *Forest Products Journal*, **55**, 132–135.
7. Arshadi, M., Nilsson, D. and Geladi, P. (2007) Monitoring chemical changes for stored sawdust from pine and spruce using gas chromatography—mass spectrometry and Visible-NIR-spectroscopy. *Journal of Near Infrared Spectroscopy*, **15**, 379–386.
8. Arshadi, M., Gref, R., Geladi, P. *et al.* (2008) The influence of raw material characteristics on the industrial pelletizing process and pellet quality. *Fuel Processing Technology*, **89**, 1442–1447.
9. Arshadi, M., Geladi, P., Gref, R. and Fjällström, P. (2009) Emission of volatile aldehydes and ketones from wood pellets under controlled conditions. *The Annals of Occupational Hygiene*, **53**, 797–805.
10. Tumuluru, J.S., Sokhansanj, S., Hess, J.R. *et al.* (2011) A review on biomass torrefaction process and product properties for energy applications. *Industrial Technology*, **7**, 384–401.
11. Prins, M.J., Ptasinski, K.J. and Janssen, F.J.J.G. (2006) Torrefaction of wood, part 1. Weight loss kinetics. *Journal of Analytical and Applied Pyrolysis*, **77**, 28–34.
12. Li, H., Liu, X., Legros, R. *et al.* (2012) Pelletization of torrefied sawdust and properties of torrefied pellets. *Applied Energy*, **93**, 680–685.
13. Larsson, S. H., Rudolfsson, M., Nordwaeger, M. *et al.* (2013) Effects of moisture content, torrefaction temperature, and die temperature in pilot scale pelletizing of torrefied Norway spruce. *Applied Energy* **102**, 827–832.
14. Stelte, W., Clemons, C., Holm, J.K. *et al.* (2011) Pelletizing properties of torrefied spruce. *Biomass Bioenergy*, **35**, 4690–4698.
15. Robson, R. (2001) Biodiversity of hydrogenases. In *Hydrogen as a Fuel: Learning from Nature* (eds Cammack, R., Frey, M. and Robson, R.). Taylor & Francis, London and New York.
16. Adams, M.W.W. (1990) The structure and mechanisms of iron hydrogenases. *Biochimica et Biophysica Acta*, **1020**, 115–145.
17. Vignais, P.M. and Billoud, B. (2007) Occurrence, classification, and biological function of hydrogenase: an overview. *Chemical Reviews*, **107**, 4206–4272.
18. Melis, A., Zhang, L., Forestier, M. *et al.* (2000a) Sustained photobiological hydrogen gas production upon reversible inactivation of oxygen evolution in the green algae Chlamydomonas reinhardtii. *Plant Physiology*, **122**, 127–136.

19. Melnicki, R.M., Eroglu, E. and Melis, A. (2009) Changes in hydrogen production and polymer accumulation upon sulfur-deprivation in purple photosynthetic bacteria. *International Journal of Hydrogen Energy*, **34**, 6157–6170.

20. Mohapatra, A., Leul, M., Mattsson, U. and Sellstedt, A. (2004) A hydrogen evolving enzyme is present in Frankia R43. *FEMS Microbiology Letters*, **236**, 235–240.

21. Redondas, V., Gomez, X., Garcia, S. *et al.* (2012) Hydrogen production from food wastes and gas post-treatment by CO2 adsorption. *Waste Management*, **32**, 60–66.

22. Martins, M. and Pereira, I.A.C. (2013) Sulfate-reducing bacteria as new microorganisms for biological hydrogen production. *International Journal of Hydrogen Energy*, **38**, 12294–12301.

23. DeVrije, T. and Claasen, P.A.M. (2003) Dark hydrogen fermentations, in *Bio-methane & Bio-hydrogen. Status and Perspectives of Biological Methane and Hydrogen Production* (eds J.H. Reith, R.H. Wijffels and H. Barten), Dutch Biological Hydrogen Foundation, Smiet Offset, The Haag.

24. Argun, H. and Kargi, F. (2011) Bio-hydrogen production by different optional modes of dark and photo-fermentation: an overview. *International Journal of Hydrogen Energy*, **36**, 7443–7459.

25. Lütke-Eversloh, T. and Bahl, H. (2011) Metabolic engineering of Clostridium acetobutylicum: recent advances to improve butanol production. *Current Opinion in Biotechnology*, **22**, 634–647.

26. Branduardi, P., Longo, V., Berterame, N.M. *et al.* (2013) A novel pathway to produce ethanol and isobutanol in Saccharomyces cerevisiae. *Biotechnology for Biofuels*, **6**, 1–12.

27. Atsumi, S., Hanai, T. and Liao, J.C. (2008) Engineering the isobutanol biosynthetic pathway in Escherichia coli by comparison of three aldehyde reductase/alcohol dehydrogenase genes. *Applied Microbiology and Biotechnology*, **85**, 651–657.

28. Huang, H.-J., Ramaswamy, S., Al-Dajani, W. *et al.* (2009) Effect of biomass species and plant size on cellulosic ethanol: a comparative process and economic analysis. *Biomass and Bioenergy*, **33**, 234–246.

29. Hamelinck, C.N., Hooijdonk, G.V. and Faaij, A.P.C. (2005) Ethanol from lignocellulosic biomass: techno-economic performance in short-, middle- and long-term. *Biomass and Bioenergy*, **28**, 384–410.

30. Jun, A., Tschirner, U.W. and Tauer, Z. (2012) Hemicellulose extraction from aspen chips prior to kraft pulping utilizing kraft white liquor. *Biomass and Bioenergy*, **37**, 229–236.

31. Sedlak, M. and Ho, N.W.Y. (2004) Production of ethanol from cellulosic biomass hydrolysates using genetically engineered Saccharomyces yeast capable of cofermenting glucose and xylose. *Applied Biochemistry and Biotechnology*, **113–116**, 403–416.

32. Kudahettige, R.L., Holmgren, M., Imerzel, P. and Sellstedt, A. (2012) Characterization of bioethanol production from hexoses and xylose by the white rot fungus Trametes versicolor. *Bioenergy Research*, doi:10.1007/s12155-011-9119-5.

33. Holmgren, M. and Sellstedt, A. (2005) Fermentation process, starter culture and growth medium. Patent WO 0548487.

34. Cristi, Y. (2007) Biodiesel from microalgae. *Biotechnology Advances*, **25**, 294–306.

35. Sharma, Y.C., Singh, B. and Korstad, J. (2011) A critical review on recent methods used for economically viable and eco-friendly development of microalgae as a potential feedstock for synthesis of biodiesel. *Green Chemistry*, **13**, 2993–3006.

36. Perego, C. and Ricci, M. (2012) Diesel fuel from biomass. *Catalysis Science & Technology*, **2**, 1776–1786.
37. Yadvika, S., Sreekrishnan, T.R., Kohli, S. and Rana, V. (2004) Enhancement of biogas production from solid substrates using different techniques – a review. *Bioresearch and Technology*, **95**, 1–10.
38. Quaak, P., Knoef, H., and Stassen, H. (1999) Energy from biomass, a review of combustion and gasification technologies. World Bank Technical Paper no. 422. The International Bank for Reconstruction and Development, Washington (DC).
39. Bridgwater, A.V. (2003) Renewable fuels and chemicals by thermal processing of biomass. *Chemical Engineering Journal*, **91**, 87–102.
40. Natarajan, E., Ohman, M., Gabra, M. *et al.* (1998) Experimental determination of bed agglomeration tendencies of some common agricultural residues in fluidized bed combustion and gasification. *Biomass and Bioenergy*, **15**, 163–169.
41. Bahng, M.K., Mukarakate, C., Robichund, D.J. and Nimlos, M.R. (2009) Current technologies for analysis of biomass thermo-chemical processing. *Analytica Chimica Acta*, **651**, 117–138.
42. Dry, M.E. (1999) Fischer-Tropsch reactions and the environment. *Applied Catalysis A: General*, **189**, 185–190.
43. Hamelinck, C.N., Faaij, A.P.C., Uil, H.D. and Boerrigter, H. (2004) Production of FT transportation fuels from biomass; technical options, process analysis and optimisation, and development potential. *Energy*, **29**, 1743–1771.
44. Qi, Z., Jie, C., Tiejun, W. and Ying, X. (2007) Review of biomass pyrolysis oil properties and upgrading research. *Energy Conversion and Management*, **48**, 87–92.
45. Yaman, S. (2004) Pyrolysis of biomass to produce fuels and chemicals feedstocks. *Energy Conversion and Management*, **45**, 651–671.
46. Bridgwater, A.V. and Peacocke, G.V.C. (2000) Fast pyrolysis processes for biomass. *Renewable and Sustainable Energy Reviews*, **4**, 1–73.
47. Goyal, H.B., Seal, D. and Saxena, R.C. (2008) Bio-fuels from thermochemical conversion of renewable resources: a review. *Renewable and Sustainable Energy Reviews*, **12**, 504–517.
48. Chiaramonti, D., Oasmaa, A. and Solantausta, Y. (2007) Power generation using fast pyrolysis liquids from biomass. *Renewable and Sustainable Energy Reviews*, **11**, 1056–1086.
49. Demirbas, A. (2002) Gaseous products from biomass by pyrolysis and gasification: effects of catalyst on hydrogen yield. *Energy Conversion and Management*, **43**, 897–909.
50. Toor, S.S., Rosendahl, L. and Rudolf, A. (2011) Hydrothermal liquefaction of biomass: a review of subcritical water technologies. *Energy*, **36**, 2328–2342.
51. Demirbas, A. (2000) Mechanism of liquefaction and pyrolysis reactions of biomass. *Energy Conversion and Management*, **41**, 633–646.
52. Demirbas, A. (2001) Biomass resource facilities and biomass conversion processing for fuel and chemicals. *Energy Conversion and Management*, **42**, 1357–1378.
53. Zhang, L., Xu, C. and Champagne, P. (2010) Overview of recent advances in thermo-chemical conversion of biomass. *Energy Conservation and Management*, **51**, 969–982.
54. McKendry, P. (2002) Energy production from biomass (part 2): conversion technologies. *Bioresource Technology*, **83**, 47–54.
55. Lim, J.S., Manan, Z.A., Alwi, S.R.W. and Hashim, H. (2012) A review on utilisation of biomass from rice industry as a source of renewable energy. *Renewable and Suitable Energy Reviews*, **16**, 3084–3094.

56. Semelsberger, T.A., Borup, R.L. and Greene, H.L. (2006) Dimethyl ether (DME) as an alternative fuel. *Journal of Power Sources*, **156**, 497–511.

57. Zheng, S., Kates, M., Dube, M.A. and McLean, D.D. (2006) Acid-catalyzed production of biodiesel from waste frying oil. *Biomass and Bioenergy*, **30**, 267–272.

58. Saha, B.C. and Woodward, J. (1997) *Fuel and Chemicals from Biomass*, American Chemical Society, Washington, DC, pp. 172–208.

59. Wyman, C.E. (1996) *Handbook on Bioethanol: Production and Utilization*, Taylor & Francis, Washington, DC, pp. 1–424.

# 8

# Policies and Strategies for Delivering a Sustainable Bioeconomy: A European Perspective

## David Turley

*NNFCC – The Bioeconomy Consultants, The Biocentre, York Science Park, UK*

## 8.1    Introduction

Biomass-derived chemicals are an important part of what has become defined as the 'bioeconomy', and the drivers encouraging the development of biobased chemicals are in many ways tightly bound up with those designed to support the wider development of a biobased economy.

The term bioeconomy has emerged to describe the concept of a biological-resource-fuelled economy that parallels the fossil-fuelled economy in its diverse range of products and in its mutual chemical, material and energy linkages. This concept is commonly combined with that of resource and energy efficiency plus reduced environmental impacts as a means of delivering what is seen as a more sustainable economic model, more suited to a planet with finite fossil resources and a burgeoning population with its associated material and energy demand.

*Introduction to Chemicals from Biomass*, Second Edition. Edited by James Clark and Fabien Deswarte.
© 2015 John Wiley & Sons, Ltd. Published 2015 by John Wiley & Sons, Ltd.

The current reality is that the bioeconomy is commonly evidenced in piecemeal developments, often acting as an adjunct to, or in tandem with, the fossil economy. A whole range of different products, materials, fuels and energy technologies are represented.

In some cases, biobased products may offer new functionalities or solutions to problems. In this case the argument for developing such products is based on the specific advantages offered that provide the appropriate economic stimulus to support its commercial development. In other cases, the same material product can be produced from biobased or petrochemical resources; in this case, where is the additional value? As an example, polyethylene (PE) can be made from sugar- and starch-derived dehydrated bioethanol or from fossil feedstocks, typically naphtha. If the former fails to offer any inherent cost advantage over its fossil-derived counterpart, then to create a market demand we depend on the additional 'social or public good' value placed on the material in terms of what is seen as its preferential environmental credentials, something that is much more difficult to quantify and value. In current market conditions, market observers identified that biobased PE was commanding a premium of between 15 and 20% on prices for fossil PE [1].

It is likely that biobased chemical, material and energy sectors will continue to be closely linked, and successful commercial development will depend on this where multiple product streams are generated in biorefinery approaches. For example, the development of biomass energy supply chains will support the logistics of feedstock supply to other outlets, or the residues from chemical extraction will be used as a low-carbon fuel.

The impacts of individual biobased developments can have a range of market impacts depending on the relative technical merits, costs of production and costs of market introduction. In the case of the latter, this may entail costs for re-tooling or modification to manufacturing processes.

Some technical developments need additional support or incentives to stimulate their commercial development. This is where political support comes into play to encourage uptake in areas which are seen to be of wider benefit to society but currently suffer, for one reason or another, from 'market failure', that is, where bio-derived chemicals or materials are not competitive with current alternatives.

From a policy and social perspective, the development of a bioeconomy is seen as delivering greater sustainability, delivering more environmentally friendly and socially acceptable products, ideally at little or no extra cost, while providing an opportunity to stimulate innovative high-tech employment as well as securing employment opportunities in the rural economy. Delivering a sustainable outcome depends on a careful balancing and weighing of the respective economic, social and environmental impacts, influenced by a multitude of associated considerations (Figure 8.1).

Delivering sustainability is a challenge, and there can be many unforeseen impacts. For example, in attempting to curb the development of transport biofuels derived from food crops, environmental organisations argue that increased biofuel use results in additional land being cleared to grow the food crops that are diverted

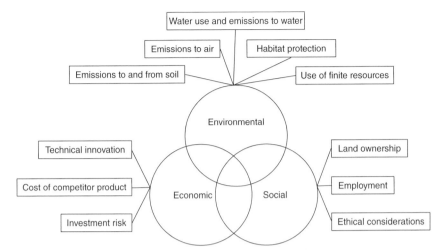

**Figure 8.1**   *Sustainable production is seen to occur when environmental, social and economic considerations are addressed and means are sought to reduce negative impacts to the greatest extent possible. In reality this represents the best compromise that can be achieved when faced with a multitude of competing impacts.*

to biofuel production, hence negating any net greenhouse gas (GHG) savings. However, this is a much more complicated issue than suggested by this simple argument which is typically based on the assumption that production per unit area is inelastic to demand; this whole issue is an extremely volatile area of debate.

The above argument is an ideal illustration of the potential difficulty faced in developing these new market opportunities, particularly where the environmental credentials of a product are the main rationale for its development. However, even in the case of land-use change, industry has responded by developing sustainability schemes and approaches to build both consumer and policymaker confidence in order to avoid or minimise such effects.

If society sees a general advantage in developing a bioeconomy, then the question is raised of how this should be encouraged and developed to deliver what society values as its most desirable outcomes. Clearly this would be impossible to achieve with one all-encompassing policy mechanism, but more likely by incremental steps and supporting mechanisms to emphasise good policy and practices that deliver the wider common goal. Not all potential supply chains and developments will contribute to this, but the aim is to set the direction of travel, agree the metrics and supporting measures and monitor progress in delivery.

## 8.2   Drivers for Change

The issues identified as rationale for supporting the development of a bioeconomy are varied and include (1) social and political; (2) economic; and (3) technical as defined in the following.

- *Social and political*: increasing energy and material self reliance; preservation of finite fossil resources; reducing fossil carbon emissions as part of a range of actions agreed in support of international climate change agreements and/or national aspirations on greenhouse gas reduction targets; and supporting disadvantaged rural economies.
- *Economic*: fostering innovation and high-value job creation; development of added-value market opportunities; development of novel cheaper means of production of high-value chemicals through application of biotechnology; and insulating from potential increasing future fossil feedstock costs and fluctuations in fossil fuel costs.
- *Technical*: new functionalities (e.g. biodegradability); and novel biosynthesis of difficult-to-synthesise chemicals.

The social and political issues and job-creating opportunities identified above provide the rationale for government interest, involvement and significant direct investment to encourage the development of the bioeconomy sector. In contrast, the economic and technical drivers influence industry interests in the sector. It can however be difficult to decouple industry incentives from policy incentives that also drive industry in the search for biobased solutions to particular problems, such as reducing waste and reducing GHG emissions, where pressure is placed directly on industry to deliver government, environmental or other objectives.

## 8.3   The Starting Point: Strategies for Change

The starting point for the development of measures to support the transition to a biobased economy is the drafting of political or corporate strategies to address the concerns highlighted in the previous section as drivers for change. More concrete measures and drivers can evolve from these, and this is where impact can be more effectively monitored to assess progress towards the goals articulated in aspirational strategies.

For example, in 2012 the EU adopted a strategy and action plan entitled *Innovating for Sustainable Growth: a Bioeconomy for Europe*, with the aim of shifting Europe towards greater and more sustainable use of renewable resources including for industrial purposes. This was driven by the stated aim of transitioning the EU from a fossil-based to a biobased economy, using research and innovation as the key drivers. The actions are focused on three key themes: (1) developing new technologies and processes for the bioeconomy; (2) developing markets and competitiveness in bioeconomy sectors; and (3) encouraging policymakers and stakeholders to work more closely together.

In a market currently estimated to be worth €2 trillion in Europe and employing 22 million people, each euro invested in EU-funded bioeconomy research and innovation is expected to return €10 of value by 2025 (EU press release, February 2012). The EU therefore clearly sees this sector as an engine for growth and prosperity.

Similarly, in the US in 2009 the National Research Council published the report *A New Biology for the 21st Century*. The report stresses the potential of biological research, among other things, to decrease US dependence on petro-leum-based products while increasing domestic manufacturing of biobased fuels and chemicals, and the need for the US to capitalise on developments in the sector as an economic imperative. Following this, in April 2012 the Whitehouse published its *National Bioeconomy Blueprint* [2]. This was designed to encourage the development of the bioeconomy for what was seen as its ability to harness innovation in the biological sciences to drive economic growth, particularly in rural America, while delivering in areas of public good. The blueprint focus on high-tech biotechnology applications, a sector currently valued at $100 billion. Here there is less emphasis on reducing GHG emissions (although biofuels do feature), and more on harnessing the value proposition offered by exploiting and nurturing leading-edge technology and innovation in biological sciences. Key actions include: supporting research and development (R&D) investments in underpinning science and technology areas; facilitating the transition from research to implementation; reducing barriers (particularly regulatory barriers and development costs); knowledge building and training development in higher education establishments; the identification of, and support for, public-private partnerships; and pre-competitive collaborations to encourage joint working.

There is clearly a lot of common ground between the US and EU objectives that stem from such initiatives, and these reflect the issues commonly afflicting developing technologies where risks are high for investors looking at new and relatively unproven technologies and recipient markets may need considerable education and development to stimulate interest.

This is the scenario of classic 'market failure' that allows governments to intervene to assist, support and nurture business and technology over the early stages of development, effectively providing market support, diminishing as technologies become established and/or technology costs fall and markets develop naturally.

Many different types of initiatives can flow from such strategies, for example ring fencing of funding within programmes for research and innovation (e.g. Horizon 2010 in Section 8.5.1.1) or measures to shape public purchasing decisions (e.g. BioPreferred Programme in Section 8.4.2.1).

Governments have the ability to create new markets though policy actions in support of strategic policy objectives. The development of biofuel and biobased heat and power markets are obvious examples; they would not exist at such scales without efforts to reduce the cost differential between renewable and fossil options or to mandate the use of renewable materials and fuels, undertaken as part of a package of measures to reduce GHG emissions.

## 8.4  Direct Measures

This section identifies the rationale adopted for such positions and the types of policy measures that can and have been adopted to support the development of biobased chemical, materials and energy.

### 8.4.1    Integrated Development

The wide range of sectors and technologies involved in the development of a biobased economy and the myriad of different possible drivers means that a broad range of policy domains are impacted. Policy solutions in the bioeconomy sector derived in response to identified needs commonly involve working across traditional government policy groupings and departments working in agriculture, environment and energy and those dealing with business, innovation and education. This does not always run smoothly, for example when there are tensions over how land should be most productively utilised.

This necessary development of integrated working within governments also mirrors what is happening in the commercial sector, where traditionally fossil-focused chemical companies in the sector have developed partnerships with agricultural commodity suppliers and biotechnology companies to take forward commercial development and demonstration projects. Current examples of such partnerships include: BP, AB Sugar and DuPont established a joint venture to develop first-generation bioethanol production through Vivergo fuels; US biotech company Amyris joined with the French oil giant Total to form Total Amyris Biosolutions, a 50–50 joint venture that will produce and market renewable biodiesel and bio jet fuel; and DSM (chemicals and healthcare) and Roquette (a leading starch manufacturer) have formed a joint venture to develop the market for biobased succinic acid.

In cases where technologies are well developed and proven, then policies may only require measures to open up suitable market opportunities, addressing the so-called 'demand side'. Where technologies are less well developed, stimulation of development and demonstration may be required to encourage industry uptake and de-risk private investment, addressing issues of 'supply side'. Typically, a mix of both supply-side and demand-side measures are required.

Ideally, any policy should work in tandem with existing market structures and operators to reduce development costs. But why are policies required at all?

This is summarised simply by Londo and Meeusen [3] who identified that the development of a biobased economy will require support for a significant period for two main reasons. First, new technologies need to be developed that will initially be more expensive than their conventional counterparts that have had decades to be refined and developed, making them as cost efficient as possible. It is expected that the new biobased options would follow a similar technology cost reduction pathway in the medium to long term, reducing the need for public support. Second, some biobased technologies will not become competitive as their wider social benefits and the real external costs reflecting the environmental impacts of their fossil-derived competitors are not recognised in their current market value. In the latter case, additional support measures are required.

Governments commonly rely on estimates of the Social Cost of Carbon (cost per tonne of $CO_2$ emitted) which represents the anticipated costs of dealing with the effects of climate change (e.g. increased flood protection) to weigh-up the benefits of taking any particular action. The EPA in the US and DECC in the UK commonly use such approaches to benchmark the cost benefit of policy support for low-carbon-oriented legislation.

The difficulty in both cases is that the resulting markets are dependent, at least in the short to medium term, on the decisions of policymakers and changes in government on a relatively regular basis, compared to the medium- and long-term timescales associated with technology development and deployment timeframes. Policymakers have to recognise this and provide sufficient safeguards.

This uncertainty significantly increases the risk for investment in such sectors. However, as is seen in Section 8.5, not all commercial developments in the biobased sector rely exclusively on policy support to promote their development; in some cases, a market advantage can be developed.

Policy drivers also differ globally depending on regional political priorities. In the US the primary diver has been to reduce dependency on fossil fuel imports from unstable regions. In the EU, responding to commitments made as part of International Climate Change Agreements has been the key driver for change. Transport accounts for around a quarter of the US energy demand and most of its oil imports. In the EU as a whole, transport accounts for a quarter of GHG emissions. It is therefore clear to see why the development of biofuels has been supported in both the US and EU, driven by supportive policy measures to deliver what has become a global commodity market in bioethanol and biodiesel fuels. Policies to support the greening of power and heat generation have followed, focusing on GHG reduction priorities.

Although there are strategies, policies and incentives to support the development of biofuels and bioenergy, with very definitive targets and measures to ensure delivery, there has been little or no direct support to encourage the development of biobased chemicals or materials widely in Europe. These sectors have had to rely on indirect support, development of increased environmental awareness in corporate decision-making, and technical development and cost reduction to build an increasing market share.

In contrast to Europe, the US has been much more pro-active and adopted an approach to stimulate market development for chemicals and materials through its BioPreferred Programme (see Section 8.4.2.1).

### 8.4.2    Policy Mechanisms

Primarily driven by the transport and power sector, a number of policy instruments such as those listed in the following have been developed to directly encourage and promote adoption of biobased alternatives.

*Feed-in tariffs*: typically used to support renewable heat and power production where a regulated, minimum guaranteed unit price is paid to private producers of renewable power fed into the grid supply. Different tariffs can be linked to specific technologies and revised relatively quickly, though such measures can also be used to provide long-term support to early adopters. Widely used internationally. Costs are generally passed directly to the end-user in higher energy bills.

*Taxes*: both incentives and penalties can be used to influence consumer purchasing habits when considering similar alternative products. Can be implemented as tax forgone (e.g. duty reduction), in which case tax revenues are then impacted which may have to be recouped through general taxation. In contrast, higher taxes can be imposed on fossil competitors (e.g. carbon taxes) to levelise total costs between fossil and low carbon alternatives; the additional costs are typically passed onto end-users. The downside to this is that it can be difficult to modify taxation levels quickly if required to temper the market response.

*Guaranteed markets and mandates*: the creation of voluntary or mandatory targets can stimulate market development, either by specifying the volume of biobased material (alone or as part of a blended material) that suppliers must deliver. Again, end-users are required to pick up any additional cost. Such measures can also provide more indirect drivers. For example, where pressure is placed on certain sectors to reduce the use of or phase out what are seen as undesirable chemicals (e.g. volatile organic solvents), this can create market opportunities for alternatives which in some cases can benefit biobased materials. Most 'total loss' lubricant fluids (e.g. chainsaw oils) are now biobased to reduce environmental pollution risks.

In implementing such policy actions, policymakers rely on evidence and advice to help identify the most appropriate actions to take to deliver change. Policymakers also require monitoring of the impacts of any interventions to assess the need for change and refinement, in order to ensure the policy is delivering as intended.

The specialist advice required in support of policy and strategy development is also likely to be multi-functional, for example focusing on greenhouse gas reductions exclusively could fail to recognise the impact of other burdens. As an example, it was soon questioned what the impact of biofuel development was on land-use change and food prices and the level of impact that could be tolerated. As a result, social impacts were also quickly recognised to be important in assessing the overall impacts of the imposition of policies to support biofuel development.

When developing market interventions, policymakers also need to control the amount of government intervention to avoid destabilising what would otherwise be stable and potentially viable markets. All of the above policy measures can have a very significant impact on the market but at very different level of cost. While feed-in tariffs, tax incentives and mandatory actions have a very significant impact, they are bureaucratic and therefore costly to manage and police. Typically, they are also limited in the products and materials to which they relate, primarily to make the overarching legislation and management simpler and more efficient to police.

### 8.4.3   Preferential Purchasing Policies

The implementation of *preferential purchasing policies* can have a much wider impact across a range of different product sectors, while being relatively cheap and simple to implement. Governments have significant spending power and typically a large estate to manage. As a result they can dictate or encourage purchasing policies and purchasing decisions, taking greater account of environmental issues in balancing the costs of different options. Large corporations, particularly well-known brand owners, operate similar policies in support of their corporate sustainability policies. These approaches have the advantage of working across a wide range of commodities. Decisions can be rapidly implemented and devolved to local level and have a direct impact.

#### *8.4.3.1   BioPreferred Programme*

The US Environmental Protection Agency (EPA) created the Environmentally Preferable Purchasing Program (EPPP) in 1993 to help federal officials meet sustainable purchasing requirements. Federal agencies are expected to consider products or services that have a lesser or reduced effect on human health and the environment when buying products and services. In doing so, agencies are using the federal government's enormous buying power (about $350 billion) to stimulate market demand for green products and services. They must consider the costs and benefits of various green products and services, using information provided by the EPA in the form of web-mounted databases and other resources to identify better practices and potential service providers. 'Environmentally preferred' does not necessarily mean biobased, but it provides an important framework to focus attention on such products and thereby stimulate uptake.

As an adjunct to the EPPP, to increase the purchase and use of biobased products the US BioPreferred program was created by the Farm Security and Rural Investment Act of 2002 (2002 Farm Bill, and expanded in the later 2008 Farm Bill). The BioPreferred program is managed by the US Department of Agriculture (USDA) and involves: (1) the certification and award of labels to qualifying products to increase consumer recognition of biobased products; and (2) supporting Federal procurement by identifying and designating categories (currently 97 product categories are covered) of biobased products that should be afforded preference when making purchasing decisions (under EPPP requirements).

Around 10,000 products are currently listed by the BioPreferred program, with the potential for many more. According to the USDA [4], federal agencies and the US Department of Defense spent approximately US$ 500 billion (£316 billion) on biobased products up to 2012, increasing awareness of biobased products and their manufacturers and creating 100,000 jobs.

The BioPreferred programme was the first example of a government procurement programme for biobased products, which was a key influence on the development of European strategies such as the Lead Market Initiative.

### 8.4.3.2    Lead Market Initiative

In its efforts to develop policies to support the development of a bioeconomy, the EU established a Lead Market Initiative for biobased products. This initiative sought to bring together sector and policy interests to develop actions to drive the bioeconomy forwards. One of the actions was to establish an Ad-hoc Advisory Group for Biobased Products, which in turn published a number of what it saw as priority actions for the EU to help develop the bioeconomy [5]. One of these was to encourage contracting authorities in all EU Member States to give preference to biobased products in tender specifications for public procurement exercises. The EU responded and issued the Green Public Procurement Guidelines, though its impact has been relatively low key to date.

## 8.5    Supporting Measures

The previous section identified the direct measures that policymakers can deliver to stimulate change towards a socially oriented public-good objective and to influence purchasing policies. But there are other actions that are also required to stimulate effective and reliable markets for biobased materials, some of which are dependent on government and some on actors in the market place.

On the supply side, drivers include: funding research and innovation; demonstration projects; and supporting early investment. Drivers of the demand side include: the development of corporate social responsibility (CSR) schemes; and the development of consumer confidence through (1) certification and standards and (2) assurance and labelling. These are all discussed further in the following sections.

### 8.5.1    Supply-Side Drivers

#### 8.5.1.1    Research and Innovation

As a new and emerging sector, significant research and development is required to optimise the use of biomass for chemical production. The expansion of the bioeconomy as a whole has been enabled by an increase in scientific knowledge and growing technical experience gained from pilot-scale development. The development of biotechnology in all forms, and in thermal processing and catalytic or biochemical reformation of syngas, has opened up the potential for a rapid expansion in the palate of chemicals and materials that can feasibly be derived from biomass resources at a competitive cost.

Most bulk biobased chemical production currently arises from existing food crop supply and processing chains, such as starch for food applications, vegetable oil refining and latterly bioethanol and biodiesel production. Typically, only part of the crop feedstock material is used in such processes (carbohydrates and oils) and further development is required to unlock the potential of the untapped cellulosic and lignin residues that offer potential for both sugar and aromatic monomer production. The technology for this is currently at pilot plant scale. Investigating means of biomass and biomass component separation and segregation may provide further useful chemical building blocks.

New forms of biomass, including algal biorefinery platforms, also offer the potential to increase the range of chemical materials produced at a globally exploitable scale.

The development of synthetic biology, fermentation technologies and bio-catalytic processes will further add to the range of tools and products available to industry to utilise base sugar feedstocks to support production of a range of different chemicals and materials.

All of the above requires significant investment by industry and public funding, both in basic and applied research and in the building and sharing of knowledge. Investment in so-called green technologies by governments is seen as a means of developing high-tech industries and associated high-value manufacturing and technology jobs. The ongoing current integration of chemical and biotechnology corporate interests also demonstrates the strength of commercial interest in and commitment to developing the biobased chemical sector.

In the EU, Horizon 2020 is the main financial instrument supporting research and innovation (from 2013 to 2020); €4.7 billion (£4 billion) has been allocated to address the challenges facing the development of biobased industries and biotech-nology. Much of this will have to be levered from industry to generate successful projects. To support this, the Biobased Industries Initiative has been established as a public-private partnership between the European Commission and the Biobased Industries Consortium (BIC). Its aim is to support research and innovation to encourage uptake of renewable biological resources. It aims to achieve this by encouraging partnerships between researchers and the private sector to fund and provide technical resources to address the challenges facing commercialisation in the biobased sector. The BIC brings more than 60 European companies, clusters and organisations from technology, industry, agriculture and forestry together to support the development of partnerships.

### 8.5.1.2  Demonstration Projects

In addition to the above support for strategic and near-market research, there is also a need for pre-competitive support to aid the demonstration of new technologies, essential for building confidence and evidence of proof of concept at a significant pre-commercial scale.

Renewable chemicals and materials that are going to be traded at significant global volumes require a large investment. For example, world-scale cellulosic ethanol plants are likely to involve capital expenditure of $100–300 million depending on scale, and large-scale syngas and thermochemical plants potentially considerably more. This is a very significant risk to investors for technologies that have limited experience of long-term operation at a reasonable commercial scale. Where governments are keen to see such technologies develop, supporting demonstration projects can help increase the evidential proof of delivery for concepts to support investment plans.

### 8.5.1.3    *Supporting Early Investment*

Measures to support significant investment in the biobased sector are relatively piecemeal and have mainly focused on support for renewable-energy-related activities. Since a number of potential developments in the biobased chemical sector can flow from such development initiatives (such as biobased derivatives from bioethanol), this is not all bad news. In Europe, the NER300 programme has been established to specifically support low-carbon demonstration projects, in this case supporting carbon capture and storage and renewable energy technologies. With the exception of support for biobased renewable energy, there is little in the way of any specific support for the demonstration of biobased chemicals and materials at an EU level. Such support has to vie with other calls on the budget for technological innovation.

In the US, the USDA runs the Biorefinery Assistance Program. This provides loan guarantees for the development, construction and retrofitting of commercial-scale biorefineries of up to $250 million. The program was established to assist the development of new and emerging technologies for the development of advanced biofuel and to increase self-reliance in fuels. To be eligible for the program, a technology must be adopted in a viable commercial-scale operation or have demonstrated technical and economic potential for commercial application in a biorefinery that produces an advanced biofuel. Loans of up to 80% can be made and the Biorefinery Assistance Program will provide loan guarantees in the range 60–90% of the loan value, depending on the scale of the project. Such measures can significantly improve the confidence of other investor partners and boost the chances of projects being realised.

The Green Investment Bank undertakes a similar role in the UK, providing finance for commercial projects where the injected capital is 'additional' to available private sector finance. In this case, the key areas of relevant interest are renewable energy; in this way it acts in a similar fashion to the US Biorefinery Assistance Program.

At smaller scales there are typically a myriad of smaller grants, loans and initiatives to support developments where solid business cases can be demon-strated and the levels of risk are much lower. All such support is provided on a case-by-case basis where there is a well-proven commercial opportunity, and such schemes are typically open to a wide range of sectors. Biobased technologies must therefore compete for their share of support.

## 8.5.2 Demand-Side Drivers

### 8.5.2.1 Corporate Social Responsibility

The business decisions of brand owners and key chemical and material manufacturers are increasingly influenced by internal CSR programmes, designed to help protect brand market shares and corporate image. CSR programmes drive the development of environmentally sustainable solutions to ensure compliance with environmental legislation and ethical standards, often with the aim of exceeding the minimum legal requirements. CSR programmes influence corporate purchasing and other decisions in order to minimise any negative environmental impact of its activities.

The actions underpinning such strategies often include environmental auditing of production procedures and analysis of energy use, which can highlight particular risks. Key issues of concern include potential future high cost and volatility of fossil fuel costs and associated feedstocks. Increasingly, companies are also examining the carbon footprint of their activities to examine how energy and materials can be used more efficiently. Other issues affecting brand owners include the impacts of tightening legislation on issues such as reducing packaging and waste and increasing recycling. Companies have looked to see how biobased materials can assist with meeting their environmental and efficiency objectives, particularly in the packaging materials sector.

Policy drivers to reduce waste volumes going to landfill have pushed up the cost of waste disposal. In turn this has helped to boost the market for biobased biodegradable polymers, particularly for items with a short life, including shopping carrier bags, disposable catering utensils and packaging materials. This has favoured the development of starch-based and latterly lactic acid and other microbial fermentation polymers to provide an increasing range of potential polymer materials with different technical characteristics to suit a variety of market needs.

There is also growing interest in biobased durable polymers because of better GHG credentials or other exploitable advantages. The falling price of biopolymers such as polyethylene terephthalate (PET; a widely recycled unique biopolymer) and biobased PE (a ubiquitous polymer in fossil form) has also increased company interest in such alternatives, as well as opportunities for examining alternative end-of-life opportunities to disposal by landfill. For example, biobased PE and other bio- and partially biobased polymers could effectively be used as a green fuel once recycling opportunities have been exhausted, possibly reaping financial rewards where such renewable energy generation is supported.

Such issues are encouraging companies to look much more carefully at the materials used and how these impact on the business more widely. This includes consideration of closed-loop systems of material management from production to end-of-life disposal to extract the maximum value and most positive environmental outcomes.

#### 8.5.2.2    Customer and Market Confidence

Where markets are driven by sustainability, environmental or other green credentials, then how can customers be sure that what they are purchasing is delivering on its promises? It is here that standards and certification play a role in protecting consumers and building market confidence, helping to protect the market position for the leading developers and the investments made in the support.

Standards and certification also relate to technology definitions, particularly important in a market that can suffer from 'green wash', that is, unsubstantiated or irrelevant claims. Such claims are probably unintentional and due to a lack of awareness [6]. As disposable incomes increase, so-called 'ethical spending' increases; such consumers can be vulnerable to marketing messages from businesses keen to exploit this. Several governments including the UK recognise this and have published *Green Claims Guidance* [7], which provides guidance on how marketing should be phrased and presented to avoid flouting trade description and advertising guidelines and legislation.

## 8.6    Bioeconomy Definitions

This article adopts the following definitions (as per British Standard PAS 600:2013):

- biobased: derived from biomass;
- biobased content: the proportion of a product that is derived from biomass;
- biobased material/product: a material wholly or partly derived from biomass; and
- biodegradable polymer: material that will primarily break down under the action of microorganisms.

### 8.6.1    Biobased Content

In the growing green economy, quantifying the impacts of products and services is of increasing commercial relevance. This is particularly important where products are marketed or rewarded on their environmental credentials. For example, this could be in terms of acquisition of tradable carbon credits through a reduction in GHG emissions. In such cases, being able to determine the biobased content of a product of both fossil and biomass origin is of increasing relevance.

In the USDA's BioPreferred Programme, biobased is defined in the covering legislation (2002 Farm Bill) as commercial or industrial products (other than food or feed) that are composed in whole, or in significant part, of biological products, renewable agricultural materials (including plant, animal and marine materials) or forestry materials, and includes biobased intermediate ingredients or feedstocks. The USDA then goes further to establish minimum biobased content standards for a whole range of product categories. In order to be listed, a product or package must meet or exceed the minimum biobased content percentage in its given category in order to use the *Certified Biobased Product* label. Where minimum biobased content standards for a product category have not been established, companies may apply for the Certified Biobased Product label if the product contains a minimum of 25% biobased content.

The emerging biobased manufacturing industry is producing products that include materials containing mixtures of both biobased and petroleum-derived components. For example, PET is one of the most commonly sold plastics worldwide, most commonly recognised as food packaging PET bottles. PET is a polyester produced by the polymerisation of a diol (ethylene glycol) and a diacid (terephthalic acid). While both monomers can be produced from biological fermentation, the former is much easier than the latter. Currently, most PET is a combination of bio- and fossil-derived monomers, and such materials can be classed as biobased. In some cases PET will be 100% biobased showing that, even within the same material, the biobased content can vary; such differences may need to be clarified to maintain customer assurance.

This issue is also important where refuse-derived wastes are combusted, where $CO_2$ of both fossil and biological carbon is released and, through gasification and fermentation technologies, can be converted into ethanol for use as a fuel or feedstock for PE or other material manufacture. In this case there is no technical difference in the product but, if a premium is demanded for biobased materials, the consumer needs assurance that the product they purchase can at least be 'deemed' to be of biological origin (though assurance procedures).

There are two approaches to determine the biobased content of a material or product: through either carbon isotope analysis or based on mass calculations. The pros and cons of each approach are discussed in the following sections.

### 8.6.1.1 Carbon Isotope Ratio Analysis

The Earth's atmosphere contains isotopes of carbon including the main stable isotope $^{12}C$ and an unstable isotope $^{14}C$ with a half-life of around 5,730 years. These are present in a relatively constant proportion. Plants absorb both forms as carbon dioxide during photosynthesis. When an organism dies, the $^{14}C$ in the residue decays over time. In the resulting fossil coal, oil and gas deposits, which are millions of years old, there is no $^{14}C$. The measurement of the ratio of $^{12}C$ : $^{14}C$ in a product in comparison to reference atmospheric ratios can be used to provide an estimate of the proportions of fossil and biogenically derived carbon in a product or material.

Such approaches provide a means of verifying the origin of a product, though this is relatively costly and time consuming. This approach is adopted in many standards, including *ASTM D6866 Standard Test Methods for Determining the Biobased Content of Solid, Liquid and Gaseous Samples Using Radiocarbon Analysis*, used to determine and independently audit the biobased content of listed products in the US BioPreferred Programme. This approach is also adopted in assessing the biobased content of polymers (CEN/TS 16137).

While this approach may enable quantification of bio- and fossil-derived ethanol produced from mixed fossil and biomass feedstock streams for example, this approach is not universal in its applicability (even within the energy sector). In Europe for example, the use of carbon ratio isotope analysis has been advocated for measuring the carbon isotope ratio in $CO_2$ emissions from power plant stacks fuelled by mixed waste. These would then be used to reward the energy generation deemed to be derived from biomass combustion on a similar proportionate basis. However, biomass materials tend to be partially oxygenated compared to fossil feedstocks such as polymers. Most biomass-derived feedstocks therefore tend to have lower calorific values and contribute less to power generation for the same level of carbon throughput. Such approaches therefore have to be supported by additional analyses.

In addition, changes in product formulation may have no (or in contrast, a very significant) impact on the biobased content of a material (assessed on a biogenic carbon basis) as an example from Beta Analytics [8] demonstrates:

A hypothetical fibreboard composite containing 30% silica with 70% cellulose will have a bio-based content of 100%. If the formula is modified to contain 20% silica and 80% cellulose, the fibreboard will still be 100% bio-based. All the carbon in the fibreboard is still coming from the cellulose.

If the silica is replaced with graphite fibres derived from petrochemicals (100% fossil carbon), the bio-based content will decrease significantly. The cellulose (which itself is only 44% carbon) now represents a significantly smaller proportion of the total carbon in the fibreboard. The same composite, now made with 70% cellulose and 30% carbon fibres will only be 51% bio-based.

### 8.6.1.2   Mass Calculations

Reporting on biobased content by weight percentage follows directly from the 'recipe' the producer uses. For example, a copolyester blend of 30% cellulose by weight and 70% fossil-based copolyester by weight would be classed as having 30% biobased content. The carbon isotope approach would show a much lower biobased value in this instance, as a much higher proportion of the cellulose element comprises oxygen.

Given the difference in outputs, some companies use both methods [9]. For example, for its biobased Sorona product description, DuPont quotes both the biobased carbon content (28%) and the renewable content by weight (37%).

Individuals may be tempted to choose an approach that gives a higher number, perceived by the customer as 'better'. However, given the risks of 'green washing', it is better for companies to report scientific, unambiguous information as DuPont does in the example above.

Given that such ambiguity can arise, when considering the issues of environmental impact a detailed life-cycle analysis (see Section 8.7) may be required to better understand the situation in question.

### 8.6.2 Biodegradability

Biodegradability is an important parameter in some sections of the biopolymer sector but also where materials are exposed to the environment, for example lubricants and greases. It is important to define what is meant by degradability in relation to biobased materials, which can be different to that for fossil- or partially biobased 'biodegradable' polymers. In the latter case, additives can be incorporated into materials such as PE, polypropylene, polystyrene and PET, accelerating the degradation of the plastic in the environment. These additives (pro-oxidants) use a salt of a transition metal such as cobalt, iron, manganese or nickel to drive the oxidation process which, under the action of heat or light, will reduce the molecular weight of the polymer to a level where bacteria and fungi in the soil environment can further reduce the material into water and carbon dioxide. These products are known as oxo-biodegradable, though the degradation process can take many months.

This has led to calls to differentiate between such materials by applying terms such as 'biodegradable' only to those materials capable of being biologically broken down completely (under specific conditions) in the absence of any other stimulus, and 'degradation' to the breakdown of materials such as oxo-biodegradable materials. Understandably this is an area of intense lobbying between manufacturers in each camp, as manufacturers of oxo-degradable polymers seek to be treated and recognised in the same way as other fully biodegradable polymers. The outcome of such debates and lobbying is commercially important and will affect the market share of any material that becomes disadvantaged by widespread adoption of any particular standard definition.

A material can only be called biodegradable with respect to the specific environmental conditions it is likely to encounter at the point of end-of-life disposal, and by determination of the degree of degradation achieved over a specific duration measured in those conditions (determined using a standard test method). As a result of the potentially different end-of-life options available (e.g. landfill, composting or anaerobic digestion) there are numerous standards specifying the test methods for different applications or end-of-life scenarios. In the UK for example these include [10] the following.

- BS EN ISO 14855-1 designed to simulate typical aerobic composting conditions for plastics in the organic fraction of solid mixed municipal waste (see also ASTM D5338).

- BS EN ISO 17556 specifies a method for determining biodegradability of plastic materials in soil (see also ASTM D5988).
- BS ISO 15985 specifies a method for the evaluation of the ultimate anaerobic biodegradability of plastics based on organic compounds under high-solids anaerobic-digestion conditions (see also ASTM D5511).

Standard tests for biodegradability have also been developed for other sectors including biolubricants (ASTM D6731 (biodegradability in soils) and ASTM D5864 (aquatic biodegradability)).

Definitions and standards help to differentiate between materials, as the associated standards are designed to provide assurance and guarantee effective performance and compatibility of the materials in question to the process being considered.

### 8.6.3  Composting Standards

The process of composting is an important parameter in the biopolymer sector, as this ranks as recycling rather than disposal and is therefore credited as being higher on the waste hierarchy that promotes reduction and reuse options over disposal.

Compostability is particularly important for materials used for packaging operations that commonly appear in the waste stream. In these cases it is important to ensure that these materials will not interfere with biological waste treatment options, such as composting and anaerobic digestion systems.

Composting standards for materials are designed to protect the status of high-quality material outputs from composting and anaerobic digestion. Commonly quoted standards for compostable plastics are ASTM D6400-04 and EN 13432 that relate to the performance of plastics in a commercially managed compost environment. Both standards were developed for hydro-biodegradable polymers (e.g. aliphatic polyesters plus modified starch) where the mechanism inducing bio-degradation is based on reaction with water. In order for a product to be composta-ble according to these standards, the following key criteria need to be met: (1) the polymer must be able to fragment into non-distinguishable pieces after screening and safely support bio-assimilation and microbial growth; and (2) they must be biodegradable, whereby 60% and 90% respiration of carbon to $CO_2$ is achieved over a period of 180 days for ASTM D6400 and EN 13432, respectively.

Break-down products should not pose any toxicological or environmental threat. In the case of composting, the time taken to degrade is the key parameter defining suitability for commercial composting operations, to which most of the composting standards relate. Composting and anaerobic digestion (AD) plant operators working to quality standards for compost and digestate (e.g. BSI PAS 100 Compost specification) can only accept materials that comply with compost-ing standards agreed within each composting or digestate standard.

Obtaining compliance with appropriate standards is therefore an important issue for material producers interested in securing a market share of the green economy.

### 8.6.4   Material Recycling

In relation to end-of-life options, clear labelling of material types and biobased products in particular can improve the rate of recycling or composting by aiding separation, preferably at source. Biodegradable and non-biodegradable bioplastics need to be very clearly labelled to avoid contamination of waste recycling streams. Failure to ensure materials reach the desired end-point may undermine the rationale for selecting a particular material in the first place. Design and labelling of materials can help and this is backed up by agreed standards defining specific types of materials, but it also requires a significant amount of consumer education to increase effective recycling of materials.

Distinguishing between materials is likely to become more difficult where the same material can be produced from fossil or biomass resources, for example PE. In such cases the material product streams will become increasingly intermingled as they approach end of life. It then becomes increasingly difficult to ascertain any added-value proposition for biobased materials other than through combustion, where the carbon isotope balance can be assessed to differentiate the contribution from fossil and biobased materials.

## 8.7   Life-Cycle Analysis

It may be necessary to provide quantified data in support of any specific environmental claim for a material or product. In such cases, life-cycle analysis (LCA) has emerged as an important standardised approach to assess the environmental impacts of a given product, where criteria or a range of criteria are examined in a defined life-cycle perspective (i.e. from cradle to grave or another defined perspective) according to a defining set of international standards (ISO 14040 and ISO 14044).

The use of a standard approach to such analysis (i.e. to agree system boundaries, units, fossil comparators and partitioning of impacts between main and co-products, etc) enables a comparison of products and approaches to production on a common basis. Most commonly, it has been used to compare energy and material use and associated GHG emissions for comparable products, including biobased and fossil alternatives. Despite this, differences exist from one team to another due to differences in units, raw material allocations or approaches adopted.

Inventories, common tools and calculators are being developed to (1) overcome and reduce such problems and (2) provide common datasets that can be incorporated into further analysis, while helping to reduce analysis costs. Examples of such tools and databases include the following.

- US Life Cycle Inventory Database [11]
- Tool for Reduction and Assessment of Chemical and other Environmental Impacts or TRACI (US EPA) [12]
- European Reference Life Cycle Database or ELCD (EU Joint Research Centre) [13]
- GREET Life Cycle Model (Argonne University) [14]

Many businesses also use life-cycle approaches to examine their own process performance and to help reduce the overall environmental burdens of their products and services. LCA is used in strategic decision-making as a tool to examine the impact of material, technology and end-of-life options. It provides insights into the different environmental trade-offs associated with product manufacture and the consumption of resources to compliment social and economic assessments. It is a useful means of examining the impact of new and novel processing systems, for example to compare chemical and industrial biotechnology approaches to deliver the same product.

The public sector also makes use of life-cycle analysis in policy implementation. This has been most strongly developed when considering the use of biomass in transport fuels and in heat and power generation, where public subsidy (in some form or other) is required to drive the market and costs per tonne of carbon saved is an important metric in determining where public funds are directed.

## 8.8   Ecolabels

In terms of informing consumers, labelling and development and promotion of recognisable logos are an important tool in informing consumer choice. Again, there are many different logos and schemes that relate to different environmental aspects of products in use. While most relate to delivery of specific attributes (i.e. low volatile organic carbon or VOC content or energy saving ratings) that help to differentiate a product, a few relate more broadly to a wider range of environmental impacts.

An example is the EU Ecolabel (Figure 8.2), developed to help identify products and services that have a reduced environmental impact throughout their life cycle from the extraction of raw material through to production, use and disposal. The EU Ecolabel is a voluntary label promoting environmental excellence and is recognised across Europe. Applications for accreditation are reviewed by a panel of independent experts and representatives of environmental non-governmental organisations. Compliance checks are also undertaken. In support of applications, companies have to submit a technical dossier containing declarations, data sheets and test data relevant to the sector in question. Currently there are around 37,000 Ecolabel products on the market. Those awarded Ecolabels must continuously review and work to improve their environmental performance. The scheme is not however limited to EU-produced materials and products; non-EU products can be approved through the same process.

**Figure 8.2**   *EU Ecolabel logo.*

**Figure 8.3**   *Compostable logo EN13432.*

The label covers services as well as products, so is not specifically aimed at supporting biobased products, chemicals and materials; biobased products may however be suited to meeting the objectives of the scheme and benefit from the awareness-raising that such affiliation provides.

The award of the EU Ecolabel is an important part of the communications message associated with promoting products. Compliance with standards can also be recognised by the owning accreditation bodies, awarding permission to use associated logos on approved materials. The compostable logo (Figure 8.3) is an example of this, controlled in its use by compliance with standard EN 13432.

To be credible and provide the highest level of assurance, such labels need to be trusted and ideally supported by third-party audit and accreditation. The development of broader schemes such as the EU Ecolabel is an attempt to provide an overarching scheme to halt the development of a plethora of individual schemes

**Figure 8.4**　*Blue Angel ecolabel.*

in different sectors, helping to reduce costs of certification for companies whose products may cross a range of sectors and avoiding confusion among consumers.

There are however a number of well-established ecolabels that have been developed by individual EU Member States. This includes the Blue Angel (Blauer Engel) Ecolabel, one of the first and most well-known ecolabels, covering over 10,000 products and services in 80 product categories. Surveys by the Federal Environment Agency of awareness of the Blue Angel ecolabel showed widespread recognition of the label in Germany, with 39% of consumers basing purchasing decisions on it [15] (Figure 8.4). The Blue Angel Label works by providing more specific information to the consumer. The text wrapped round the circular central image provides an indication of the main environmental properties of the product, relevant to the sector of use, while the text below the circular emblem provides a description of the protection goal of the positive environmental attribute of the product.

The value of such labelling will no doubt continue to raise as awareness and understanding of such schemes increases and as consumer environmental concerns increasingly affect their purchasing decisions. The development of energy efficiency ratings on electrical appliances and labelling of food in relation to its fat, salt and sugar content are all indicative of how the consumer is being educated in their purchasing decision to make more informed choices.

The chemicals industry is broad and complex, covering markets ranging from niche to commodity scale, meaning it is difficult to develop overarching schemes to promote renewable chemicals by themselves. It is more likely that development will be encouraged through individual sector developments. In the absence of any overarching policy driver to promote the development of biobased chemicals and products, developing an informed consumer base is an important task for the renewable chemical industry.

## 8.9    Concluding Remarks

The previous section has demonstrated that in Europe, with the exception of the energy sector, there is no direct incentive promoting the use of biobased products including biobased chemicals and materials. Given the sheer complexity and scale of the chemicals sector, this should not be surprising. Policymakers have used what limited tools are available to encourage development, however; support for development of a bioeconomy is strong and, where possible, monies and support is targeted.

Most development in the biobased chemical and materials sector to date has had to rely on exploitation of technical advantages in niche sector opportunities. Opportunities still exist in these sectors for expansion and growth, for example in biobased polymers. Outside of the traditional oleochemicals and starch industries, much of the biobased chemicals industry is still in its infancy; for example, the number of world-scale biobased polymer production facilities is limited to a handful of developments.

Other levers would help. In the EU, the Ad-Hoc Advisory Group within the Lead Management Initiative for biobased products has called for measures that included a call to set indicative or binding targets for biobased content in products. To date this idea has gained little traction within the EC and seems unlikely to be supported widely. The demand drivers for biobased materials will therefore continue to be opportunities that arise across a wide range of sectors, driven by environmental policy initiatives encouraging reduced waste, increased material recycling rates, reduced energy use and reduced environmental impacts of materials in production and use.

Consumer education is also required to build recognition of the value of biobased chemicals and resulting products, but also in building confidence in the labels and logos used to promote such materials when faced with an array of competing choices.

Continued innovation and research will also be a key requisite, along with investment support for scaled-up facilities. Separate measures will also be required to encourage and develop the skilled technologists and practitioners that are required to deliver biobased technologies.

There remain key issues to address regarding the sustainability of feedstocks. Policymakers will hesitate to introduce or maintain strong supportive measures for the bioeconomy if there is a lack of social acceptance or direct public opposition for the expanding use of biobased resources over fears that this will impinge on food security. The move to utilise non-food cellulosic biomass rather than starch and sugar food crops as material sources should help allay such fears. Public concern also remains over the role that biotechnology may play in developing feedstock crops and resources that improve system efficiencies and the wider acceptance of such technologies. Responding to such public concerns and demonstrating sustainable good practice in all aspects of raw material production, processing and manufacturing will remain key to development of the sector and maintenance of public confidence.

The issues that have been encountered in the development of the biofuel and bioenergy sectors have shaped the development of policy thinking in the biobased sector. The benefit in this case is that the issues are clearer now and, with associated development of sustainable sourcing and through communication of environmental messaging, consumers should have greater confidence in what they are purchasing and what their purchases should be delivering.

## References

1. ICIS Chemical Business (2013) *Bioplastics Surge Towards Commercialization*. Available at http://www.icis.com/resources/news/2012/07/02/9573828/bioplastics-surge-towards-commercialization/ (accessed 2 September 2014).

2. The Whitehouse (2012) *National Bioeconomy Blueprint*. Available at http://www.whitehouse.gov/sites/default/files/microsites/ostp/national_bioeconomy_blueprint_april_2012.pdf (accessed 2 September 2014).

3. Londo, H.M. and Meeusen, M.J.G. (2010) Policy making for the biobased economy, in *The Biobased Economy, Biofuels, Materials and Chemicals in the Post-oil Era*, vol. **12** (eds J. Sanders, M. Meeusen and H. Langerveld), Earthscan, London, pp. 203–213.

4. De Guzman, D. (2012) *Boosting Economy With Biobased Products*. ICIS Green Chemicals. Available at http://www.icis.com/blogs/green-chemicals/2012/02/boosting-economy-with-biobased/ (accessed 2 September 2014).

5. Ad-hoc Advisory Group for Bio-based Products (2014) *Lead Market Initiative Ad-hoc Advisory Group for Bio-based Products Priority Recommendations*. Available at http://ec.europa.eu/enterprise/policies/innovation/policy/lead-market-initiative/files/bio-based-priority-recommendations_en.pdf (accessed 2 September 2014).

6. Futerra (2014) *The Greenwash Guide*. Available at http://www.futerra.co.uk/downloads/Greenwash_Guide.pdf (accessed 2 September 2014).

7. Defra (2011) *Green Claims Guidance*. Available at https://www.gov.uk/government/uploads/system/uploads/attachment_data/file/69301/pb13453-green-claims-guidance.pdf (accessed 2 September 2014).

8. Beta Analytics (2014). Understand BioBased Content. Available at http://www.biobasedtesting.com/biobased.htm (accessed 2 September 2014).

9. Bioplastics Council (2012) *Understanding Biobased Content*, Society of the Plastics Industry, Bioplastics Council. Available at http://www.plasticsindustry.org/files/about/BPC/Understanding%20Biobased%20Content%20-%200212%20Date%20-%20FINAL.pdf (accessed 15 September 2014).

10. British Standards Institution (2013) *Bio-based Products – Guide to Standards and Claims*, BSI Standards, PAS-600:2013.

11. NREL (2013) *US Life Cycle Inventory Database*. Available at https://www.lcacommons.gov.nrel/search (accessed 4 September 2014).

12. EPA (2013) *Tool for the Reduction and Assessment of Chemical and Other Environmental Impacts (TRACI)*. Available at http://www.epa.gov/nrmrl/std/traci/traci.html (accessed 2 September 2014).

13. JRC (2013) *European Reference Life-Cycle Database*. Available at http://elcd.jrc.ec.europa.eu/ELCD3/ (accessed 2 September 2014).

14. Argonne National Laboratory (2013) *GREET Life Cycle Model*. Argonne National Laboratory. Available at http://greet.es.anl.gov/ (accessed 2 September 2014).

15. The Blue Angel (2014) *Great Brand Awareness – The World's Most Renowned Eco-label*. Available at http://www.blauer-engel.de/en/company/index.php (accessed 2 September 2014).

# Index

References to figures are given in *italic* type; references to tables are given in **bold** type.

*Introduction to Chemicals from Biomass*, Second Edition. Edited by James Clark
and Fabien Deswarte.
© 2015 John Wiley & Sons, Ltd. Published 2015 by John Wiley & Sons, Ltd.